Birds of
SENEGAL AND THE GAMBIA

HELM FIELD GUIDES

Birds of SENEGAL AND THE GAMBIA

Nik Borrow and Ron Demey

CHRISTOPHER HELM
LONDON

HELM
Bloomsbury Publishing Plc
50 Bedford Square, London, WC1B 3DP, UK

BLOOMSBURY, HELM and the Helm logo are trademarks
of Bloomsbury Publishing Plc

First published in the United Kingdom 2012

A catalogue record for this book is available from the British Library

Library of Congress Cataloguing-in-Publication data has been applied for

ISBN: 978-1-4081-3469-6; eBook: 978-1-4081-7101-1

6 8 10 9 7

Design by Mark Dando at Fluke Art
Printed and bound in India by Replika Press Pvt. Ltd.

FSC www.fsc.org MIX Paper from responsible sources FSC® C016779

CONTENTS

ACKNOWLEDGEMENTS

It is with pleasure that we take this opportunity once again to thank all those who, in a major or minor way, directly or indirectly, contributed to our *Birds of Western Africa* (2001) and *Field Guide to the Birds of Western Africa* (2004), on which this book is based. For the status and distribution of bird species in Senegal and The Gambia we have drawn extensively upon the work of Gérard and Marie-Yvonne Morel, and the more recent field guide by Clive Barlow and Tim Wacher, and are happy to acknowledge the debt.

Nigel Redman, the commissioning editor, is thanked for having initiated the project and for his assistance in the production of this book.

Mark Adams and Robert Prys-Jones at the Natural History Museum in Tring, UK, and Michel Louette at the Royal Museum for Central Africa, Tervuren, Belgium, are thanked for continuing to facilitate access to skins.

Once again, Julie and Marc Dando have expertly executed the digital manipulation of images, production of maps, and design and layout of the plates and text.

NB would like to thank his mother for continued support and the many Birdquest clients, friends and companions in the field, in particular Didi Abdel-Haviz, Mark Beaman, Sering Bojang, Dick Forsman, Solomon Jallow, James Lidster, Idrissa Ndiaye, Dembo Sonko and Wandifa Touray.

As always, RD wishes to express his deep gratitude to Rita Swinnen for her love, her patience and her enthusiastic and unstinting support.

Authors' note Many interesting observations remain buried in personal notebooks or unpublished reports. It is therefore recommended that relevant data be submitted to a refereed journal for publication, e.g. *Malimbus* or *Bulletin of the African Bird Club* (see page 23).

Future fieldwork will certainly add to our knowledge of the region's avifauna and the information presented in this guide, especially concerning species' ranges, can (and will) become out of date. The authors (c/o the publishers, Christopher Helm) would therefore be pleased to receive any comments and information which updates and corrects the text, maps or illustrations.

As was the case for *Birds of Western Africa*, NB produced the artwork and the maps, whilst RD wrote the text. However, as we worked in close collaboration with each other and critically examined each other's work, this book is once again a joint project.

INTRODUCTION

Despite a dramatic human population growth over the past few decades and the concomitant widespread habitat degradation and loss of biodiversity typical of so many African countries, Senegal and The Gambia still hold a remarkable and diverse avifauna. The two countries, often collectively referred to as Senegambia, have a combined bird list of 668 species, comprising both Afrotropical residents and migrants, and species of Palearctic origin. Over 30 of these are of global conservation concern, from Critically Endangered to Near Threatened. Lying immediately south of the Sahara, Senegambia is particularly important as a stopover and wintering place for Palearctic migrants after their crossing of the desert. The many coastal and inland wetlands attract large numbers of Palearctic and intra-African migrant waterbirds, whilst the dry and seemingly poor, but actually complex, Sahel ecosystem provides food and shelter for multitudes of Palearctic passerines throughout the northern winter, coinciding with the region's dry season. Offshore, seabird passage can be impressive, and the Cap Vert peninsula, Africa's westernmost point, is probably the best seawatching site of the eastern Atlantic seaboard. Both countries offer an excellent introduction to Afrotropical birding, whilst some Sahel specialities can more easily be seen in northern Senegal than anywhere else in West Africa.

This book primarily aims to enable observers identify any bird that may be encountered in Senegal and The Gambia. To be as user-friendly as possible, species accounts, accompanied by a small distribution map, face each plate, as in the majority of modern field guides. Although most of the illustrations have been taken from our *Birds of Western Africa* (2001), as amended in the subsequent *Field Guide to the Birds of Western Africa* (2004), they have been rearranged into new plates and many amendments and new figures have been included. In addition, some plates have been completely re-painted. The present format, with fewer species per plate, has enabled us to enlarge the illustrations and expand the text. The distribution of all species has been updated and the maps have been redrawn. Although the maps are necessarily small and often rather sketchy, we believe it is far better to have such maps than none, as they permit an understanding of a species' known or inferred occurrence at a glance. We would, however, once more like to emphasise that they do not and cannot provide a detailed reflection of actual distributions and should therefore be used with common sense. New information, corrections and comments are most welcome (c/o the publishers) and will, as far as possible, be incorporated in future editions of the book.

We hope that this field guide will serve its purpose and further stimulate interest in and study of Senegambia's birds by both foreign and local observers, and will thus ultimately contribute to the conservation of the region's vital biodiversity.

SENEGAL
THE GAMBIA
GUINEA-BISSAU
PARC NATIONAL DES OISEAUX DU DJOUDJ
Senegal River
Richard Toll
Dagana
Lac de Guier
Djeuss
Lampsar
NDIAEL FAUNAL RESERVE
St Louis
PARC NATIONAL DE LA LANGUE DE BARBARIE
16°
Lagbar
Vallé du Ferlo
Louga
Lompoul
Tiângol Kosses
Revnérou
Dara
Linguér
Fas Boye
Mboro
Mékhé
Darou Mousti
15°
Velingara
Rufisque
Tivaouane
Touba
Tiel
Gassane
Mbaké
Pikine
Tiaroye
Rout
Thiès
Bambey
Bargny
Diourbel
PARC NATIONAL DES ILES DE LA MADELEINE
Gossas
Mbabane
Tiadiave
Tataguine
Mbour
Nianing
Fatick
Guinguineo
Joal-Fadiot
Foundiougne
Kaolack
14°
Sokone
Nianija Bolon
PARC NATIONAL DU DELTA DU SALOUM
Nioro Du Rip
Kolar Bolon
Bao Bolong
Karang
Nja Kunda
Farafenni
Georgetown
Kerewan
Salikene
Bakau
Banjul
Mansa Konko
Brikama
Bintang Bolon
Bwiam
13°
Soungrougrou
Kolda
Casamance
Bignona
Marsassoum
Sèdhiou
RÉSERVE KALISSAYE
Casamance
Ziguiinchor
Oussouye
PARC NATIONAL DE BASSE-CASAMANCE
16°
15°

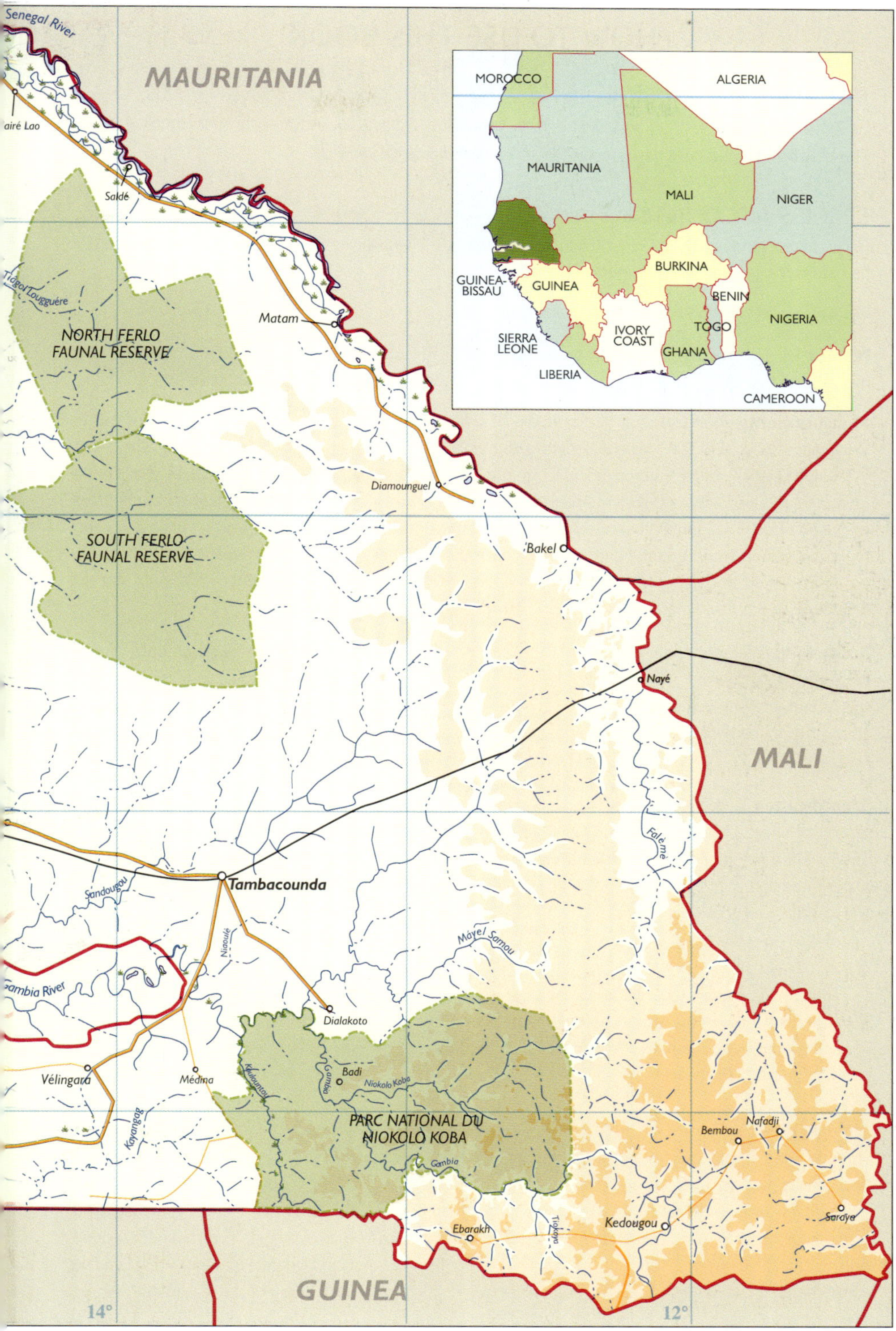
Senegal River
MAURITANIA
Saldé
Matam
NORTH FERLO
FAUNAL RESERVE
SOUTH FERLO
FAUNAL RESERVE
Diamounguel
Bakel
Nayé
MALI
Falémé
Tambacounda
Sandougou
Niaoulé
Mayel Samou
Gambia River
Dialakoto
Vélingara
Médina
Badi
Gambia
Niokolo Koba
PARC NATIONAL DU
NIOKOLO KOBA
Gambia
Bembou
Nafadji
Saraya
Ebarakh
Kedougou
GUINEA
14°
12°
MOROCCO
ALGERIA
MAURITANIA
MALI
NIGER
BURKINA
GUINEA-
BISSAU
GUINEA
BENIN
NIGERIA
TOGO
SIERRA
LEONE
IVORY
COAST
GHANA
LIBERIA
CAMEROON

HOW TO USE THIS BOOK

This book illustrates and briefly describes all species definitely recorded from Senegal and The Gambia ('Senegambia') up to 2010, as well as some that have been claimed but whose occurrence requires proof. The design and layout have been executed with field identification in mind.

SPECIES ACCOUNTS

Many species groups have a short introduction highlighting the key features. This information is intended to be read in conjunction with the species accounts.

The species accounts comprise the following elements.

Names The first line of each species account contains the English name followed by the scientific (binomial) name (see pages 15–17 for an explanation of the taxonomy and nomenclature used in this book).

Measurements To indicate the bird's relative size, **length** (in cm) is mentioned after the species name. **Wingspan** (WS) and tail-streamer length are mentioned where relevant. Measurements are taken from authoritative sources, complemented by our own mensural data obtained from specimens. Some measurements from *Birds of Western Africa* (2001) and *Field Guide to the Birds of Western Africa* (2004) have been corrected. Total-length measurements represent the length of museum skins lying on their backs and measured from bill tip to tail tip. Note that direct comparison of these measurements can be quite misleading because they do not take into account other aspects of physiognomy, such as relative bill, neck and tail lengths, bulk, etc. It is therefore most useful only when comparing related species.

Subspecies If the species has a single subspecies occurring in Senegambia, its name is placed at the beginning of the species account. If more than one subspecies occurs, these are normally succinctly described within the species account. It should be borne in mind that differences between subspecies are often subtle (e.g. colour tones), or based on measurements, and are thus often very hard or impossible to discern in the field.

Identification Each species account typically begins with the key features. Principal characteristics of the different plumages follow, in the order adult male, adult female, juvenile and, if relevant, immature. We have attempted to include as much information as space permitted, without repeating what can be seen on the plates—the text is intended to supplement the illustrations and to highlight the most relevant identification features. Diagnostic or important identification features are placed in ***semi-bold italics***. Distinguishing features of similar species are given where relevant. (For more detailed species descriptions, please refer to our *Birds of Western Africa*).

The term 'immature' (used generally for a non-adult bird) has normally not been used to indicate juvenile plumage (the first plumage of true, non-downy, feathers). For young passerines of Palearctic origin wintering in our area, 'immature' indicates 'first-winter' plumage. In some cases 'immature' is used when the available information prevents greater precision. The terms 'winter' and 'summer' refer to northern hemisphere seasons.

The terms 'small', 'medium-sized' and 'large' indicate only the relative size of a species compared to its close relatives. For the sake of brevity, and when not indicated otherwise (or obviously different), the term 'top of head' includes forehead and crown, 'throat' includes the chin, and 'legs' includes legs and feet. 'Above' generally encompasses head and upperparts, and sometimes also tail; 'below' the entire underparts; in some cases, certain parts are excluded, but this should be obvious from the context. 'Flight feathers' include primaries and secondaries (but not tail feathers). 'Wing feathers' include flight feathers and wing-coverts. 'Upperparts' often includes scapulars and wing-coverts (if concolorous). Study of the illustrations should preclude any confusion.

Colour names have deliberately been kept as simple as possible and most will be readily understood. Where two colours are combined (compound colours), the last named is dominant; in other words, they should be interpreted as the second colour tinged with the first; e.g. reddish-brown is brown tinged red. Where the suffix -ish is added, this indicates a weaker or less distinct shade of that colour.

Frequently used colours include buff (pale or dull yellowish with a brownish or beige tinge), chestnut (dark reddish-brown), olive (dull yellowish-green, like the fruit) and horn (pale brownish-yellow; used solely for bills). Dusky (dirty greyish or brownish) is often used for rather indistinct darkish markings. The term dark (opposite pale) is used for dark plumage markings that lack any obvious colour.

Habitat and habits (HH) The habitat in which the species normally occurs is briefly indicated. Some typical behaviour relevant to identification is mentioned wherever space permitted. Other relevant information given in this section includes the species' status and relative abundance.

Status The status of each species is denoted as follows:

Resident: a species that resides within its range throughout the year and breeds; the opposite of a migrant.

Intra-African migrant: a species that breeds in one part of Africa and spends the post-breeding season in a different area, or appears only seasonally in another part.

Palearctic migrant: a species that breeds in the Palearctic region (Europe, North Africa and part of Asia) and spends the boreal winter in sub-Saharan Africa.

Vagrant: a species outside its normal range.

More than one category may be applicable. It is sometimes difficult to distinguish between a genuine vagrant and a species with only a few records. Some 'vagrants' may prove to be more or less regular, albeit rare or scarce, given increased observer coverage.

Abundance Our aim has been to be as user-friendly as possible. Therefore we use only five, easily understood, abundance categories:

Common: invariably encountered singly or in significant numbers within its normal habitat.

Fairly common: usually, but not invariably, encountered within its normal habitat (= 'frequent' or 'not uncommon').

Uncommon: relatively frequently, but not regularly, encountered within its normal habitat.

Scarce: only irregularly and infrequently encountered within its normal habitat.

Rare: rarely encountered, often implying fewer than c.10 records.

Note that relative abundance is somewhat subjective and can vary across the species' range and according to season.

Globally threatened species Species included in *Threatened Birds of the World* (BirdLife International 2000) and its most recent update (see www.birdlife.org) are indicated at the end of the HH section by 'Red List category' followed by one of the four categories occurring in Senegambia:

Critically Endangered: species facing an extremely high risk of extinction in the wild in the immediate future.

Endangered: species facing a very high risk of extinction in the wild in the near future.

Vulnerable: species facing a high risk of extinction in the wild in the medium-term future.

Near Threatened: species coming very close to qualifying as threatened.

For a comprehensive discussion of the above categories, see www.birdlife.org.

Voice Only the most characteristic vocalisations are (briefly) given. In some cases, where the species is mainly silent or the vocalisations are unimportant for identification purposes, these have been omitted when space was limited. Transcriptions of calls and songs are placed in *italics*. References to Claude Chappuis' (2000) outstanding collection of 15 CDs of *African Bird Sounds* – an invaluable tool – are presented within square brackets, with CD and track number. It should be noted that these CDs inevitably contain certain errors (see for example the review in *Bull ABC* 9: 74–78, 2002). Discrepancies between Chappuis' accompanying booklet and this book are therefore due to the fact that references of vocalisations known to us as having been misidentified have been omitted or placed under the correct species.

As is widely acknowledged, transcribing bird sounds so that they can be unambiguously interpreted by others is almost impossible. Phonetic renditions and verbal descriptions are necessarily highly subjective and open to misinterpretation, but there appears to be no convenient alternative. The user of this book is therefore advised to listen to sound-recordings and compare them with the transcriptions, in order to understand the authors' interpretation.

In the transcriptions of vocalisations the following conventions apply:

single vowels are pronounced short (thus a as in 'apple', e as in 'extra', i as in 'it', u as in 'full')

ee as in 'see', 'be'

iiiiii is higher pitched than *eeee*

k as in 'cat' (c is not used for the hard 'k' sound)

ch as in 'check'

sh as in 'sheep'

CAPITAL LETTERS indicate that the component in question is considerably louder than the others.

Pauses between notes or syllables are denoted as follows (after Alström *et al.* 1991):

seesee no discernable pause

see-see very short pause

see see normal pause, as in ordinary conversation

see, see longer pause (at least *c.*1 second)

see...see pause of more than 2 seconds

PLATES

All species positively recorded and several species only claimed from Senegambia are illustrated in colour. Our aim has been to illustrate as many distinct plumages as space would permit. Thus, distinctive male, female and immature plumages are depicted, subject to the availability of representative specimens or personal field notes. This is also the case with races that are sufficiently distinctive to be separated in the field. Only in a few cases, however, has the juvenile plumage of passerines been illustrated, as this appears unnecessary and potentially confusing. Indeed, this plumage is usually retained for only a short period and identification is normally facilitated by the presence of the parents.

Wherever possible or desirable, care has been taken to respect family groupings. On most plates, the species are arranged in order of resemblance and are thus not necessarily in taxonomic order. The order of the families, however, largely follows taxonomic sequence. A checklist of the birds of Senegal and The Gambia in systematic order is given on pages 316–333.

A few plates depict birds painted to different scales; in such cases the various birds or groups are divided by a narrow line on the plate.

The majority of the illustrations are those from our *Birds of Western Africa* (2001), as amended in the subsequent *Field Guide to the Birds of Western Africa* (2004). They have, however, been rearranged into new plates and many additional amendments and new figures have been included.

MAPS

Distribution maps are provided for all species. The maps are based on published data up to 2010. Unpublished data known to us have been taken into account, when considered reliable. The maps reflect the known or inferred distribution of a species in areas of suitable habitat and should not necessarily be taken as providing a comprehensive and detailed reflection of actual distributions. They must therefore be used with caution and common sense.

Key to the maps:

- resident
- mainly resident but partially migratory or erratic within range
- non-breeding visitor (main range)
- non-breeding visitor (sparse occurrence)
- breeding visitor
- X vagrant or isolated record
- **?** uncertain record or range

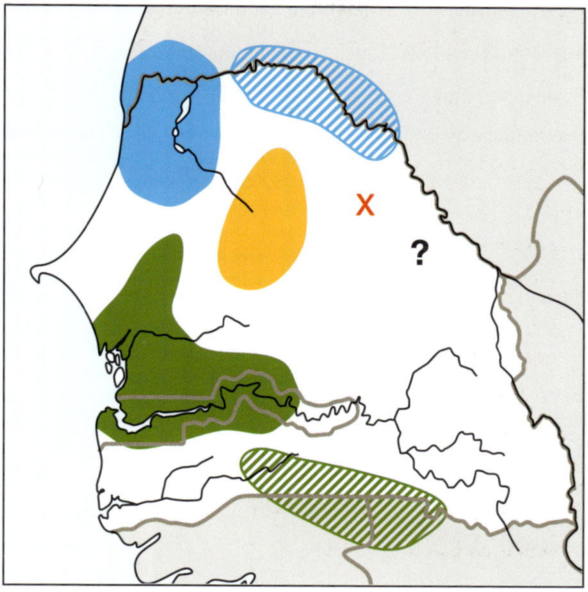

TAXONOMY AND NOMENCLATURE

In general, we have followed *The Birds of Africa*, vols. 1–7 (Brown *et al.* 1982; Urban *et al.* 1986, 1997; Keith *et al.* 1992; Fry *et al.* 1988, 2000; Fry & Keith 2004), although in some cases we have preferred other recent authors, where these adopt what we consider a more advanced or consistent view. The names used by the African Bird Club (see the checklist of African birds at www.africanbirdclub.org) is largely similar to ours, although there are a few differences.

In some cases, where a taxon has been variably treated as a subspecies or a species by different authors, the specific name under which it is also often treated is placed within parentheses, to indicate that this does not imply an undisputed taxonomic decision. This has been done in cases where there is evidence to suggest that the taxon may preferably be treated as a separate species, e.g. *Larus (cachinnans) michahellis*, *Bubo (bubo) ascalaphus*, and *Hippolais (pallida) opaca*, although a few others, e.g. *Ficedula (hypoleuca) speculigera* and *Parus (leucomelas) guineensis*, are more controversial. It has also been done in one case, *Acrocephalus (scirpaceus) baeticatus*, where evidence suggests that the form often treated as a full species may actually best be 'lumped' again.

Although we are aware that recent molecular studies have proposed an increasing number of changes, we have basically retained the more familiar names used in our *Field Guide to the Birds of Western Africa* (2004). As explained in the latter work, these may in some cases differ from those in *Birds of Western Africa* (2001; hereafter abbreviated as *BOWA*). Those differences are given again below, together with a few additional ones introduced in this book.

Changes in the scientific name resulting from the correction of the gender

Scientific names have been corrected following David & Gosselin (2002a,b, 2011; the two corrections from 2011 are indicated with *). These are, in alphabetical order: *Alopochen aegyptiac***a**, *Amaurornis flavirostr***a**, *Bleda canicapill***us**, *Butorides striat***a**, *Chlidonias hybrid***a**, *Cinnyris coccinigastr***us**, *Columba livia gymnocycl***a**, *Delichon urbic***um**, *Euplectes macrour***a**, *Heliolais erythropter***us**, *Leptoptilos crumenif***er***, *Numida meleagris galeat***us**, *Phalaropus fulicari***us**, *Tchagra senegal***us**, *Tigriornis leucoloph***a**, *Treron calv***us**, *Turnix sylvaticus lepuran***a***.

New scientific names resulting from taxonomic changes

Name used in *BOWA* (2001)	Name used in this book	References
Phoenicopterus ruber	*Phoenicopterus roseus*	Knox et *al.* (2002)
Hieraaetus fasciatus	*Aquila fasciata*	Helbig et *al.* (2005)
Hieraaetus spilogaster	*Aquila spilogaster*	Helbig et *al.* (2005)
Aquila wahlbergi	*Hieraaetus wahlbergi*	Dickinson (2003), Helbig et *al.* (2005)
Eupodotis savilei	*Lophotis savilei*	del Hoyo et *al.* (1996), Dickinson (2003)
Eupodotis melanogaster	*Lissotis melanogaster*	del Hoyo et *al.* (1996), Dickinson (2003)
Catharacta skua	*Stercorarius skua*	Dickinson (2003), Sangster et *al.* (2004)
Catharacta maccormicki	*Stercorarius maccormicki*	Dickinson (2003), Sangster et *al.* (2004)
Larus sabini	*Xema sabini*	Dickinson (2003)
Myrmecocichla cinnamomeiventris	*Thamnolaea cinnamomeiventris*	Dickinson (2003)
Melaenornis pallidus	*Bradornis pallidus*	del Hoyo et *al.* (2006)
Cyanomitra obscura	*Cyanomitra olivacea*	Dickinson (2003)
Malaconotus sulfureopectus	*Chlorophoneus sulfureopectus*	Dickinson (2003)
Ploceus superciliosus	*Pachyphantes superciliosus*	*Birds of Africa*
Amandava subflava	*Sporaeginthus subflavus*	*Birds of Africa*
Lonchura cucullata	*Spermestes cucullatus*	*Birds of Africa*
Lonchura bicolor	*Spermestes bicolor*	*Birds of Africa*
Lonchura fringilloides	*Spermestes fringilloides*	*Birds of Africa*
Lonchura cantans	*Euodice cantans*	*Birds of Africa*
Miliaria calandra	*Emberiza calandra*	Dickinson (2003), Sangster et *al.* (2004)

African Masked Weaver *Ploceus velatus*, treated as a single species in *BOWA* (2001), is now considered to consist of two species, Vitelline Masked Weaver *P. vitellinus* and Southern Masked Weaver *P. velatus* (*Birds of Africa*).

Changes in the taxonomic order resulting from a change at family level

The Cisticolidae, previously included in the Sylviidae, have been raised to family level, following Dickinson (2003) and del Hoyo et *al.* (2006).

New English names In a few cases we have replaced the English name of a species used in *BOWA* (2001) by that used in *Birds of Africa*'s last volume or another recent source, when there appeared to be a good reason for doing so. For example, *Circus aeruginosus* (previously Eurasian Marsh Harrier) has become Western Marsh Harrier, as there is an Eastern Marsh Harrier in Asia. *Rhinopomastus aterrimus* (previously Black Wood-hoopoe) has become a scimitarbill, like the other members of the genus.

Name used in *BOWA* (2001)	Name used in this book	References
Short-toed Eagle	Short-toed Snake Eagle	del Hoyo *et al.* (1999)
Eurasian Marsh Harrier	Western Marsh Harrier	Dickinson (2003)
Little Buttonquail	Common Buttonquail	Dickinson (2003)
European Scops Owl	Eurasian Scops Owl	Dickinson (2003)
Black Wood-hoopoe	Black Scimitarbill	Dickinson (2003)
Cliff Chat	Mocking Cliff Chat	Dickinson (2003)
Black-winged Red Bishop	Black-winged Bishop	*Birds of Africa*
White-cheeked Oliveback	Grey-headed Oliveback	*Birds of Africa*
African Quailfinch	Black-faced Quailfinch	*Birds of Africa*
Cut-throat	Cut-throat Finch	*Birds of Africa*
Parasitic Weaver	Cuckoo Finch	*Birds of Africa*
Baka Indigobird	Barka Indigobird	*Birds of Africa*

English names and their spelling A set of unique English-language names for the extant species of the birds of the world has been provided recently by a committee appointed by the International Ornithological Congress (IOC; see worldbirdnames.org). We think the IOC list is best considered as a work in progress. As several names proposed for Afrotropical species differ from those used in western Africa, we have not adopted these.

Regarding the spelling of English bird names, we mostly approve of the IOC rules, as these largely reflect ours (cf. *BOWA*). However, we note one or two inconsistencies in the use of hyphenation and capitalisation of compound names. First, if the second word of a compound name is a type of bird and the taxon is in that bird family, we agree that the name should be spelled with two words, e.g. Night Heron, Ant Thrush, Swamp Warbler, Forest Flycatcher. However, when both nouns are names of birds or bird families the IOC advises to insert a hyphen 'to signify that the taxon belongs to the family of the second word, not the first', e.g. Eagle-Owl (but Scops Owl), Sparrow-Lark (but Bush Lark). Since the rationale given is weak or even invalid – it is indeed hard to believe that anyone would ever make the mistake of thinking that an Eagle Owl is an eagle – we see no reason to deviate from the general rule by inserting a hyphen in these cases. We therefore continue to employ Cuckoo Hawk, Eagle Owl, Robin Chat, Sparrow Lark, etc.

Second, we also accept the rule that if the second word is a kind of bird but the taxon is not a member of the bird family named, the compound name should be spelled as a single word (e.g. Cuckooshrike, previously Cuckoo-shrike, as this is not a shrike), except where it would be hard to pronounce or look odd (e.g. Painted-snipe). Following this rule, we continue to write e.g. Wood-hoopoe (as this is not a hoopoe).

GEOGRAPHY, CLIMATE AND HABITATS

The West African countries of Senegal and The Gambia are located on the Atlantic coast between latitudes 12°20' and 16°30'N and longitudes 11°20' and 17°30'W. The Gambia, mainland Africa's smallest country with a land area of 11,147 km^2, encloses the lower 475 km of the Gambia River and is entirely surrounded by Senegal except for its coastline. It extends for *c.*320 km inland and is mostly only *c.*25–30 km wide. Senegal covers an area of 196,720 km^2, extending inland for *c.*500 km, and is bordered by Mauritania to the north, Mali to the east and Guinea and Guinea-Bissau to the south.

TOPOGRAPHY

Senegal and The Gambia are predominantly flat or gently undulating, lying almost entirely below 100 m. Only in the south-eastern corner of Senegal, which is part of the foothills of the Fouta Djallon massif in Guinea, a few hills attain 300–400 m with the maximum altitude along the border with Guinea reaching *c.*500 m.

The coast is generally low and sandy, with four major estuaries of (from north to south) the rivers Senegal, Sine-Saloum, Gambia and Casamance. A few rocky outcrops occur along the coast, with the highest, at Cap de Naze on the 'Petite Côte', south of Dakar, reaching a maximum height of 74 m. In The Gambia, the highest laterite cliffs, at Fajara and Brufut, are only 8–12 m in height. There are a few rocky offshore islands and reefs, including the Parc National des Îsles de la Madeleine, opposite the Cap Vert peninsula, Africa's westernmost point, on which Dakar lies. In The Gambia, the small, low and unstable offshore Bijol Islands are formed by a reef. The Senegal River, West Africa's second-largest river, flows north out of the Fouta Djallon to form, over a length of more than 1,350 km, the eastern and northern border of Senegal.

CLIMATE

The principal feature of the climate is the alternate wet and dry seasons, which are governed by the movement of the Intertropical Convergence Zone (ITCZ), a zone of low pressure towards which blow winds from the northern and southern high-pressure belts. South-westerly winds from the Atlantic Ocean are warm and moist, while north-easterly winds from the Sahara, known as Harmattan, are hot, dry and dusty. The ITCZ annually oscillates north and south following, with a lag of 1–2 months, the position of the sun. At the northern summer solstice it lies near the Tropic of Cancer and wet maritime winds produce a rainy season north of the equator. When the ITCZ moves south and the sun is over the Tropic of Capricorn, most of West Africa comes under the influence of the continental, hot, dry Harmattan. Overall, the dry season lengthens and rainfall decreases with increasing latitude.

Senegambia experiences a single rainy season, which is longest at the coast in the south-west (May–September/October, with August generally the wettest month; mean annual rainfall *c.*1,500 mm) and shortest in the north and north-east (August–September; mean annual rainfall *c.*300 mm). However, the onset and duration of the dry and rainy seasons and the annual rainfall all vary considerably from year to year.

Temperatures average around 28°C on the coast, but are higher inland. Mean monthly maximum temperatures range from 24–31°C at the coast to 32–41°C from *c.*350 km inland, whilst mean monthly minimum temperatures are 16–23°C at the coast and 12–25°C inland. The hottest months are those just preceding the rains; coldest months usually are December/January.

MAIN HABITATS AND VEGETATION ZONES

The northern part of Senegal, north of *c.*15°15'N, lies within the Sahel biome and is characterised by dry thorn-bush savanna dominated by *Balanites aegyptiaca* and *Acacia* species. The rest of the country and The Gambia lies mainly within the Sudan-Guinea Savanna biome and was originally covered by various types of wooded grassland, with variable tree cover and tree height (both gradually decreasing northward, reflecting the decline in rainfall) and grasses growing under and between the trees. Coastal Casamance and the adjoining area in The Gambia, in the south-west, fall within the Guinea-Congo lowland forest transition zone and patches of semi-evergreen forest still occur, together with a small fragment of the avifauna typical of that habitat.

Throughout both countries, much of the natural vegetation has been extensively modified by man, even in the relatively sparsely populated north. In the Sahel, the once widespread *Acacia nilotica* forests along the lower Senegal River, an important habitat for Palearctic migrants, have largely disappeared due to the combined effects of drought and destruction by man for firewood, charcoal and clearance for agriculture. Further south, vast areas are now devoid of large trees, which have been cut down to make way for food crops and, in the Sudan savanna zone, ground-nuts. In some areas, however, large trees remain, forming a kind of cultivated parkland. Former agricultural plots are overgrown by farmbush and second growth. Domestic stock is numerous and roams widely, resulting in large areas being affected by overgrazing. Fires regularly rage through the savannas in the dry season. Although a natural phenomenon to which taxa of the zone are well adapted, these fires are now almost always deliberately started by man.

In the south-east (including the Parc National du Niokolo-Koba), larger trees and gallery forests of various width occur, with denser patches of woodland forming thickets, occasionally with more or less large stands of bamboo, especially along small temporal streams. Open areas of dry edaphic grassland, called bowals (also spelled 'boval'), occur where the rocky substrate is covered by only a thin layer of soil, allowing the growth of short grass, but no trees.

Mangroves occur on all major estuaries, the most important being those on the northern bank of the Casamance River delta and along the Saloum and Gambia rivers; along the latter they extend up to c.200 km upstream. Extensive mudflats are exposed at low tide. Behind the mangroves are seasonally flooded areas, including saltmarshes, shallow lakes and sparsely vegetated mudflats. Freshwater wetlands, most of them seasonal, occur further inland.

The mainly sandy coastal zone is dominated by dunes, which are often unstable. Between Saint Louis and Dakar a strip of c.150 km long and 5 km wide behind the dunes comprises a string of permanent freshwater lakes and temporarily wet depressions, called 'niayes', which are important for a wide variety of breeding and wintering waterbirds and also raptors.

The hydrological regime of the Senegal River delta and adjacent wetlands has been considerably altered by the construction of dams and dykes in an attempt to increase the area of irrigated cultivation (principally of rice). The natural cycle of an annual flood followed by falling water-levels and the intrusion upriver of salt water from the estuary no longer occurs and has resulted in the drying up of large areas. The Senegal floodplain nevertheless continues to attract huge concentrations of waterbirds, for which it remains of vital importance.

Offshore, pelagic seabirds pass in large numbers during migration; these can best be observed from the Cap Vert peninsula. Inshore, resident and passing or wintering species may find resting and feeding opportunities.

Throughout, various artificial habitats which may attract certain bird species have been created by man, including reservoirs, sewage farms, urban areas with gardens, artificial saltpans, bridges and culverts, and cultivated areas (especially ricefields).

IMPORTANT BIRD AREAS

Senegal has 17 Important Bird Areas (IBAs), The Gambia has 13 (Coulthard 2001, Robinson 2001). They are listed numerically by their IBA number and mapped here. IBAs have been identified against globally agreed criteria; for more information see Fishpool & Evans (2001) and BirdLife International's website. Map numbers below refer to IBA codes listed opposite (e.g. number 1 on Senegal map = SN001 etc).

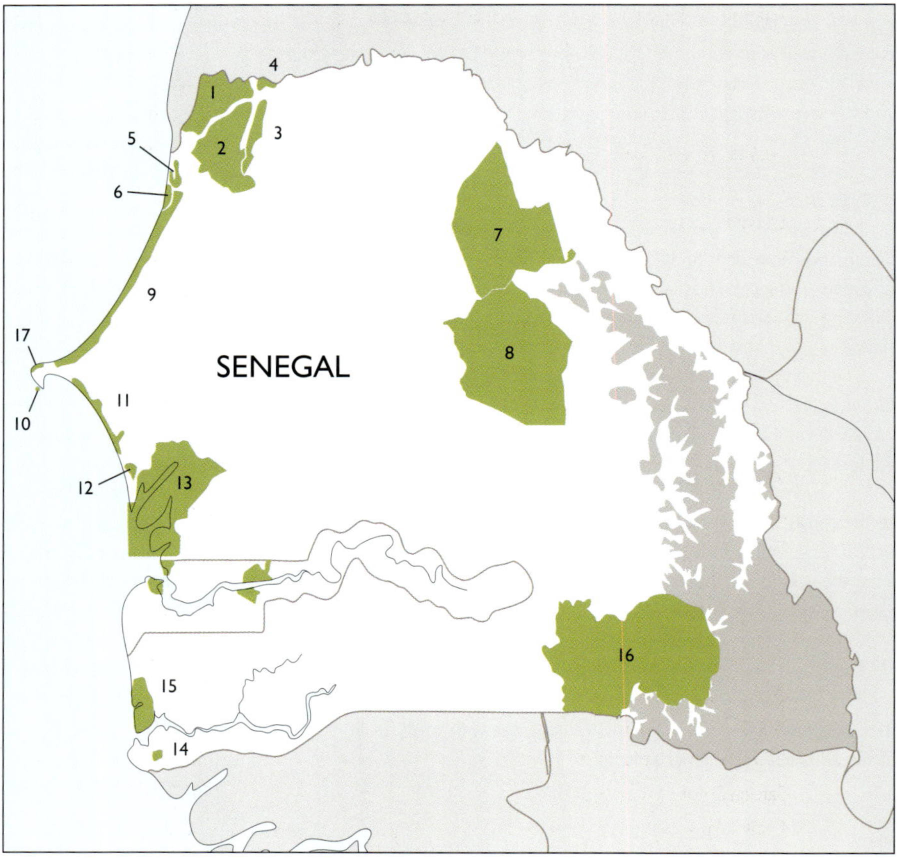

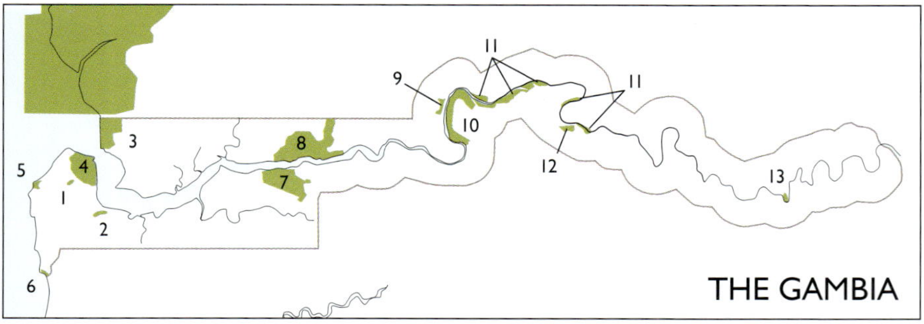

SITE NAME	IBA CODE
Senegal	
Djoudj wetlands	SN001
Ndiaël basin (including the 'Trois Marigots')	SN002
Lac de Guiers	SN003
River Senegal (Ntiagar to Richard-Toll)	SN004
Guembeul Avifaunal Reserve and St Louis lagoons	SN005
Parc National de la Langue de Barbarie	SN006
Ferlo North	SN007
Ferlo South	SN008
Niayes (from Dakar to St Louis)	SN009
Parc National des Iles de la Madeleine	SN010
La Petite Côte	SN011
Joel-Fadiouth	SN012
Delta du Saloum	SN013
Parc National de Basse Casamance	SN014
Kalissaye Avifaunal Reserve	SN015
Parc National du Niokolo-Koba	SN016
Cap Vert	SN017
The Gambia	
Abuko Nature Reserve	GM001
Pirang Forest Park	GM002
Niumi National Park	GM003
Tanbi wetland complex	GM004
Tanji River (Karinti) Bird Reserve	GM005
Allahein to Kartung coast	GM006
Kiang West National Park	GM007
Bao Bolon Wetland Reserve	GM008
Samba Sotor to Kaur wetlands	GM009
Dankunku wetlands	GM010
Islands of Central River Division	GM011
Jakhaly ricefields	GM012
Prufu–Darsilami area	GM013

TAXONOMY: SOME DEFINITIONS

Taxonomy is the science of classification and naming of living organisms. The classification of organisms attempts to reflect relationships between them and works on a hierarchical system by which an organism is placed in categories of decreasing level, from 'kingdom' to 'species' and 'subspecies'. These categories are known as taxa (singular: taxon; hence 'taxonomy'). The main categories useful to the field ornithologist are the following.

Vertebrates are divided into **classes**: mammals, birds, reptiles, amphibians and fish. Birds are vertebrates characterised by the possession of feathers and belong to the class Aves, which is the scientific name for 'birds'.

Class Aves contains approximately 29 large groups called **orders**, with names ending in '-iformes'. Birds of the order Passeriformes (passerines, or songbirds) are placed together because they share certain morphological characters. Other orders include, for example, Ciconiiformes (storks, herons and ibises), Strigiformes (owls) and Piciformes (woodpeckers, honeyguides and barbets).

Orders are divided into c.180 **families**, with names ending in '-idae', following the same logic of shared characters. Examples of families within the order of Passeriformes are: Hirundinidae (swallows and martins), Pycnonotidae (bulbuls), Ploceidae (weavers) and Estrildidae (Estrildid finches).

Within families, birds are clustered into more than 2,000 **genera** (singular: genus). Examples of different genera within the family Estrildidae are: *Estrilda*, *Uraeginthus*, *Pytilia*, *Lagonosticta*, *Spermestes*, etc.

A genus comprises one or several **species**. For example, three species belonging to the genus *Estrilda* occur in Senegambia: *Estrilda caerulescens*, *Estrilda melpoda* and *Estrilda troglodytes*. According to the Biological Species Concept, which is followed here, a species can be defined as a population, or group of populations, of actually or potentially interbreeding individuals, reproductively isolated from all other such populations. Members of a species should be able to interbreed freely and produce fertile offspring.

The scientific name of a species is based on an internationally accepted, binomial system and consists of a two-part, latinised name, which is conventionally written in *italics*. The first part is the generic name (with first letter capitalised), the second the specific name (all lower case).

Finally, a species may be divided into different **subspecies** (also called races). Subspecies are groups of similar-looking individuals, slightly different from other groups, but belonging to the same species. Members of subspecies can or could still interbreed with other members of the same species. A subspecies' name is added as a third part to the species' scientific name, e.g. *Cuculus clamosus gabonensis*. The first population to be described becomes the **nominate subspecies** and carries the same subspecific and specific names, e.g. *Motacilla flava flava*. If there is no doubt as to which species or genus is involved, the name can be shortened to *Motacilla f. flava* or *M. f. flava*.

ORGANISATIONS

INTERNATIONAL

BIRDLIFE INTERNATIONAL

Wellbrook Court, Girton Road, Cambridge CB3 0NA, UK.
Tel. +44 (0)1223 277318 Fax. +44 (0)1223 277200
E-mail: birdlife@birdlife.org Website: www.birdlife.org

BirdLife International is a global partnership of conservation organisations that strives to conserve birds, their habitats and global biodiversity. It is the official Red List authority on birds for the International Union for Conservation of Nature (IUCN). A network of Important Bird Areas (IBAs) and Endemic Bird Areas (EBAs) has been identified with the aim to conserve the species within these areas.

Publications: *World Birdwatch* magazine (quarterly) and, with Cambridge University Press, *Bird Conservation International* (quarterly).

AFRICAN BIRD CLUB (ABC)

c/o BirdLife International, Wellbrook Court, Girton Road, Cambridge CB3 0NA, UK.
E-mail: info@africanbirdclub.org Website: www.africanbirdclub.org

The ABC aims to provide a worldwide focus for African ornithology and encourage an interest in the conservation of the birds of the region. Its website and journal are valuable sources of information on African birds.

Publication: *Bulletin of the African Bird Club* (bi-annual).

REGIONAL

WEST AFRICAN ORNITHOLOGICAL SOCIETY / SOCIÉTÉ D'ORNITHOLOGIE DE L'OUEST AFRICAIN

Website: malimbus.free.fr

The Society aims to promote scientific interest in the birds of West Africa and to further the region's ornithology, principally through publication of its journal.

Publication: *Malimbus* (bi-annual).

NATIONAL

WEST AFRICAN BIRD STUDY ASSOCIATION (WABSA)

c/o Atlantic Hotel, P.M.B 676 Serrekunda, Banjul, The Gambia.
Tel. +220 4228601/+220 4223637
E-mail : mcamara@post3.tele.dk Website: www.gambia.dk/nature/wabsa.html

The association aims to create local awareness of birds and biodiversity and help create, protect and maintain local sanctuaries for wildlife.

BIRD TOPOGRAPHY

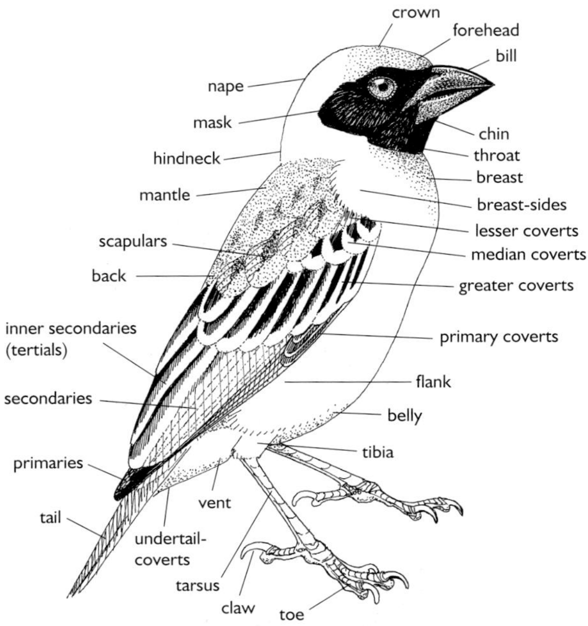

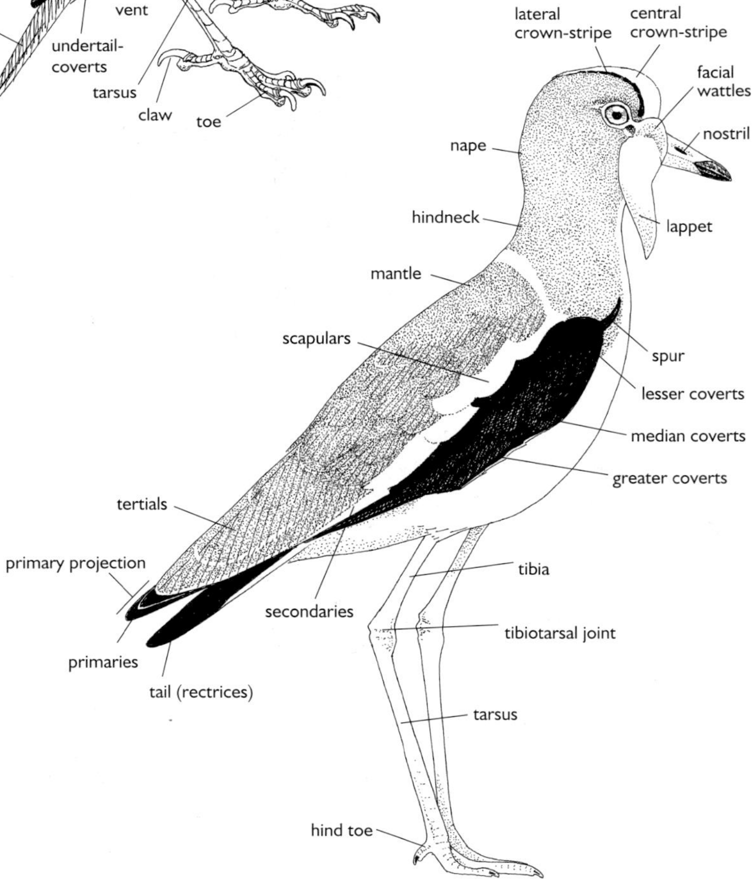

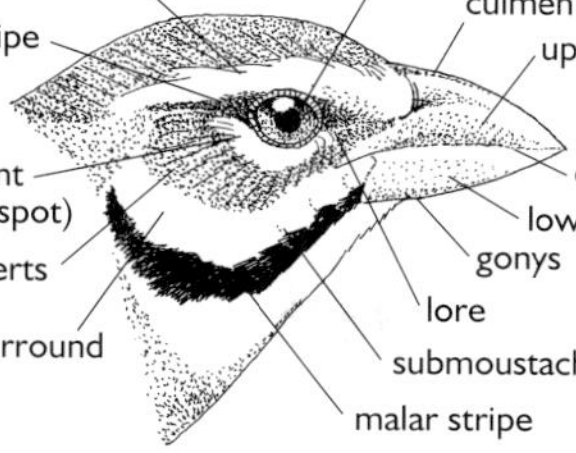
supercilium
eye-stripe
eye-crescent
(subocular spot)
ear-coverts
ear-coverts surround
iris
culmen
upper mandible
cutting edge
lower mandible
gonys
lore
submoustachial stripe
malar stripe

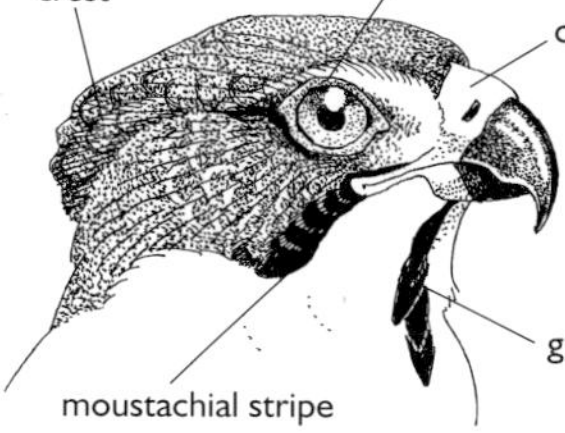
crest
orbital ring (bare skin)
cere
gular stripe
moustachial stripe

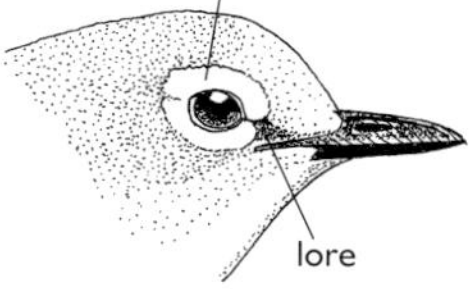
eye-ring (feathering)
lore

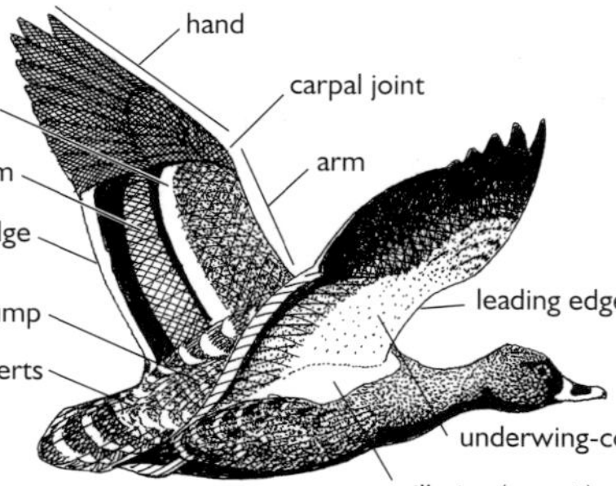
hand
carpal joint
arm
wingbar
speculum
trailing edge
rump
uppertail-coverts
leading edge
underwing-coverts
axillaries (armpit)

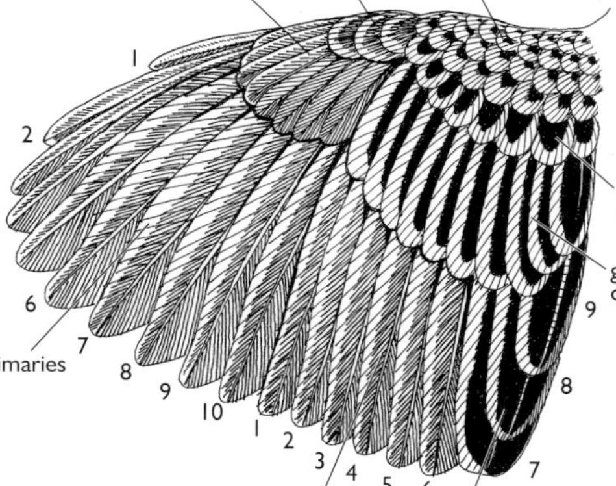
alula
lesser coverts
primary coverts
1
2
6
7
primaries
8
9
10
1
2
3
4
5
6
secondaries
inner secondaries (tertials)
7
8
9
median
coverts
greater
coverts

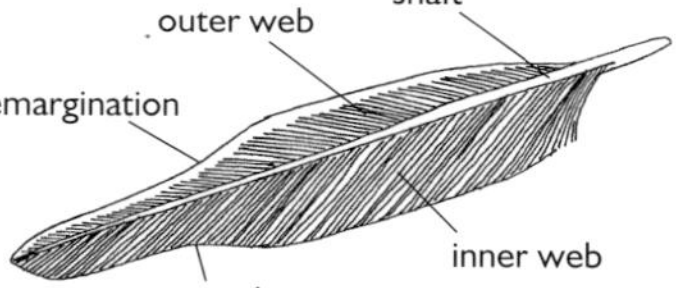
shaft
outer web
emargination
inner web
notch

GLOSSARY

adult A bird in final plumage (plumage no longer changing with age).

aerial Air-frequenting (e.g. an aerial feeder).

Afrotropical Referring to the biogeographical region that comprises sub-Saharan Africa.

allopatric Mutually exclusive geographically. Applied to (taxonomically) closely related forms (populations) whose ranges do not overlap. Opposite of sympatric. See also parapatric.

antiphonal Referring to precisely timed alternating singing or calling by two birds, usually a mated pair.

arm Inner part of wing, between the body and the carpal joint (wing-coverts and secondaries).

axillaries The feathers at the junction of the underwing and the body; the 'armpit' of a bird.

bare parts Those parts not covered by feathers (including bill, cere, eyes, wattles, orbital ring and feet).

bib A contrasting, usually dark area, on the throat and/or upper breast.

biome A major regional ecological community, characterised by a distinctive flora and fauna, e.g. Guinea-Congo Forests and Sudan-Guinea Savanna.

brood parasite A species that lays its eggs in the nest of another (the 'host') and plays no parental role in raising its young, e.g. cuckoos, honeyguides and indigobirds.

call Brief vocalisation, used mainly to maintain contact with conspecifics or to alert to danger. Often consists of a single, simple note (cf. song).

cap A contrasting patch on the top of the head.

carpal The bend of the wing between 'hand' and 'arm', or carpal joint.

carpal bar A contrasting, dark bar on the upperwing, running diagonally from the carpal joint towards the body. Exhibited in certain plumages by some gull species.

carpal joint The joint at the bend of the wing.

carpal patch A contrasting mark or area at or near the carpal joint.

casque An enlargement on the upper surface of the bill, in front of the head, as on hornbills.

cere Bare and often brightly coloured skin at the base of the upper mandible, containing the nostrils.

cheek Loosely applied to the area on the side of the head.

cline A gradual change of a species' morphological characters across its geographical distribution.

collar A band of contrasting colour on the neck.

colony An assemblage of birds within a discrete area to nest (hence colonial species).

congeneric Belonging to the same genus (hence congeners).

conspecific Belonging to the same species.

crepuscular Active at dawn and dusk.

crest A tuft of elongate feathers on the head.

cryptic Colours or markings aiding concealment (as in nightjars).

culmen The ridge of the upper mandible.

deciduous A tree that loses its leaves during set periods of the year, remaining leafless for some time.

dimorphic Having two distinct morphs or forms.

dissonant Not in harmony, harsh-toned.

distal Furthest from centre of body or point of attachment; opposite of proximal.

disyllabic Consisting of two syllables.

diurnal Active during the day / in daylight.

duetting Male and female of a pair singing simultaneously or antiphonally in response to each other.

eclipse plumage A female-like plumage acquired by males of some species (e.g. ducks and some sunbirds) during or following breeding.

edgings/edges In relation to feather patterns, indicates outer feather margins. Edgings can result in distinct paler or darker panels of colour on wings or tail.

emergents Trees whose crowns are conspicuously taller than the surrounding canopy.

endemic Native or confined to a particular area (e.g. turacos are endemic to Africa).

epiphyte A plant that uses another for support, not for nutrients.

extralimital Outside a given area.

eye-crescent A crescent above or below the eye; often occurring both above and below (forming broken eye-ring).

eye-ring A ring of tiny feathers surrounding the orbital ring.

eye-stripe A usually dark stripe that extends back from the bill over the lores and through the eye (post-ocular stripe).

face Anthropomorphic term, encompassing the forehead, lores, supercilium, ear-coverts and upper throat, or any combination of these.

feral A free-flying bird derived from domesticated or captive stock.

filoplume A thin, hair-like feather.

first-winter An immature plumage stage acquired after juvenile plumage and recognised by the presence of retained, and more worn, juvenile feathers in wings and tail (refers to winter season in northern hemisphere and is therefore used for Palearctic migrants).

flight feathers Main wing feathers (primaries and secondaries), but excluding tail feathers.

forest A continuous stand of trees at least 10 m tall, their crowns interlocking.

forest outliers Patches of forest away from main forested areas.

form A general term for distinguishable entities, including species, subspecies, morphs, etc.

fringes In relation to feather tracts, indicates complete feather margins (compare edgings). Contrasting fringes can result in a scaly appearance to body feathers or wing-coverts.

frontal shield An area of bare, fleshy or horny skin on the forehead, usually brightly coloured (e.g. on gallinules)

frugivorous Fruit-eating (hence frugivore, fruit-eater).

gallery forest Forest along rivers.

gape The fleshy interior and corners of the bill.

genus (plural **genera**) A taxonomic category between family and species, representing one or more species with a common ancestor (phylogenetic origin); the genus name forms the first part of the two-part scientific name.

gliding Flight on a direct course without, or between, wingbeats.

gorget A distinctively coloured or streaked band across the throat or upper breast ('necklace').

graduated referring to a tail in which the central rectrices are longest, the others becoming progressively shorter toward the sides.

granivorous Feeding on grains or seeds (hence granivore, seed-eater).

grassland Land covered by grasses and other herbs, either with woody plants, or the latter not covering more than 10% of the ground.

gregarious Commonly assembling in groups.

gular Related to the throat. A gular pouch is a loose area of skin extending from the throat (e.g. on pelicans). A gular stripe is a stripe extending on the centre of the throat (usually narrow and dark).

hand The outer part of the wing, between carpal joint and tip.

herbaceous Of or similar to herbs.

Holarctic Biogeographical region that includes the Palearctic and Nearctic regions, i.e. most of the temperate zone of the northern hemisphere.

hood A contrasting area covering all or most of the head and neck.

hybrid The product of a cross between individuals of unlike genetic constitution; usually used for a cross between individuals of different species.

immature A general term for a non-adult bird.

inner wing Inner part of the wing; also called arm.

insectivorous Insect eating (hence insectivore, insect-eater).

inselberg An isolated hill or mountain rising abruptly from its surroundings.

iridescent The glossy or 'metallic' effect of changing colours caused by reflected light from specially structured feathers, e.g. on glossy starlings and sunbirds.

iris (plural **irides**) The round, coloured membrane surrounding the pupil of the eye.

juvenile A bird in its first feathered, non-downy, plumage.

jizz An overall impression of the appearance a bird in the field based on a combination of characters.

leading edge The front edge of the wing. Generally referred to when it is marked with a contrasting (dark or pale) band.

littoral Situated near a (sea) shore.

local Occurring within a small or restricted area.

mangrove Open or closed stands of salt-adapted evergreen trees or bushes occurring on shores between high- and low-water mark.

mask A dark area of plumage surrounding the eye, usually extending from the base of the bill to the ear-coverts.

melanistic A blackish morph.

midwing panel A contrasting pale area in the middle of the inner wing on some gull species.

migratory Making regular geographic movements.

mirror A subterminal white spot on the wingtip of a gull. Not to be confused with the white primary tips (these, if present, are always visible on a gull at rest; mirrors are usually not visible on the closed wing).

monotypic A biological group having a single representative. E.g. 'monotypic genus' (a genus with a single species), 'monotypic species' (a species with no subspecies). See polytypic.

montane Growing or living in mountainous areas.

morph A normal but distinct plumage variant which is not related to sex, age or season.

morphological Pertaining to form and structure.

mottled Plumage marked with coarse spots or irregular blotches.

moult The process of replacement of old feathers by new. Moult may be complete (all head, body, wing and tail feathers replaced during the same period) or partial (involving the renewal of all or most contour feathers, except those of wings and tail). In smaller species moult is usually repeated during a set period each year; in larger species (e.g. raptors) it may be spread over several years with overlapping moults.

Nearctic Biogeographical region comprising N America south to the tropics.

nocturnal Active at night.

nomadic Referring to a wandering or erratically occurring species with no fixed territory when not breeding.

nominate (**race** or **subspecies**) The first described and named form of a species, typified by its subspecific name being the same as the specific e.g. *Sylvietta virens virens*.

non-passerines All orders of birds except the passerines.

notched Referring to a tail in which the central feathers are slightly shorter than the outer ones, forming a very shallow fork or notch.

nuchal Relating to the hindneck or nape (used with reference to a crest, patch or collar).

orbital ring Ring of bare skin immediately surrounding the eye (not to be confused with eye-ring).

outer wing Outer part of the wing; also called hand.

Palearctic Biogeographical region that includes Europe, N Africa, the Middle East and N Asia south to the Himalayas and Yangtze River.

parapatric Occupying different but contiguous geographical areas. Applied to taxonomically related populations whose ranges are closely contiguous but do not overlap. See also allopatric and sympatric.

partial migration Migration by part of a population.

passage migrant A migrant that occurs regularly but only briefly at a locality during migration to and from its breeding and wintering grounds.

passerines Members of the large Order Passeriformes, often referred to as 'perching birds' or 'songbirds', characterised by perching with three toes pointing forward and one toe back. In Senegambia, includes all species from larks onward.

pectoral tufts Coloured tufts at each side of the breast.

pelagic Of the open sea. A pelagic species spends most of its life at sea, far from land.

pied Patterned black and white.

polyandrous A female animal that mates with more than one male.

polygamous An animal that has two or more mates of the opposite sex at the same time.

polygynous A male animal that mates with more than one female during the breeding season.

polymorphic Having more than two distinct plumage morphs.

polytypic A biological group having more than one representative. E.g. 'polytypic genus' (a genus with two or more species), ' polytypic species' (a species with two or more subspecies). See monotypic.

post-ocular stripe A short, usually pale, stripe, which extends back from the eye. Sometimes reduced to a post-ocular spot.

primary forest Forest in a virgin or undisturbed state.

primary projection The distance that the tips of the primaries project beyond the tertials on a closed wing.

proximal Nearest to centre of body or to point of attachment; opposite of distal.

race Synonymous with subspecies.

rainforest Closed-canopy forest in areas of high rainfall.

range Geographical area in which a taxon is distributed.

raptor Bird of prey (generally refers to diurnal species, not owls).

rectrices (singular **rectrix**) Main tail feathers.

rictal bristles Sparse, though often prominent, bristles at the base of the bill.

remiges (singular **remex**) Flight feathers (primaries and secondaries).

resident A species or population that occurs year-round in the same area and breeds there, even though some individuals may not remain in the same area throughout the year.

riparian Bordering water.

riverine Living or growing on a river bank.

roost A sleeping or resting place.

saddle Generally used to indicate a part of the upperparts contrasting in colour with the rest of the upper surface (e.g. in juvenile marsh terns)

Sahel Semi-arid zone between savanna and desert, characterised by scattered, thorny vegetation.

savanna Habitat dominated by grasses with a varying proportion of trees and shrubs.

secondary forest Forest regenerating after a greater or lesser degree of disturbance, often by selective logging or agriculture. It is characterised by a lack of large trees and a significant proportion of coloniser species.

secondary bar A contrasting dark bar on the secondaries, as in immature plumages of some gull species.

sedentary A species that remains in the same site throughout the year, individuals wandering no more than a few kilometres at the most.

serrated With a fine, saw-like edge (as in the outer edge of the outer primary in saw-wings).

shorebird Another term for wader.

shoulder patch An area of contrastingly coloured wing-coverts.

soaring Circling flight (often in thermal of warm air).

song A more complex pattern of vocalisations, mainly uttered in the breeding season (cf. call).

speculum A usually iridescent panel on the secondaries in dabbling ducks.

stratum (plural **strata**) Level (used to indicate levels within forest, e.g. middle strata = mid-levels).

streamer An exceptionally long, slender tail feather, as in some bee-eaters, paradise flycatchers and sunbirds.

square Referring to a tail in which all feathers are of equal length, forming a straight border.

subadult An imprecise term indicating a bird in nearly adult plumage; often used when precise age is difficult to establish, e.g. in large raptors.

subspecies A population of a given species that differs more or less obviously in appearance from one or more other populations of the same species. Subspecies are assumed to be (at least theoretically) capable of interbreeding but to be relatively isolated from each other, with interbreeding limited to areas of contact. The border between species and subspecies is often arbitrary. The nominate subspecies is the one first named: its scientific name has the second and the third terms identical, e.g. *Tauraco persa persa*.

subterminal Near the end. A subterminal band is a contrasting (dark or pale) band, usually broad, situated near the tip of a feather or feather tract (used particularly in reference to the tail).

supercilium A usually pale stripe, which extends from the base of the upper mandible, above the lores and the eye (extending behind the eye; compare supraloral stripe).

superspecies A group of closely related and largely or entirely allopatric species.

supraloral stripe A short, usually pale, stripe, which extends from base of upper mandible, over lores, to just above or in front of eye (not extending behind the eye; if it does it becomes a supercilium).

sympatric Occurring in the same geographical area. Opposite of allopatric.

taxon (plural **taxa**) A named form (this unit of biological classification can refer to a group of organisms of any taxonomic rank, e.g. family, genus, species, subspecies, etc.)

taxonomy The science of classification and naming of all life forms.

terminal At the end. A terminal band is a contrasting (dark or pale) band, usually broad, situated at the tip of a feather or feather tract (used particularly in reference to the tail).

terrestrial Living or occurring mainly on the ground.

thermal Rising current of heated air (used by e.g. raptors to gain height).

thicket A closed stand of bushes and climbers usually between 3 and 7 m tall.

top of head Includes forehead and crown.

trailing edge The rear edge of the wing, often contrasting with the rest of the wing (cf. leading edge).

understorey The lowest stratum in forest or woodland.

vent The area around the cloaca (anal opening), just behind the legs (not to be confused with the undertail-coverts).

vermiculations Narrow, often wavy, bars that generally create an overall effect and are visible only at close range.

wader General term used, in the plural, for members of the families Rostratulidae (painted-snipes) to Scolopacidae (sandpipers and allies); another term for shorebird.

wattle Naked, fleshy, usually brightly coloured, skin on head (e.g. around eye), base of bill or throat (e.g. as in wattle-eyes, African Wattled Plover and Black-casqued Hornbill).

window A contrasting pale panel in the outer wing of some raptors and gulls.

wingbar Generally a narrow and well-defined dark or pale bar on the upperwing, and often referring to a band formed by pale tips to the greater or median coverts (or both, as in 'double wingbar').

wing feathers Includes flight feathers and wing-coverts.

wing panel A contrasting, usually pale band on the wing; broader and generally more diffuse than a wingbar (often formed by pale edges to the remiges or coverts). At rest, usually refers to a panel on the secondary coverts, secondaries and/or tertials; in flight, usually refers to a panel on the primary bases.

wingspan The distance as measured from wing-tip to wing-tip.

wooded grassland Land covered by grasses and other herbs, with woody plants covering 10–40% of the ground.

woodland An open stand of trees at least 8 m tall with a canopy cover of 40% or more. The field layer is usually dominated by grasses.

zygodactyl Form of foot with the outer and inner toes pointing backward, the middle toes forward, as in turacos, cuckoos, parrots, owls and woodpeckers.

ABBREVIATIONS AND SYMBOLS

♂	male
♀	female
ad	adult
juv	juvenile
imm	immature
br	breeding
non-br	non-breeding
c.	circa (about)
C	central
esp.	especially
incl.	including
N, NE, E, SE, S, SW, W, NW	compass directions (northern, north-eastern, eastern, etc.)
WS	wingspan
Jan, Feb, etc	months (January, February, etc.)
[CD2:67]	CD and track number *African Bird Sounds* (Chappuis 2000)

SHEARWATERS (Plates 1–2) Medium-sized to large pelagic seabirds with long narrow wings and horny-plated, hook-tipped bills with external tubular nostrils ('tubenose'). Fly low over waves on stiff wings, long glides alternating with a few fast flaps. Spend most of their lives at sea and are normally only seen on land at breeding colonies or as storm-driven waifs. Most species breed in burrows or crevices in sea cliffs on oceanic islands. Normally silent away from breeding colonies. Feed mostly on small marine organisms, predominantly squid, fish and crustaceans. Many may gather around fishing boats to retrieve offal. Occurrence in W Africa inadequately known.

Cory's Shearwater *Calonectris diomedea* 45–56 cm; WS 110–125 cm

Rather featureless large shearwater (about size of Lesser Black-backed Gull) with ***ashy-brown upperparts*** and ***unmarked white underparts***. At close range shows diagnostic ***pale yellowish bill*** and variable amount of white in uppertail-coverts. Ashy-brown extends onto head- and neck-sides. Flight appears lazy with several deep wingbeats followed by long, low glide. Wings typically held bowed downwards. ***C. d. diomedea*** (Scopoli's Shearwater) has white wedges on inner webs of otherwise black outer primaries, resulting in narrower black band on outer wing and more angled white underwing panel. ***C. d. borealis*** is rather heavier-looking than nominate, with heftier bill, rather darker head and mantle. Primaries wholly dark, creating solid, well defined, broad black band on outer wing and rounded white underwing panel. **HH** Offshore visitor from Palearctic (both taxa); seasonally fairly common to common off Dakar (mainly Sep–Nov and Mar–Jul). Breeds on Mediterranean (nominate) and Atlantic islands (*borealis*), dispersing south and west in non-breeding season. Separation of races at sea often complicated by birds moulting primaries. **Voice** [CD1:2b, 3a]

Cape Verde Shearwater *Calonectris edwardsii* 40 cm; WS 90–110 cm

As Cory's Shearwater but usually smaller (wings 10–15% shorter) and ***slimmer***; head smaller, more angular; tail relatively longer; head and upperparts slightly darker and greyer brown; bill ***slimmer, grey with dark subterminal band*** (looking black-tipped at distance). Underwing pattern and flight as Cory's. **HH** Visitor off Senegal (common in Oct, e.g. >4,400 Oct 2007). Breeds Cape Verde, presumably dispersing south in non-breeding season. Red List category: Near Threatened. **Voice** [CD1:3b]

Great Shearwater *Puffinus gravis* 43–51 cm; WS 105–122 cm

Large shearwater, with ***distinctive dark cap***, white collar, brown-grey upperparts and white uppertail-coverts. Underwing largely white with ***dark diagonal band on underwing-coverts*** (appearing smudgy at distance). White underparts with diffuse dark patch on belly (often difficult to see) and grey-brown undertail-coverts. Bill dark and relatively fine. At long range, ***gleaming white blaze on neck-sides contrasts strongly with dark cap and upperparts***. Flight strong and powerful, stiff wingbeats interspersed by glides. Wings normally held straight. **HH** Offshore visitor from Southern Ocean; rare off Dakar. **Voice** [CD1:4a]

Northern Fulmar *Fulmarus glacialis* 43–52 cm; WS 101–117 cm

Rather stocky seabird with relatively large head and short neck and tail. Even at great distance easily distinguished from gulls by flight on stiff, straight, relatively narrower and shorter wings. Head and underparts typically white; upperparts pale grey with diffuse white patch at base of primaries; tail paler grey. Wings lack white trailing edge. Small dark spot in front of eye. Flight consists of rapid wingbeats followed by long periods of gliding low over waves. Floats high when swimming. Shearwaters have more slender appearance, with longer, narrower wings, narrower tail and thinner bill, and are darker above, incl. on head. Large gulls have different jizz, with buoyant flight on long, narrow, rather pointed, black-tipped wings; upperwings have white trailing edge. **HH** Palearctic vagrant, Senegal (single record, the first for the Afrotropics: one freshly dead off Dakar, early Sep 1999, ringed as nestling in Ireland).

borealis
diomedea
Cory's Shearwater
Cape Verde Shearwater
Northern Fulmar
Great Shearwater

Sooty Shearwater *Puffinus griseus* 40–50 cm; WS 95–110 cm

Large, ***all-dark*** shearwater ***with diagnostic silvery underwing-coverts*** (esp. immediately behind carpal joint; often difficult to see at distance). Long, slim body; ***long, narrow and pointed wings often swept back***. Flight strong, direct with stiff-winged flaps followed by descending glides. Tends to arc rather higher than other shearwaters, esp. in strong winds. **HH** Offshore visitor from Southern Ocean. Possible at any season, though most likely Sep–Jan (one of the commonest shearwaters off Dakar, e.g. c.1,000–9,700 in Oct, 1995–2008). Red List category: Near Threatened. **Voice** [CD1:6a]

Bulwer's Petrel *Bulweria bulwerii* 25–29 cm; WS 67–73 cm

Medium-sized, long-winged, ***all-dark*** petrel with pale diagonal band on upperwing. ***Long, narrow, graduated tail*** (usually held closed and appearing pointed) diagnostic. Flight buoyant and erratic, circling low over waves. Sooty Shearwater is much larger, with different flight action and square-ended tail. **HH** Breeds on Cape Verde and other Atlantic islands. Disperses south in non-breeding season (Nov onwards). Scarce offshore Mauritania–Gulf of Guinea. **Voice** [CD1:2a]

Manx Shearwater *Puffinus puffinus* 30–35 cm; WS 71–83 cm

Medium-sized shearwater, ***black above and white below***. Black of cap and upperparts extends to below level of eye and onto breast-sides. Wings rather straight and not angled backwards from carpal joint. Typical flight a series of rapid stiff-winged strokes followed by shearing over waves, banking from side to side, alternately showing black upperside and white underside. **HH** Offshore visitor from Palearctic; regular but rather scarce off Dakar (Aug–Oct). Transequatorial migrant, wintering mainly off east coast of S America; also S Africa. **Voice** [CD1:4b]

Balearic Shearwater *Puffinus mauretanicus* 30–40 cm; WS 76–93 cm

Resembles Manx Shearwater, but slightly larger and longer-bodied and lacks striking contrast between upper- and underside, having brownish (not black) upperparts merging into dirty greyish-buff (not pure white) underparts and grey-brown vent, flanks and axillaries. Dark rear body can show conspicuously when banking. Plumage variable, darkest birds may resemble Sooty Shearwater (but latter has more uniform head and underparts). Majority of palest birds have brown 'shawl' across upper breast and more uniform-looking head, nape and upperparts. More typical, darker birds have more clearly defined chocolate-brown cap (encompassing crown, lores and ear-coverts) which contrasts with paler nape. **HH** Vagrant from W Mediterranean. Recorded off Mauritania and, rarely, Senegal. Red List category: Critically Endangered. **Voice** [CD1:5a]

Macaronesian Shearwater *Puffinus baroli* 25–30 cm; WS 58–67 cm

Recalls compact, miniature Manx Shearwater with smaller bill. Underwing margins and tips of wings have less black. Stiff-winged flight usually low, consisting of several rapid wingbeats followed by brief glide. ***P. b. baroli*** Black of cap does not usually extend to level of eye, giving bare-faced appearance; undertail-coverts white. Breeds Atlantic islands. ***P. b. boydi*** has more dark on head-sides and dark undertail-coverts. Breeds Cape Verde. **HH** Both taxa probably scarce but regular offshore to at least 9°S (e.g. 0–56 in Oct off Dakar, 1995–2008; occasionally to mouth of Gambia River). **Voice** [CD1:5b]. **Note** Previously included in Little Shearwater *P. assimilis* but the three forms breeding in tropical and subtropical parts of Atlantic Ocean (these and *lherminieri*) appear not to be closely related to the other forms of the complex.

Sooty Shearwater
Bulwer's Petrel
Balearic Shearwater
Manx Shearwater
baroli
Macaronesian Shearwater
baroli
boydi

STORM-PETRELS *Small, swallow-like pelagic seabirds. Largely dark with white rump. Solitary or gregarious at sea. Identification often difficult; flight action is an important clue. Feed mostly on small crustaceans, squid, fish and other animal matter. Breed colonially in burrows or crevices on oceanic islands. Silent away from breeding colonies. Occurrence in W Africa inadequately known.*

European Storm-petrel *Hydrobates pelagicus* 15–16 cm; WS 37–41 cm

Smallest and darkest storm-petrel in West Africa. All black except contrasting square white rump and diagnostic, but variable, ***white band on underwing-coverts***. Feet do not project beyond square tail in flight. Narrow, pale diagonal bar sometimes visible on upperwing, esp. in fresh plumage. Flight weak and fluttering, bat-like, with almost continuous wing strokes interspersed by short glides. Often patters surface with wings raised. **HH** Fairly common to uncommon offshore visitor from Palearctic (recorded year-round). Regular in small numbers off Dakar, Sep–Nov and Feb–May; occasionally to mouth of Gambia River. Probably offshore throughout the year. **Voice** [CD1:7b]

Wilson's Storm-petrel *Oceanites oceanicus* 15–19 cm, WS 38–42 cm

O. o. oceanicus/exasperatus. Small, with ***pale diagonal upperwing bar and U-shaped white rump which wraps round to sides of undertail-coverts***. Short, rounded wings lack obvious bend at carpal joint. Legs project beyond ***square tail*** in flight (sometimes creating illusion of forked tail). Underwing mostly dark, sometimes with faint greyish band on underwing-coverts. Flight direct and purposeful with rapid wingbeats and short glides, often recalling a swallow. 'Walks' on water when feeding, pattering over surface with high-raised wings. Wings typically held rather straight when gliding. **HH** Fairly common to common offshore visitor from Southern Ocean (mainly Apr–Oct). Most abundant storm-petrel off Dakar in Sep–Oct; occasionally to mouth of Gambia River. Probably regular off entire W African seaboard throughout the year. **Voice** [CD1:6b]

Leach's Storm-petrel *Oceanodroma leucorhoa* 19–21 cm; WS 43–48 cm

O. l. leucorhoa. Relatively large storm-petrel, with long angular wings showing obvious bend at carpal joint and ***rather long, forked tail*** (fork visible from certain angles only). ***White rump V-shaped***, less conspicuous and with much less lateral extension than other storm-petrels; ***usually has diagnostic dark central division*** which, though hard to see, gives rump a smudgy appearance. Upperwing has ***prominent dirty grey diagonal band*** across coverts. Underwing all dark. Feet do not project beyond tail in flight. Flight buoyant and bounding with deep tern-like wingbeats followed by short shearwater-like glides and characteristically sudden changes of speed and direction. Unlike other storm-petrels does not habitually foot-patter. Wings held angled when gliding. **HH** Fairly common offshore visitor from Palearctic (recorded year-round). Regular in small numbers off Dakar in autumn and May–Jun; occasionally to mouth of Gambia River. Probably regular off entire W African seaboard. **Voice** [CD1:8b]

Madeiran Storm-petrel *Oceanodroma castro* 19–21 cm; WS 43–46 cm

Relatively large, with white rump (extending to lateral undertail-coverts) and slightly forked tail (often appearing square). Wings rather angular with obvious bend at carpal joint and pale diagonal bar on upperwing-coverts; underwing all dark. Feet do not project beyond tail in flight. Flight typically buoyant and zigzagging, with deep wingbeats and low shearwater-like glides with wings held horizontally or bowed downwards. Often patters surface with wings held horizontally; frequently sits on water. Difficult to separate from Wilson's and Leach's Storm-petrels. Wilson's has long, projecting legs and more noticeable wingbar; Leach's has rather similar jizz, but less extensive white rump and more forked tail. **HH** Breeds on Cape Verde and other Atlantic islands. Recorded in small numbers off Dakar in autumn. Probably regular offshore. **Voice** [CD1:8a]

White-faced Storm-petrel *Pelagodroma marina* 19–21 cm; WS 41–44 cm

P. m. eadesi. Distinctive. Rather large, greyish storm-petrel with ***white underparts*** and white face with dark patch ('mask') on ear-coverts. Wings broad in centre. Rump grey. Very long legs project beyond slightly forked tail. Sails low over sea on stiff wings. Flight erratic, with legs dangling and pendulum-like swinging from side to side, giving impression of bouncing off wave tops. **HH** Breeds on Cape Verde and other Atlantic islands. Scarcely recorded off Dakar. Probably regular offshore Mauritania to at least Liberia. **Voice** [CD1:7a]

European
Storm-petrel
Wilson's
Storm-petrel
Leach's
Storm-petrel
Madeiran
Storm-petrel
White-faced
Storm-petrel

TROPICBIRDS *Medium-sized, graceful, highly aerial seabirds. Adults largely white; tail wedge-shaped with two greatly elongated central feathers. Sexes similar. Young heavily barred black on upperparts, lacking long tail streamers. Buoyant, rather tern-like, purposeful flight, fluttering wingbeats alternated with long glides. Feed by hovering, then plunging vertically. Breed colonially on islands.*

Red-billed Tropicbird *Phaethon aethereus* 45–50 cm; WS 100–115 cm

Adult Mainly white with ***red bill***, black streak through eye and long white tail streamers (46–56 cm). Upperparts narrowly barred black (sometimes difficult to see in strong sunlight); outer primaries and their coverts black, forming diagnostic band on upperwing. **Juvenile/immature** Closely barred upperparts appear grey at distance; inner secondaries and upperwing-coverts barred; black nuchal collar; black-tipped yellowish bill. **HH** Small population breeds on islets off Senegal (Îles de la Madeleine). Offshore vagrant, Gambia. **Voice** Shrill *keek* or *karreek*. Also loud, piercing calls, only likely to be heard at breeding colonies. [CD1:12]

FRIGATEBIRDS *Large seabirds with extremely long, angular wings and very long, deeply forked tails. Bill long and hooked. Very buoyant flight; soar effortlessly for long periods; never settle on sea. Forage in flight by swooping down on prey or by harassing other seabirds, forcing them to disgorge their catch. Readily take offal. Plumages variable, differing according to age (period needed to acquire adult plumage probably 4–6 years) and sex.*

Magnificent Frigatebird *Fregata magnificens* 90–110 cm; WS 215–245 cm

Unmistakable. **Adult male** Entirely black; red gular pouch (inflated in display, but unlikely to be seen in a vagrant). **Adult female** All-dark underwings; white breast. **Immature** Head, breast and belly white; all-dark underwing; broken breast-band. Seven transitional immature plumages described. **HH** Breeds Cape Verde (rare and declining). Offshore vagrant, Gambia (Mar 1965, Oct 1980). A frigatebird photographed in coastal Gambia in 2005 was either this species or extralimital Ascension Frigatebird *F. aquila*. **Voice** Silent at sea. [CD1:20]

GANNETS AND BOOBIES *Large, conspicuous seabirds with stout, tapering bills, cigar-shaped bodies, long, narrow wings and wedge-shaped tails. Flight steady and purposeful with shallow flaps followed by a glide. Flocks often fly in single file. Feed principally by plunge-diving from relatively great height. Breed colonially. In W Africa only likely to be encountered offshore, unless sick or storm-driven.*

Northern Gannet *Sula bassana* 85–100 cm; WS 170–192 cm

Largest seabird likely to be encountered (only vagrant frigatebirds are larger). **Adult** Predominately white with creamy-yellow head and neck, and black wingtips. **Juvenile** Entirely dark brown with small white spots on upperparts, and a whitish, V-shaped rump patch. **Immature** Becomes progressively whiter, producing confusing patterns; much individual variation. Adult plumage acquired over 4 years. **Second calender-year** Head and underparts largely white; above mostly dark, with white on uppertail-coverts and usually on lesser upperwing-coverts. Crown often dark; remnants of juvenile plumage form partial dark collar. **Third calender-year** Head and underparts as adult; flight and tail feathers mainly black; upperparts and wing-coverts mottled black and white. **Fourth calender-year** As adult, but central tail feathers black and secondaries mixed black and white. (Potential vagrant Cape Gannet *S. capensis* similar, but has black secondaries and tail in all plumages.) **HH** Fairly common to scarce non-breeding visitor to at least 10°N (Sep–Jul). **Voice** [CD1:14]

Brown Booby *Sula leucogaster* 65–75 cm; WS 135–150 cm

S. l. leucogaster. **Adult** Distinctive pattern of uniformly dark brown upperparts, head, throat and upper breast contrasting strongly with white underparts, incl. underwing-coverts. Bill yellowish. **Juvenile/immature** As adult but white of underparts sullied dusky-brown. Adult plumage acquired over 2–3 years. Juvenile Northern Gannet has white crescentic mark on rump; also larger, less agile, with slower flight and relatively shorter tail. **HH** Breeds on Cape Verde, Alcatraz (Guinea) and Gulf of Guinea Is. Rare wanderer along Senegambian coast. **Voice** [CD5:2b]

juv
ad
Red-billed Tropicbird
ad ♂
Magnificent Frigatebird
ad ♂
ad ♀
imm
ad
ad
Northern Gannet
Brown Booby
juv
2nd-year
juv
ad
ad

GREBES *Small waterbirds, adapted for diving and swimming under water. Sexes similar. Feed on fish and invertebrates. Build floating nest. Chicks have stripes on head-sides and neck.*

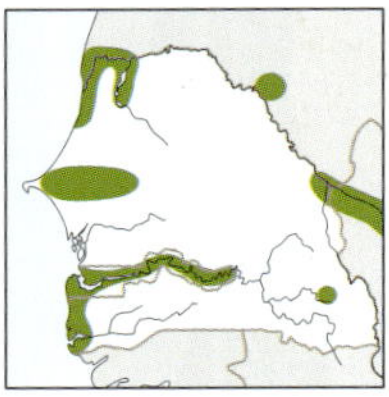

Little Grebe *Tachybaptus ruficollis* — 25–29 cm

T. r. capensis. Small, compact, mainly dark grebe, with rounded head, short neck, stubby bill and fluffy rear end. **Adult breeding** Blackish-brown above with ***bright chestnut cheeks and foreneck*** and ***conspicuous yellowish-white gape patch***. **Adult non-breeding** Duller and paler, with contrasting dark cap and reduced, more whitish gape patch. **Juvenile** Head-sides with irregular white stripes. **HH** Locally common. Singly or in pairs, occasionally in groups, on various aquatic, preferably well-vegetated, habitats. Swims buoyantly and dives frequently. Subject to local movements to utilise seasonal waterbodies. Presence of Palearctic migrants (nominate race: somewhat smaller, with slightly less white in secondaries; indistinguishable in field) suspected, N Senegal. **Voice** Loud, descending, whinnying trill *bi-i-i-i-i-i...* uttered frequently and year-round, often in duet. [CD1:10a]

Black-necked Grebe *Podiceps nigricollis* — 28–34 cm

P. n. nigricollis. Medium-sized grebe with short, slender, slightly upturned and sharply pointed bill, conspicuous red eyes and high, steep forehead with forecrown peak. In flight has white secondaries. **Adult non-breeding** Black cap diffusely extending to below eye. Ear-coverts and neck often dusky. **Adult breeding** Black neck, loose golden ear-tufts and chestnut flanks. **Juvenile** As adult non-breeding but browner overall. **HH** Palearctic migrant/vagrant to aquatic habitats. Has become more regular visitor to N Senegal (Oct–May) since 1990. Single record, coastal Gambia (Kotu Sewage Farm, Nov 1989). **Voice** [CD1:11]

Great Crested Grebe *Podiceps cristatus* — 46–51 cm; WS 59–73 cm

P. c. cristatus. Large, slim grebe with long, slender neck and long, pink, sharp-pointed bill. In flight has much white on forewing, scapulars and secondaries. **Adult non-breeding** Largely dark above and white below; triangular head with blackish on crown not reaching to level of eye. **Adult breeding** Unmistakable, with double-horned crest and chestnut and black tippets. **HH** Palearctic vagrant, Senegal (Jan 1971, Sep 1980). Could occur in any aquatic habitat, incl. sea. **Voice** [CD1:10b]

PELICANS *Huge waterbirds with characteristic massive bill and very large, distensible gular pouch. Sexes similar. Flight strong, with heavy wingbeats alternated by glides; head drawn back. Often soar. Float high on surface when swimming. Often flying in formation and foraging in groups. Feed almost exclusively on fish.*

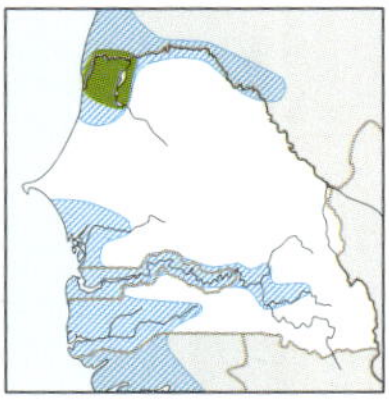

Great White Pelican *Pelecanus onocrotalus* — 140–175 cm; WS 245–310 cm

Adult non-breeding Mainly white with diagnostic ***black flight feathers***. Bill greyish with pink cutting edges, pouch yellow. Bare facial skin pink. **Adult breeding** Acquires variable pinkish tinge to plumage and small ragged crest on rear of crown; bare-part colours intensify. **Juvenile** Dull greyish-brown above, whitish below; flight feathers dark brown. Bill, pouch and legs greyish. **Immature** Paler than juvenile, whitening progressively with age, flight feathers darkening, bare parts becoming more yellow. In flight, note underwing pattern (brown on coverts progressively lost with age); contrasting white rump. Full adult plumage probably attained in 3rd or 4th year. **HH** Widespread and locally common resident or intra-African migrant in large aquatic habitats. Breeds in Senegal. Roosts and nests on ground. Often fishes in coordinated groups. Flies in V-shaped or curved skeins, often at great height. Movements in W Africa unclear. **Voice** [CD1:18]

Pink-backed Pelican *Pelecanus rufescens* — 125–150 cm; WS 215–250 cm

Adult Smaller and duller than Great White Pelican, with pale grey cast to plumage and, in flight, less contrasting wing pattern. Bill and pouch pale yellowish. When breeding, acquires short grey crest on nape, small blackish patch in front of eye, and pinkish tinge to upperparts; bare-part colours intensify. **Juvenile** Brownish above with grey bill, greenish-yellow pouch and greyish-pink legs. **Immature** Becomes progressively paler. Full adult plumage probably attained in 3rd or 4th year. In flight, underwing pattern as adult; contrasting white rump. **HH** Widespread and locally common in large aquatic habitats. Normally fishes singly. Usually roosts and nests in trees, sometimes far from water, occasionally on ground. Principally resident; local movements recorded. **Voice** [CD5:4b]

ad non-br
juv
Little Grebe
ad br
ad non-br
ad br
Black-necked Grebe
Great Crested Grebe
ad br
ad non-br

Great White Pelican
ad
imm
ad
imm
ad
imm
imm
ad
Pink-backed Pelican

CORMORANTS *Medium-sized to large waterbirds with dark plumage, strong bills hooked at tip, long necks and bodies, and long, wedge-shaped tails. Catch fish underwater principally by surface-diving. Stand upright, wings often held open to dry. Fly with outstretched neck. Flight strong and direct with regular wingbeats. Take-off from water requires run across surface, beating it with both feet at once.*

Long-tailed Cormorant *Phalacrocorax africanus* 51–56 cm; WS 85 cm

P. a. africanus. Small cormorant with short bill, short neck and long tail. **Adult breeding** Blackish with green gloss and short crest on forehead. Bill and bare facial skin yellow, becoming orangey or red at start of breeding season. Eye red. **Adult non-breeding** Duller, without crest. Some brownish like immature but with red eye. **Immature** Mainly dull brownish above, pale brownish to buffish-white below. **HH** Widespread and locally common resident in various aquatic habitats. Often favours smaller rivers and pools than Great Cormorant; also less gregarious. Swims low in water, with neck erect and bill held up at angle; often only head and neck above surface. Subject to local movements related to water levels. **Voice** [CD5:3b]

Great Cormorant *Phalacrocorax carbo* 80–100 cm; WS 125–150 cm

Large cormorant with long, thick neck. **Adult *P. (c.) lucidus* breeding** (White-breasted Cormorant) White from throat to breast; white patch on thigh. Bare skin at base of lower mandible yellow. Non-breeding adult duller, without thigh patch. ***P. c. maroccanus*** has white confined to throat and upper neck. **Immature** Brownish above, whitish below. **HH** *P. (c.) lucidus* is a locally common resident in aquatic habitats; *P. c. maroccanus* a Palearctic vagrant, N Senegal. Gregarious. Roosts in groups; breeds colonially. Usually feeds singly; flocks may form where prey abundant. **Voice** [CD1:15]

DARTERS *Large aquatic birds superficially resembling cormorants but slimmer, with long, slender bills without hook. Found mainly on fresh water. Feed singly, spearing fish with bill. Breed colonially in trees.*

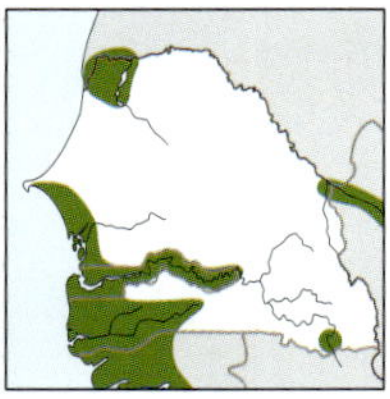

African Darter *Anhinga rufa* *c.*80 cm; WS 120 cm

Slim, with ***long pointed, dagger-shaped bill*** and ***slender neck with distinct kink*** (obvious in flight and when perched). In flight, long graduated tail often held in fan shape. **Adult male** Mainly black with chestnut foreneck, white stripe below eye extending onto neck-sides, and buffish white streaks on wing-coverts and scapulars. **Adult female** Crown and hindneck brown (not black), rest of neck buffish-brown, white neck-stripe less distinct. **Immature** As female, but even browner and paler, without white neck-stripe. Adult plumage acquired over *c.*2 years. **HH** Locally common but declining resident in various aquatic habitats; avoids sea, but frequents coastal lagoons. Swims with only head and neck above water. Perches in trees and on rocks, sometimes on shore, often with wings spread like cormorant. May soar on thermals, often at great height. Irregular, local movements occur, presumably in response to fluctuating water conditions. **Voice** [CD5:4a]

FINFOOT *Unobtrusive grebe-like waterbird with long neck and small head. Inhabits quiet forested rivers, streams, lakes and lagoons with overhanging vegetation.*

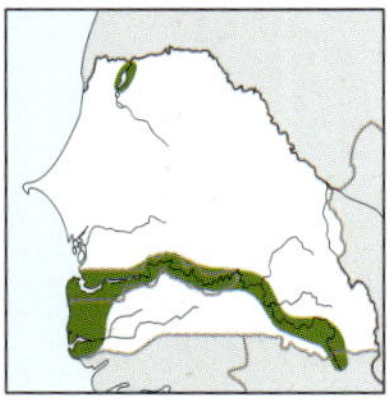

African Finfoot *Podica senegalensis* 50 cm

P. s. senegalensis. Distinctive waterbird with ***darkish upperparts, long stiff tail, red bill and bright orange-red legs and feet***. **Adult male breeding** Forehead to hindneck black with green gloss, separated from slate-grey throat and foreneck by whitish line from eye down neck-side. Upperparts dark brown glossed green, finely spotted white; lower breast and belly white; flanks barred dark brown and white. **Adult male non-breeding** Throat and foreneck white. **Adult female** Smaller, forehead to hindcrown brown, throat and foreneck white. **Juvenile** As female but paler, much browner with few white spots; lower neck, breast and flanks tawny buff. Bill dark. **HH** Locally scarce to frequent resident of well-vegetated, shaded forest and woodland streams and mangroves. Swims low in water, usually near water's edge. Pumps head and neck back and forth with each leg-stroke. Occasionally creeps with remarkable agility through undergrowth of muddy banks. Movements related to changing water levels recorded. **Voice** Calls include dry cackles and guttural notes. Usually silent. [CD6:13]

juv/imm
ad br
ad
ad non-br
Long-tailed Cormorant
ad
juv
ad br
lucidus
ad br
maroccanus
Great Cormorant
juv
ad ♀
ad ♂
ad ♂
African Darter
ad ♂
African Finfoot
ad ♀
juv

BITTERNS, HERONS AND EGRETS (Plates 7–10) *Slender, medium-sized to large wading birds with long necks and legs and long, straight, pointed bills. Flight strong with regular wingbeats and neck retracted, forming an S (unlike storks, spoonbills, ibises and cranes). Sexes similar. Adult plumage attained during 2nd, 3rd or 4th calendar year. Most species usually found in or near water. Feed on a wide variety of prey, most taking fish and other aquatic animals.*

Black-crowned Night Heron *Nycticorax nycticorax* 56–65 cm; WS 90–100 cm

N. n. nycticorax. Medium-sized, stocky heron with relatively short neck and legs. **Adult** ***Crown, mantle and scapulars black*** contrasting with ***grey wings, rump and tail***. Head-sides and underparts white. Two long white plumes on hindneck (absent in non-breeding). Eye crimson. Legs and feet pale yellow, becoming red during courtship. Feet project clearly beyond tail in flight. **Juvenile** Dark brown streaked buff with conspicuous buff spots on wings. Adult plumage takes 3 years to acquire, the white spotting on upperparts and dark streaking on underparts becoming progressively less marked with advancing age. Black crown acquired during 2nd year. In flight, note absence of white on back and distinct feet projection. **HH** Frequent in various aquatic habitats. Gregarious and largely nocturnal. Mostly roosts by day in trees or dense marshy vegetation, leaving roost at dusk. Occasionally seen feeding in daylight. Breeds in colonies in trees or bushes, usually in company of other herons, cormorants or darters. Migrants from W Palearctic occur alongside residents during northern winter (some, especially immatures, year-round); high numbers in Senegal delta. **Voice** Distinctive, low, harsh *kwok* in flight. [CD1:23]

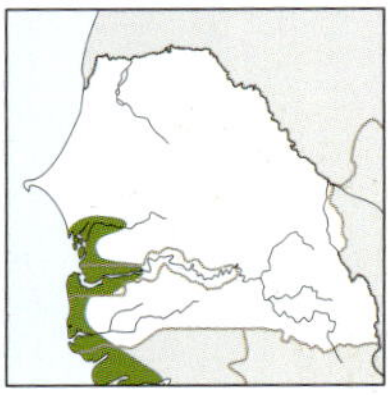

White-backed Night Heron *Gorsachius leuconotus* 50–55 cm

Adult Medium-sized, stocky heron with black head and ***huge eye surrounded by white patch***. Upperparts blackish-brown with white plume-like scapulars forming diagnostic but rather inconspicuous ***triangular patch on back***. Short nuchal crest. Throat white, neck and breast rufous, belly white with black streaking. In flight, note inconspicuous white patch on back; feet project slightly beyond tail. **Juvenile** Darker than juvenile Black-crowned with fewer wing spots and blackish forehead; eye huge. In flight, note presence of white scapulars and slight feet projection. **HH** Uncommon and local resident along well-vegetated streams, ponds and creeks. Largely nocturnal and secretive. Roosts in dense foliage or swampy vegetation. Easily overlooked; seldom seen unless flushed, or when active at dusk or at night. **Voice** Usually silent. Croaking alarm note. [CD5:6a]

White-crested Tiger Heron *Tigriornis leucolopha* 66–80 cm

Adult Cryptically patterned, barred black and buffish-brown heron with diagnostic white crest and long, dagger-shaped bill. When not raised, crest is half-concealed by nape feathers and often difficult to see. Crown may appear black. Appears all-dark in flight; primaries tipped white. **Juvenile** More strongly barred than adult. Juvenile Black-crowned and White-backed Night Herons are smaller, streaked (never barred), and have white spots on wing-coverts. **HH** Rare resident of shaded forest streams; also mangroves. Secretive and largely nocturnal, hence seldom seen. 'Freezes' in bittern-like posture when alarmed. **Voice** Far-carrying, single or double, low moaning note, regularly repeated in slow tempo, usually for brief periods before dawn and after sunset. [CD5:5b]

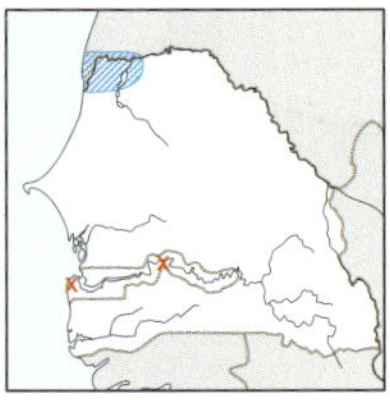

Great Bittern *Botaurus stellaris* 70–80 cm; WS 125–135 cm

Medium-sized, stocky and thick-necked with brown, cryptically patterned plumage, ***black crown and black malar stripe***. Heavy, dagger-shaped yellowish bill. Flight owl-like with deep wingbeats on broad, rounded wings strongly bowed downwards. Immature has crown and malar stripe browner. Heavier looking than White-crested Tiger Bittern, is streaked (not barred), lacks white crest, and occurs in quite different habitat. **HH** Palearctic vagrant, but may be more regular than records suggest. Frequents dense reedbeds, marshes, swamps, mangroves. Active by day and night. 'Freezes' in upright posture, bill 'sky-pointing' when threatened. Very secretive. **Voice** Silent outside breeding season. [CD1:21]

ad
juv
juv
ad
Black-crowned Night Heron
White-backed Night Heron
White-crested Tiger Heron
ad
juv
ad
Great Bittern

PLATE 8: SMALL BITTERNS AND HERONS

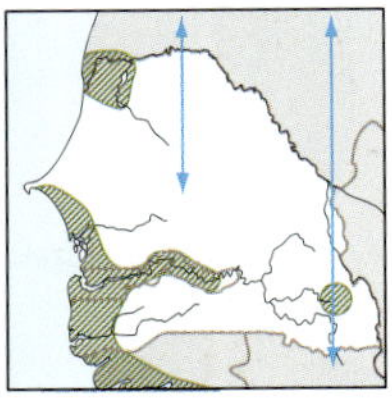

Little Bittern *Ixobrychus minutus* 27–38 cm; WS 49–58 cm

Very small, thick-necked heron with ***contrasting pale wing panel*** (especially striking in flight). **Adult male** Crown and back black, head-sides and neck rich chestnut (resident race ***payesii***) or pale buffish (Palearctic race ***minutus***). **Adult female** Streaked brownish, with duller black crown, nape and tail, and browner, less contrasting upperwing-coverts. **Juvenile** More heavily streaked than female. Following post-juvenile moult resembles adult but has darker, mottled and spotted wing panel. **HH** Uncommon to locally fairly common in reedbeds, marshes, areas of rank vegetation near water, occasionally mangroves. Secretive and shy. 'Freezes' in erect posture with bill 'sky-pointing' when threatened. Distribution of races inadequately known; nominate dominates in N Senegal. **Voice** Usually silent. Hard *ker-ek* and sharp, rapid *kekekekek*. Breeding male utters deep croaking *hogh* or *woof* at c.2-second intervals. [CD1:22]

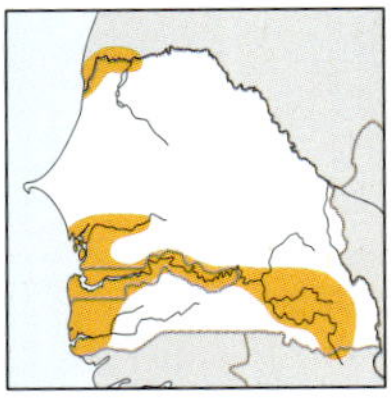

Dwarf Bittern *Ixobrychus sturmii* 25–30 cm; WS 50 cm

Adult ***Tiny heron*** with uniformly ***slate-grey upperparts*** (incl. face) and ***buffish underparts (darkest on belly) heavily streaked black***. Prominent black streak from base of bill to centre of belly. Lores and orbital ring bluish to yellowish-green. Bill blackish with variable amount of yellow at base and below. Flight slow and heavy with legs dangling. Confusion only likely with Green-backed Heron, esp. in flight, but latter has contrasting dark crown and unstreaked, pale grey neck, and is larger. **Juvenile** Upperparts duller, feathers edged buff. **HH** Locally frequent to rare at vegetated streams and ponds, marshes, reedbeds, temporary pools, seasonally inundated areas, mangroves. 'Freezes' in typical bittern stance when alarmed. Wet season migrant; perhaps resident in some areas. **Voice** Occasionally *kehk!* or a loud croak when disturbed. [CD5:5a]

Green-backed Heron *Butorides striata* 40–48 cm; WS 60–70 cm

B. s. atricapilla. **Adult** Small, stocky, short-necked heron with erectile nuchal crest, dark crown and upperparts, grey head-sides and underparts, and white throat with buff line reaching belly. Greenish tinge to upperparts only visible at close range or in good light. In flight appears largely dark with conspicuous yellow legs (reddish-orange when breeding), dangling on take-off. **Juvenile** Mainly brown above with white or buff spots on wing-coverts; head and underparts prominently streaked; distinct malar stripe. Dwarf Bittern is smaller and has heavily streaked underparts. Beware of confusing flying bird with Giant Kingfisher, which has similar size, upperpart coloration and call. **HH** Common resident in heavily vegetated margins of rivers, streams, lakes and pools, mangroves, occasionally floodplains. In tidal habitats often far from cover at low tide. Usually adopts hunched stance when feeding, with body almost horizontal and legs bent. **Voice** Loud, explosive *kyah* when flushed. [CD5:7a]

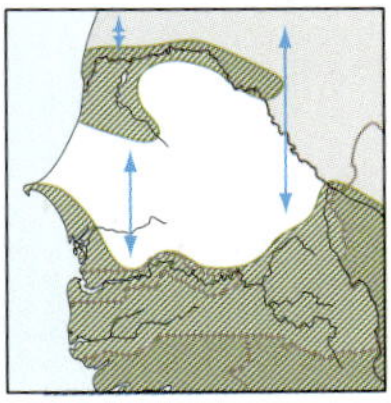

Squacco Heron *Ardeola ralloides* 42–47 cm; WS 80–92 cm

Small, stocky, short-legged heron, appearing largely ***cinnamon or buffish-brown at rest*** but with ***contrasting white wings, rump and tail in flight***. **Adult breeding** Largely cinnamon with long cream-coloured and black-edged feathers on crown and nape forming a crest; bill bright blue with black tip. **Adult non-breeding** Much duller. Earth-brown above, streaked below, with shorter crest. Bill, legs and feet greenish-yellow. **Juvenile** As adult non-breeding but darker brown, breast more streaked, wings mottled brown. **HH** Common in aquatic habitats. Often solitary, occasionally in loose groups. Palearctic migrants may occur alongside residents during northern winter (mainly Oct–Mar). **Voice** Croaking and nasal grating sounds. [CD1:24]

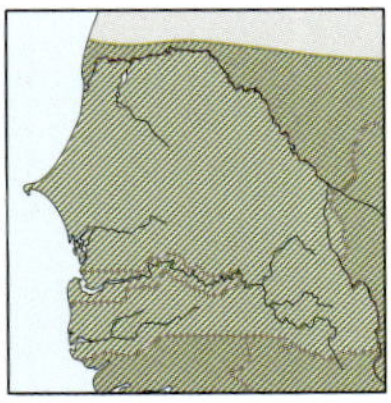

Cattle Egret *Bubulcus ibis* 48–56 cm; WS 90–96 cm

B. i. ibis. Commonest and most easily seen heron. Small, short-necked and largely white. Feathering extends below lower mandible to form distinct jowl. Typical stance is hunched with neck sunk into shoulders. In flight, appears stocky with hunched neck and rather short, heavy bill; flies with rapid, shallow wingbeats. **Adult breeding** Crown, breast and mantle rich buff. Bill and lores bright orange to coral-red. Legs coral- to dusky-red. **Adult non-breeding** All-white with yellow bill and lores; legs greenish-yellow to greyish. **Juvenile** Bill, legs and feet black. **HH** Common, tame and gregarious, occurring in open fields, grassland, inundated areas, margins of freshwater lakes and rivers, rubbish dumps and around human habitation. Will feed in drier habitats than other egrets. Often with cattle, taking insects disturbed by them. Readily perches and roosts in trees. Flies in loose flocks, often in line or V-formation. Breeds during or just after rains; then absent from many areas. **Voice** Short, gruff *kok* or *kwok*. [CD1:25]

minutus ♂
payesii ♂
Little Bittern
♀
juv
payesii ♂
ad
Dwarf Bittern
juv
Green-backed Heron
ad
juv
juv
ad non-br
ad
Squacco Heron
ad br
ad non-br
ad br
ad non-br
ad br
Cattle Egret

Black Heron *Egretta ardesiaca* 48–55 cm; WS 90–95 cm

Medium-sized, slender and ***all-black heron*** with conspicuous ***deep-orange or yellow feet*** and ***dark eyes***. **Adult** Has long, loose plumes on crown, nape and mantle. Feet may briefly turn red at start of breeding season. Eyes often erroneously stated to be yellow. **Juvenile** As adult but duller. Differs from dark morph Western Reef Egret by smaller size, darker and more uniform plumage (no white throat), darker bill and legs, and dark eye. **HH** Frequent to locally common resident of shallow waters of floodplains, marshes, margins of rivers, lakes and lagoons, mangroves and mudflats. Often in tight groups, but also solitary. Distinctive feeding posture with wings spread and held forwards over head to form shaded 'umbrella' canopy. **Voice** Usually silent. [CD5:7b]

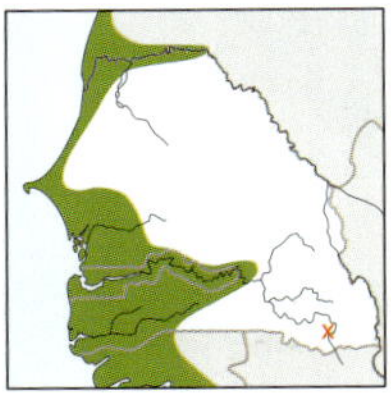

Western Reef Egret *Egretta gularis* 55–65 cm; WS 86–104 cm

E. g. gularis. Medium-sized egret with, in breeding plumage, long plumes on nape, scapulars and breast. Dimorphic. **Adult dark morph** Blackish- to slate-grey with ***white throat*** and ***variable amount of white on primary coverts***. Legs greenish-black merging into greenish-yellow feet, yellow often extending up rear edge of legs to tarsal joint or even tibia. Eye yellowish, lores olive or yellowish-olive. **Adult white morph** Very similar to Little Egret and some birds may not be safely separable in the field; note rather heavier bill with slightly drooped tip, horn-brown or grey with yellowish base to lower mandible (sometimes all black when breeding), and yellowish feet and lower legs. **HH** Common resident. Largely confined to coastal areas (mangroves, tidal pools, mudflats, saltpans, lagoons), occasionally inland at freshwater sites. Mostly solitary away from colonies, but occasionally in small, loose parties. **Voice** Throaty growl *rraaahh*. [CD1:26]

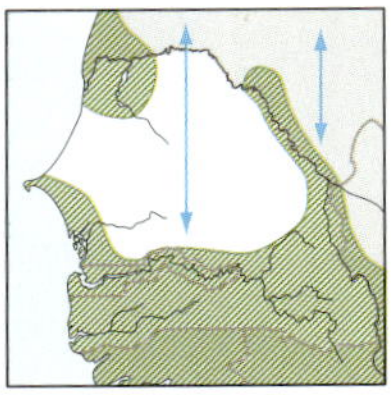

Little Egret *Egretta garzetta* 55–65 cm; WS 88–95 cm

E. g. garzetta. Medium-sized, all-white egret, distinguished from other white egrets by combination of ***slender, straight black bill, black legs*** and sharply demarcated ***yellow feet***. Compare white morph Western Reef Egret. **Adult breeding** Plumes on nape, mantle and breast; yellow lores. **Adult non-breeding** Lacks plumes; lores grey-green; bill often pale along basal half of lower mandible. **Juvenile** Bare parts dull coloured; no plumes. **HH** Fairly common in all aquatic habitats, both salt and fresh water. Feeds principally by wading through shallow water snapping at prey with quick movements. Often runs quickly on land or in shallow water, opening and closing wings as it does so. Real abundance perhaps partially obscured due to confusion with white-morph Western Reef Egret. Residents augmented by Palearctic migrants during northern winter (mainly Sep–Mar). **Voice** Harsh, throaty growl *wrraaahh*. [CD1:27]

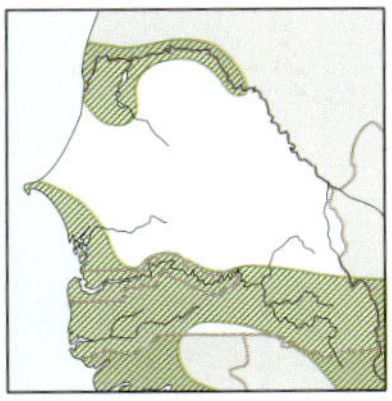

Intermediate Egret *Egretta intermedia* 65–72 cm; WS 105–115 cm

E. i. brachyrhyncha. Medium-sized, all-white egret with yellow bill and mainly black legs and feet. Resembles Great Egret, but smaller, with shorter, less kinked neck, ***shorter bill***, and ***line of gape ending level with posterior edge of eye***. Legs above tarsal joint yellow. Eye and lores yellow. **Adult breeding** Has extensive filamentous plumes on foreneck, breast and scapulars (may reach beyond tip of tail), and slight crest on rear crown, but lacks long nape plumes of Little and Western Reef Egrets. Bill becomes red with yellow tip, lores and orbital skin bright green, eye red and tibia crimson. **HH** Fairly common to uncommon in various freshwater and coastal habitats, incl. floodplains, inundated areas, margins of rivers and lakes, and mudflats. Often associated with other species of herons and egrets. Feeds by walking slowly in shallow water or on dry grassland, lunging to catch prey. **Voice** Usually silent. [CD5:8a]

Great Egret *Egretta alba* 85–100 cm; WS 140–170 cm

E. a. melanorhynchos. Large, slender, long-necked white egret, with ***line of gape extending c.1 cm beyond posterior edge of eye*** (diagnostic). Largest egret, in size similar to Black-headed Heron. In flight, and often at rest, long neck curved into distinctive S-shaped bulge. Flight slower than other egrets with legs projecting further. **Adult non-breeding** Bill long and yellow, eye yellow, lores and orbital skin olive-green. Legs and feet black. **Adult breeding** Bill becomes black, eye bright red, lores and orbital skin emerald-green. Long plumes develop on scapulars (extending well beyond tip of tail), foreneck and breast. **Juvenile** Bill yellow tipped blackish. **HH** Common to uncommon in most fresh and saltwater habitats. Mostly solitary away from breeding colonies; occasionally in small flocks. Often feeds in deep water when only head and neck visible above surrounding vegetation. Feeds by wading slowly or standing motionless. **Voice** Loud, deep, croaking *krraak*. [CD1:28]

Black Heron
Western Reef Egret
ad dark morph
ad white morph
Western Reef
Little
Intermediate
Great
juv
Little Egret
ad br
Great Egret
Intermediate Egret
ad br
ad br
ad non-br

Species on this plate are all large and readily identified. Juveniles/immatures are duller versions of adults.

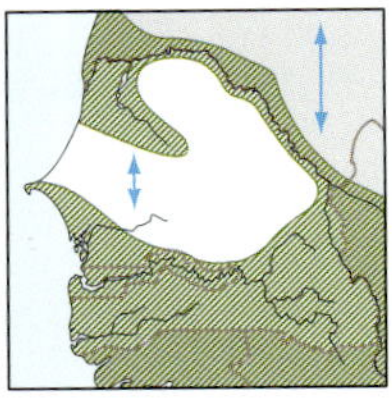

Grey Heron *Ardea cinerea* 90–100 cm; WS 155–175 cm

Large, mainly grey-and-white heron. **Adult *cinerea*** Head largely white with diagnostic black stripe from behind eye merging with black plumes on nape. In flight has prominent white carpal patches contrasting with black flight feathers; underwing uniformly dark grey. Bill and lores yellow, becoming orange to vermilion at start of breeding season. Local Mauritanian race ***monicae*** (Pallid Heron) very pale. Black-headed Heron has crown and head-sides black (dark grey in juvenile) and underwing-coverts white, contrasting strongly with blackish flight feathers. **Juvenile** Duller and plainer grey, lacking any black, with grey crown and hindneck, and greyish (not white) throat and upper neck. **HH** Common to fairly common in most aquatic habitats, incl. marshes, margins of lakes and rivers, estuaries and shallow sea coasts, coastal ricefields, occasionally in grassland. Mostly solitary, but sometimes in small groups. Feeds by waiting motionless in shallow water. Also wades and walks through shallows. Palearctic migrant (mainly Aug–Apr; some, especially immatures, year-round) and resident (small breeding population in Sine Saloum and Senegal delta). *A. c. monicae* is a rare non-breeding visitor, N Senegal (breeds Banc d'Arguin, Mauritania). **Voice** Loud, croaking *fraank* or *kraak*. [CD1:30,31]

Black-headed Heron *Ardea melanocephala* 92–96 cm; WS 150–160 cm

Adult Large, mainly black-and-white heron with black head and hindneck, contrasting white throat and foreneck, and grey belly. Upperparts darker grey than Grey Heron. Bill relatively short and deep at base, black above with greenish-yellow lower mandible. In flight, shows distinctive two-tone underwing, white underwing-coverts contrasting with blackish flight feathers. **Juvenile** Dull grey above with dark grey crown and head-sides, less black on hindneck and variable amount of rufous-buff on underparts. **HH** Uncommon to seasonally locally common. Solitary and largely terrestrial, frequenting grassland and cultivated areas rather than aquatic habitats, though does occur in marshes and margins of rivers and lakes. Often in association with man and found far from water. Feeds by walking slowly over open areas. **Voice** Raucous, nasal *kuark*. [CD5:8b]

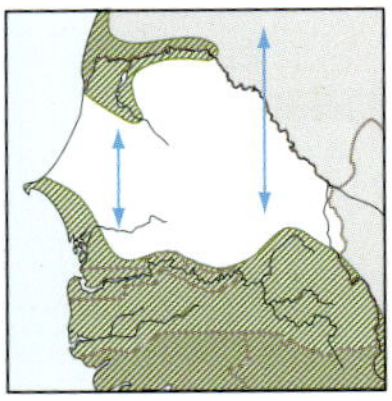

Purple Heron *Ardea purpurea* 78–90 cm; WS 120–150 cm

A. p. purpurea. Slender, lightly built heron with long thin bill and snake-like neck. **Adult** Crown black, head-sides, neck and underparts rufous striped with black. Upperparts dark slate-grey tinged purple-chestnut on shoulders. In flight, long neck kinked into very distinct bulge, upperwing dark with little contrast, underwing-coverts rufous, toes long and often held splayed. Goliath Heron is much larger, with rufous (not black) crown and heavier bill. **Juvenile** Sandy-brown above, neck indistinctly streaked, legs yellow. **HH** Uncommon to fairly common in aquatic habitats. Usually solitary. Skulks in dense aquatic vegetation. Feeds by standing motionless with body almost horizontal, catching fish with rapid strike of head. Will occasionally freeze in bittern-like posture. Palearctic migrant (mainly Sep–Apr; some year-round) and resident (very small numbers breed erratically in N Senegal). **Voice** Loud, harsh *kaark*, similar to, but higher pitched than Grey Heron. [CD1:29]

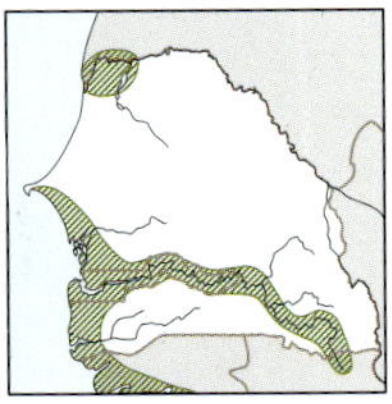

Goliath Heron *Ardea goliath* 135–150 cm; WS 210–230 cm

The world's largest heron. **Adult** Head, hindneck and underparts rufous-brown; throat white; foreneck and upper breast white streaked with black; upperparts dark grey. Blackish bill powerful and deep. In flight, brown underwing-coverts contrast with dark grey flight feathers; legs held below horizontal. **Juvenile** Basically as adult, but brown parts paler, grey areas suffused with rufous, head-sides and neck whiter. **HH** Rare to locally fairly common. Mainly in freshwater habitats, incl. lakes, marshes, mangroves and rivers; also lagoons and estuaries. Mostly solitary and rarely encountered away from water. Feeds by standing motionless for long periods, occasionally by walking and wading. **Voice** Far-carrying, raucous *kwaaark*. [CD5:9a]

ad
ad
juv
ad
monicae
juv
ad
Black-headed Heron
ad
Grey Heron
juv
juv
ad
Goliath Heron
Purple Heron

HAMERKOP Unique, medium-sized waterbird, endemic to the Afrotropical region. Sexes similar. Feeds on amphibians and small fish; also crustaceans, invertebrates, occasionally small mammals. Builds huge, domed nest of sticks, grass and mud, usually in major fork of large tree. Pairs have several nests in home territory.

Hamerkop *Scopus umbretta* 50–56 cm; WS 85–94 cm

S. u. umbretta. Unmistakable. Uniformly dark brown, with thick backward-pointing crest and stout hook-tipped bill combining to give head distinctive shape. Flight rather buoyant and owl-like, on broad, rounded wings; wingbeats often rather jerky. Glides with neck extended. In soaring flight can appear raptor-like. **Juvenile** similar to adult. **HH** Widespread and fairly common to common in any aquatic habitat within open woodland, incl. lakesides, river banks, seasonal pools and marshes, but also in drier areas. Singly or in pairs, occasionally groups. **Voice** Distinctive nasal trumpeting *yip-purr, yip-yip-yip-purr-purr-yip-yip* uttered when in groups or during display. Flight call a high-pitched nasal *yip* or *wek*. [CD5:9b]

STORKS (Plates 11–12) Large to very large wading and terrestrial birds with long bills, necks and legs. Tail short. Full adult plumage attained during 2nd, 3rd or 4th calendar year. Fly with neck outstretched (unlike herons), except Marabou Stork, and legs trailing. Normal flight action consists of soaring and gliding, often at great heights, alternated with slow wingbeats. Gregarious or solitary. Most species normally silent.

African Openbill Stork *Anastomus lamelligerus* 80–94 cm

A. l. lamelligerus. Unmistakable. **Adult** ***Blackish stork with distinctive large ivory-coloured bill with wide gap between mandibles***. Feathers of neck and upperparts glossed green, bronze and purple. **Juvenile** Dull brown with buff edges to mantle and upperwing-coverts, and shorter, almost straight bill lacking gap. **HH** Intra-African migrant, rarely recorded in Senegal. Frequents aquatic habitats, mainly freshwater. Gregarious. Feeds entirely on snails and freshwater mussels. **Voice** [CD5:10b]

Woolly-necked Stork *Ciconia episcopus* *c.*90 cm

C. e. microscelides. Unmistakable. **Adult** ***Large, mainly black stork with white neck, belly and undertail-coverts***. Black on forecrown diffusely extending onto midcrown and head-sides but leaving narrow white forehead. Tail black and slightly forked (often obscured by long undertail-coverts). Black areas strongly glossed bluish and purplish. Bill black with reddish tip and culmen. Legs black. **Juvenile** Duller, more brownish, lacking white forehead. In flight, white head, neck, belly and projecting undertail-coverts. No other stork has combination of black body and white neck. **HH** Rare to seasonally and locally fairly common in various habitats, usually near water. Mainly solitary, but may gather in small flocks. **Voice** [CD5:11b]

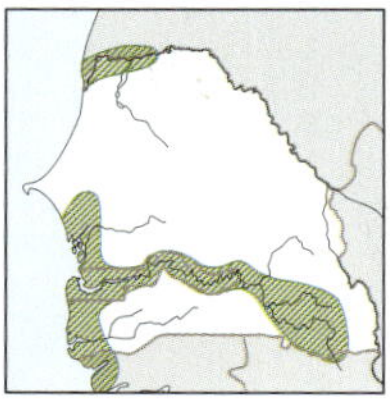

Saddle-billed Stork *Ephippiorhynchus senegalensis* 145–150 cm; WS 240–270 cm

Adult Unmistakable. ***Very large black-and-white stork with impressive black and red bill with yellow 'saddle'***. Head, neck, upper and underwing-coverts, scapulars and tail black, rest of plumage white. Legs dark with pinkish-red tarsal joints and toes. In flight, white flight feathers conspicuous. Male has brown eyes and two small yellow (sometimes red) wattles hanging from base of bill. Female has yellow eyes and usually lacks wattles. **Juvenile** Mainly greyish, with dusky bill lacking saddle. **Immature** Duller version of adult; bill duller red and black, with yellow saddle starting to appear. **HH** Rare to scarce. Singly or in pairs in any freshwater or alkaline habitat containing fish, incl. floodplains, large rivers and swamps. Normally shy and wary. **Voice** [CD5:12a]

Marabou Stork *Leptoptilos crumenifer* *c.*150 cm; WS 230–285 cm

Unmistakable. **Adult** ***Huge, heavy-bodied*** stork with dark slate-grey upperparts, white underparts, ***bare pinkish head and neck***, and ***massive pale bill***. Has two inflatable air-sacs, one long and pendulous hanging below throat, the other smaller and generally hidden between shoulders. In flight, ***uniform dark grey underwings with white axillaries diagnostic***; head tucked into shoulders. **Juvenile/immature** First-year is dark brown on upperparts with buff-edged wing-coverts. Second and 3rd year increasingly like adult. **HH** Scarce to locally fairly common, Senegal (fairly common, Niokolo Koba, Jan–Feb and May–Jun); uncommon to seasonally and locally fairly common, Gambia. In open habitats, terrestrial and aquatic. Eats virtually any animal matter. Generally gregarious. **Voice** [CD5:12b]

ad
Hamerkop
juv
ad
African Openbill Stork
ad
Saddle-billed Stork
♂
♀
juv
juv
ad
ad
Woolly-necked Stork
ad
juv
Marabou Stork
ad

Yellow-billed Stork *Mycteria ibis*

95–105 cm; WS 155–165 cm

Adult ***Mainly white with black flight feathers and tail, slightly decurved yellow bill, and red legs.*** Has bright orange-red bare facial skin and variable amount of pink suffusion on upperwing-coverts, esp. in breeding season. In flight, black flight feathers and tail. White Stork has red bill and white tail. **Juvenile** Largely greyish-brown on head, neck and upperparts; breast and belly white. Bill dull greyish-yellow, facial skin dull orange, legs brownish. By end of first year more like adult but duller. **HH** Regular in small numbers in N Senegal, more widespread and seasonally and locally fairly common in Gambia/S Senegal. In all aquatic habitats, incl. rivers, freshwater marshes, lakes, estuaries and mudflats. Gregarious, but seldom in large flocks. Often associated with other waterbirds such as herons, storks and pelicans. **Voice** [CD5:10a]

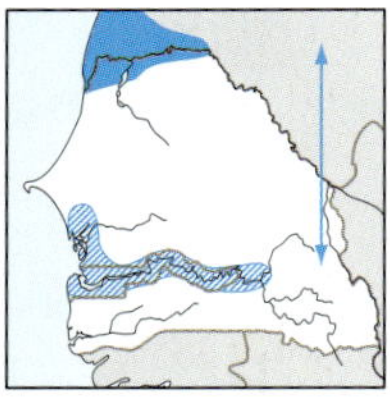

White Stork *Ciconia ciconia*

100–120 cm; WS 155–165 cm

C. c. ciconia. **Adult** ***Large white stork with black flight feathers and red bill and legs.*** **Juvenile** Duller, with off-white plumage, shorter, blackish bill with red base and dull red legs. In flight, black flight feathers, white tail. Yellow-billed Stork has yellow bill and black tail. **HH** Palearctic migrant (mainly Oct–Mar), principally N Senegal (uncommon), occasionally further south; rare but sometimes in large groups, Gambia. Mainly in grassland and open savanna where particularly attracted to swarms of locusts and bush fires. Also in wetland areas. Gregarious, migrating in flocks. **Voice** [CD1:33]

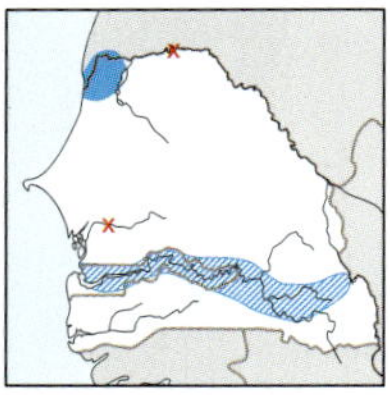

Black Stork *Ciconia nigra*

95–100 cm; WS 145–155 cm

Adult ***Large, mainly black stork with white breast and belly, and red bill, orbital skin and legs.*** **Juvenile** Sooty-brown with bill, orbital skin and legs pale grey-green. Bill and legs become red during 2nd calendar year, when c.10 months old. In flight, all-black upperparts. Compare Abdim's Stork (smaller, white rump and lower back, yellowish-green bill and legs, shorter legs projecting less in flight. Immatures more alike, but white rump and dusky-greenish legs with pink joints diagnostic of Abdim's). **HH** Palearctic migrant, frequenting aquatic and dry habitats. Regular in small numbers, N Senegal, occasionally further south; previously very rare but now more regular, Gambia. Often solitary, but may associate with other storks. **Voice** [CD1:32]

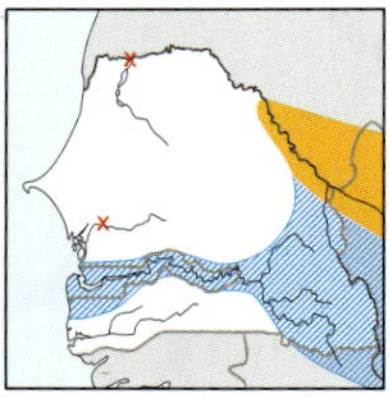

Abdim's Stork *Ciconia abdimii*

75–81 cm

Adult Rather ***small, largely black stork with white breast, belly, rump and lower back.*** Bill greenish-horn with red tip. Bare facial skin grey-blue with red patch in front of eye and under bill. Legs dull green with reddish 'knees' (tarsal joint). During breeding season bare parts become brighter. **Juvenile** Duller, more brownish above and lacks gloss of adult. Bill and legs duller and rather darker. In flight, white lower back and rump; compare Black Stork. **HH** Scarce to locally fairly common intra-African migrant, Senegal; rare, Gambia. Breeds in northern tropics prior to and during wet season. Moves south at onset of dry season, to arrive in southern tropics early in rains. Mainly terrestrial, feeding in open grasslands, pastures and cultivated areas. Gregarious. Congregates at swarms of locusts, outbreaks of army worm caterpillars and grass fires. **Voice** [CD5:11a]

juv
ad
ad
ad
Yellow-billed Stork
ad
juv
White Stork
ad
ad
Black Stork
ad
juv
Abdim's Stork
ad
juv

IBISES Medium-sized terrestrial and wading birds with rather long necks and legs and slender decurved bills. Fly with strong wingbeats and neck outstretched. Feed in dry habitats, on forest floor or in shallow wet areas by probing in soft mud.

Glossy Ibis *Plegadis falcinellus* — 55–65 cm; WS 80–95 cm

Small, dark, rather slender ibis (other dark ibises less slender). Distinctive flight with rapid wingbeats followed by short glide, ***legs projecting well beyond tip of tail***. **Adult breeding** ***Rich purplish-chestnut glossed with green on wings***. Lores pale bluish bordered above and below by narrow white line. **Adult non-breeding** Duller; head and neck with variable amount of fine white streaking. **Juvenile** Sooty-brown; head and throat variably mottled white. Following post-juvenile moult, head and neck streaked with white, but less distinctly than adult. **HH** Fairly common to scarce Palearctic visitor (Nov–Jun; some year-round) in freshwater habitats incl. marshes, floodplains, inundated areas, margins of rivers and lakes. Occasionally coastal lagoons, estuaries. Gregarious; usually in small groups. Often associated with herons, egrets and storks. Usually flies in line, sometimes in V formation. **Voice** Mostly silent. Occasionally a harsh, low *graa-graa-graa* in flight. [CD1:34]

Hadada Ibis *Bostrychia hagedash* — 76–89 cm

B. h. brevirostris. Large dark and noisy ibis with rather short legs and dark bill with red on upper mandible. **Adult** Largely dull brown with ***white malar stripe*** and metallic green gloss on wing-coverts. Flight rather irregular and jerky with one deep wingbeat followed by several shallow flaps; wings broad and rounded, bill held downwards, ***legs do not project beyond tail***. **Juvenile** Duller, lacking green gloss. **HH** Locally fairly common to scarce along streams and rivers in open woodland, also marshes, mangroves, forest edge. **Voice** Distinctive, far-carrying, nasal *haa! haa-de-dah!* frequently uttered in flight. Also a loud *HAAA!* when alarmed. [CD5:13b]

Northern Bald Ibis *Geronticus eremita* — 70–80 cm; WS *c.* 130 cm

Adult Unmistakable. Stocky, blackish ibis with shaggy crest, ***'bald' reddish head***, and reddish bill and legs. Legs do not project beyond tail in flight. **Juvenile** Head feathered; no crest; duller plumage. In 2nd year nuchal crest and violet gloss on coverts begin to appear, but head still largely feathered. Full adult bare-part coloration probably attained during 3rd year. **HH** Palearctic vagrant, Senegal (Niokolo Koba, May 1985; Matam, 30 Dec 1988 & 11 Jan 1989). Red List category: Critically Endangered. **Voice** [CD1:35]

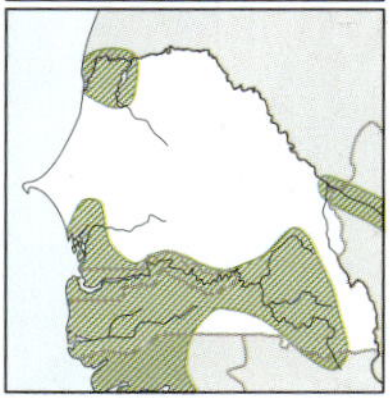

Sacred Ibis *Threskiornis aethiopicus* — 65–82 cm; WS *c.* 120 cm

T. a. aethiopicus. **Adult** Distinctive, ***mainly white*** ibis with ***bare black head and neck, long, heavy black bill, and long blue-black scapular plumes falling over folded wings and tail***. Black tips to flight feathers visible in flight. Wing edge and flanks often tinged buffish-yellow. Dull grey bare patch at breast-sides extending in line on underwing becomes blood-red when breeding. **Juvenile** Head and neck feathered, blackish with white mottling; throat and foreneck white. Scapulars less developed, primaries more extensively tipped black. In subsequent plumages ornamental scapulars develop, black on primaries and greater primary-coverts gradually decreases. Full adult plumage with completely bare head and neck attained in 3–4 years. **HH** Uncommon or scarce to locally common in variety of habitats incl. margins of rivers and lakes, marshes, floodplains, grasslands, cultivation, rubbish dumps and occasionally estuaries, lagoons and sea coast. Gregarious. Often very tolerant of man. Flocks tend to fly in V formation. **Voice** [CD5:15a].

ad br
ad non-br
ad br
juv
Glossy Ibis
ad
juv
ad
Hadada Ibis
Northern Bald Ibis
ad
juv
ad br
Sacred Ibis

SPOONBILLS *Large, white wading birds with characteristic spoon-shaped bill and long neck and legs. Fly with neck outstretched; sometimes soar. Gregarious or solitary. Feed by sweeping bill from side to side in water. Mainly silent away from breeding colonies.*

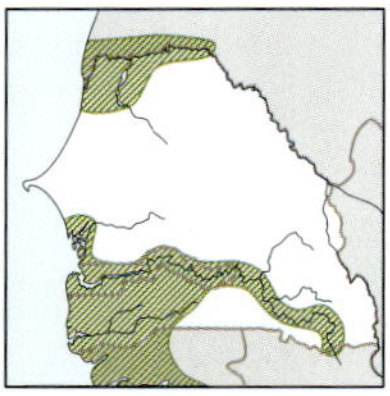

African Spoonbill *Platalea alba* — *c.*90 cm

Adult breeding All white with bare face red; bill grey with red margins; legs red; loose, fluffy nuchal crest. **Adult non-breeding** Lacks crest; bare-part coloration somewhat duller. **Juvenile** Bill yellowish; legs black; no red on face. In flight, note black tips to primaries and primary-coverts. **HH** Uncommon or scarce to locally fairly common. Mainly coastal, in shallow water of e.g. lagoons and estuaries. Occasional inland at lakes and marshes. **Voice** [CD5:15b]

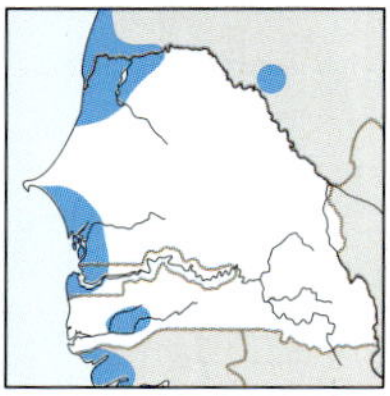

Eurasian Spoonbill *Platalea leucorodia* — 80–90 cm; WS 115–130 cm

Adult breeding *leucorodia* Bill black with yellow tip; legs black; long, loose nuchal crest; yellowish breast-band. Non-breeding adult lacks crest and breast-band. **Adult breeding *balsaci*** Bill all black. Breeds Mauritania (Banc d'Arguin; locally common). **Juvenile** As adult non-breeding but with black tips to primaries, dull pinkish bill and throat, and dull yellowish-brown legs. **Immature** Bill and legs gradually darken, black on flight feathers decreases, outer primaries retaining some black on tips. Full adult plumage probably attained in 3rd calendar year. **HH** Nominate is fairly common to scarce Palearctic visitor (mainly Sep–Mar; immatures remain in winter quarters during first summer); some *balsaci* (mainly immatures) disperse to Senegal. Mainly coastal, in shallow water of e.g. lagoons and estuaries. Occasional inland at lakes and marshes. Flocks tend to fly in line or V formation. **Voice** [CD1:36]

FLAMINGOS *Large wading birds with extremely long necks and legs, pink plumage and unique down-curved bills adapted for filter-feeding. Fly in lines or V formations with neck outstretched. Highly gregarious and quite noisy. Occur on shallow brackish, alkaline or saline lakes and lagoons.*

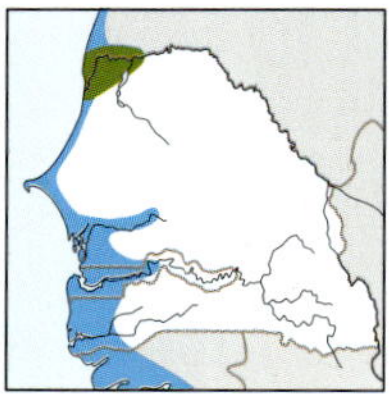

Greater Flamingo *Phoenicopterus roseus* — 127–140 cm; WS 140–165 cm

Adult Pale pink plumage (***appearing largely white at distance***); ***pink bill with contrasting black tip***. In flight, scarlet lesser and median coverts contrast strongly with black flight feathers. **Juvenile** Dull greyish plumage; note rather large, dark grey bill tipped black. In subsequent plumages, first becomes whiter, then progressively more pink. Adult plumage attained in 3–4 years. **HH** Locally common to uncommon in coastal aquatic habitats. Wades with head submerged when feeding, or swims and up-ends in deeper water. Breeds N Senegal. Mainly resident, though some dispersal. Migrants from W Palearctic occur during northern winter. **Voice** Goose-like *hank-hank* and a nasal *gnaaaa*. [CD1:37]

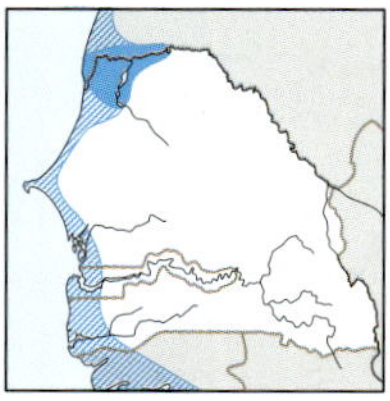

Lesser Flamingo *Phoeniconaias minor* — 80–90 cm; WS 95–100 cm

Adult ***Rather small, rose-pink flamingo*** with dark red, black-tipped bill (***appearing entirely dark at distance***); much smaller than Greater Flamingo. In flight, lesser and median coverts contrast less than in Greater Flamingo. **Juvenile** Brownish-grey; bill smaller than Greater Flamingo. In subsequent plumages, first becomes whiter, then gradually more pink. Adult plumage attained in 3–4 years. **HH** Locally fairly common to scarce in aquatic habitats (esp. Senegal delta); mainly coastal. Normally highly gregarious, but also observed singly and in small groups. Feeds mainly at night, either wading in shallow water or swimming in deeper water moving head and neck in rhythmic scything curves. Occasionally associates with Greater Flamingos. Breeds SW Mauritania. Red List category: Near Threatened. **Voice** High-pitched *chissik* or *kwirrik*. Feeding flocks utter constant murmuring *murr-err*. [CD5:16a]

juv
ad br
African Spoonbill
juv
ad br
leucorodia
ad br
balsaci
juv
Eurasian Spoonbill
ad
ad
Greater Flamingo
juv
ad
Lesser Flamingo
juv

DUCKS AND GEESE (Plates 15–18) Medium-sized to large waterbirds with plump bodies, short, robust legs, webbed feet and flat bills rounded at tip; tail short. Flight fast and direct with neck outstretched. Often gregarious.

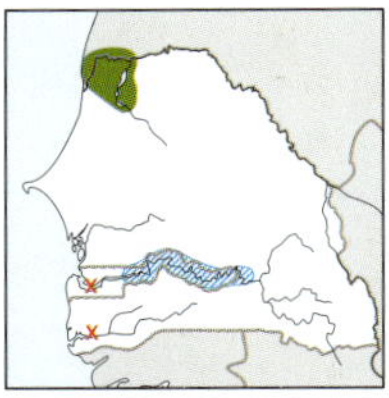

Egyptian Goose *Alopochen aegyptiaca* 63–73 cm; WS 134–154

Unmistakable. Large and bulky with ***pink bill*** and ***long pink legs***. In flight, ***white wing-coverts contrast with blackish flight feathers and form diagnostic panels above and below***. **Adult** Chocolate-brown eye patch. **Juvenile** Duller; bill and legs greyish. **HH** Fairly common in Senegal delta; rare Afrotropical visitor elsewhere. Usually in pairs or small groups in wetland habitats. Readily perches in trees. **Voice** Hoarse *taash taash*... in flight. Male utters harsh, wheezy hiss when displaying. Female gives harsh, nasal, slightly trumpeting cackle *honk-haah-haah-haah*. [CD1:46]

Common Shelduck *Tadorna tadorna* 58–71 cm; WS 100–120 cm

Unmistakable. **Adult male** Mainly black and white with rufous breast-band and red bill with knob. In flight, heavy; white with black head and black flight feathers. **Adult female** Slightly smaller and duller, often some white at bill base and no knob on bill. **HH** Rare Palearctic winter visitor, Senegal delta. Mainly on flat sandy shores, bays and estuaries, but also on fresh water near coast and in cultivated fields and grassland. **Voice** Male utters low whistling calls, female a rapid *ga-ga-ga-ga-ga-ga-gak*. [CD1:48]

Knob-billed Duck *Sarkidiornis melanotos* 56–76 cm

S. m. melanotos. Unmistakable. Large, bulky duck with mainly ***white head, neck and underparts*** and iridescent black upperparts. In flight, ***white underparts*** contrast with ***black wings and upperparts***. **Male breeding** Bill with fleshy black protuberance ('knob' or 'comb') on upper mandible (reduced in non-breeding). **Adult female** Smaller, without knob on bill. **Juvenile** Dark brown above, buffish-brown below. **HH** Uncommon to locally and seasonally common in freshwater wetlands. Seasonal movements usually related to availability of water. **Voice** Mainly silent. Calls varied; usually in display. [CD5:19a]

Spur-winged Goose *Plectropterus gambensis* 75–100 cm; WS 160–180 cm

P. g. gambensis. Huge black-and-white perching duck with long neck and long, pinkish legs. Flight slow and laboured with 'fingered' wings producing loud swishing noise. **Adult male** Black plumage glossed iridescent green and purple; red frontal knob. In flight, white forewing above and below. **Adult female** Smaller and duller; frontal knob reduced or absent. **Juvenile** Duller and browner; face entirely feathered. **HH** Uncommon to locally and seasonally common in freshwater wetlands. Usually gregarious. Feeds mainly in early mornings and at night. **Voice** Mainly silent. Male occasionally utters soft, wheezy *cheweh* or *cherwit*. [CD5:18a]

Brent Goose *Branta bernicla* 55–62 cm; WS 105–120 cm

B. b. bernicla (Dark-bellied Brent Goose). Distinctive, small dark goose with ***black head, neck and upper breast***, dark grey belly and contrasting white vent and rump. Bill and legs blackish. **Adult** ***White flash on neck-sides***. **Juvenile** Lacks white neck patch and has pale edges to wing-coverts forming bars at rest. **HH** Mainly on mudflats and estuarine waters. Palearctic vagrant, S Senegal (two photographed, Cap Skirring, Feb 1997). **Voice** Low, guttural *hrot*. [CD1:44]

ad
juv
Egyptian Goose
ad
♀
♂
Common Shelduck
♂
♂
♀
juv
Knob-billed Duck
Spur-winged Goose
♂
♂
♀
Brent Goose
ad

Fulvous Whistling Duck *Dendrocygna bicolor* 45–53 cm

Upright-standing fulvous-brown duck with ***bold white streaks on flanks***. In flight, ***white U-shaped uppertail-coverts*** contrast with blackish wings and tail and fulvous-brown body. **Juvenile** duller. **HH** Locally common, N Senegal; uncommon to rare elsewhere. Frequents well-vegetated freshwater lakes, marshes, ricefields and rivers. Feeds mainly at night. Often mixes with White-faced Whistling Ducks. **Voice** Clear, slightly nasal whistle *k-wheew* or *ksweeoo*, in flight and usually repeated. [CD5:16b]

White-faced Whistling Duck *Dendrocygna viduata* 43–48 cm

Upright-standing, mainly dark duck with long neck and legs. In flight, appears uniformly dark with chestnut forewing. **Adult** Diagnostic ***white face and throat*** contrast with black rear of head and upper neck. Lower neck, breast and mantle dark chestnut; breast-sides and flanks closely barred black and white; belly and undertail-coverts black. White face may become stained brownish. **Juvenile** (plumage retained for only a few weeks) Duller, with buffish face and upper neck, blackish crown and hindneck, and chestnut breast-band. **HH** Common in most open freshwater habitats, esp. with some emergent vegetation. Highly gregarious. Often forages in ricefields. Active at night but also feeds diurnally. Subject to much local movement. **Voice** Vocal, esp. in flight. Characteristic, clear, sibilant 3-note whistle *swee-swee-sweeoo*, usually repeated. [CD5:17a]

Black-bellied Whistling Duck *Dendrocygna autumnalis* 48–53 cm

Large whistling duck with, in adult, conspicuous ***pinkish-red bill***. Dark brown cap contrasts with pale grey-brown head and neck; conspicuous whitish eye-ring. In flight, mainly dark with long, broad white band on centre of upperwing (diagnostic). **HH** Neotropical vagrant, Gambia (Kotu sewage ponds, Dec 1998). Only record for Africa; possibility of escape cannot be eliminated. Normally in freshwater habitats. **Voice** A whistled *wee-cheew* or *weechew-weeweewheew...*

African Pygmy Goose *Nettapus auritus* 30–33 cm

Small and brightly coloured. Mainly dark glossy green above and rusty-orange below. In flight, largely dark with white face and belly; upper- and underwing dark with ***conspicuous white patch on inner secondaries***. **Adult male** White head with black crown and bright metallic green neck-sides bordered by black. Bill bright yellow with black nail. **Adult female** Duller. Head-sides and foreneck mottled greyish; narrow dusky eye-stripe. **Juvenile** Duller than adult female; eye-stripe more prominent. **HH** Uncommon to locally fairly common in freshwater wetlands rich in emergent vegetation, particularly water-lilies. Usually in pairs or small groups. Occurrence often sporadic; movements in response to changing water levels and habitat availability. **Voice** Male utters soft, rather melodious, whistled *kewheep* and *khep-khep-kheew*, usually repeated. Female a soft quack. [CD5:19b]

White-backed Duck *Thalassornis leuconotus* 38–40 cm

T. l. leuconotus. Small, buff-and-brown barred duck with contrasting white loral patch. Sloping forehead, large-headed and hump-backed appearance, and short tail held flat on water produce distinctive shape. In flight, entirely brown and buff except ***white lower back*** (inconspicuous at distance); feet protrude beyond tip of very short tail. **HH** Regular in small numbers, N Senegal. Favours vegetation-rich lakes and pools. Sits low in water and spends most of day resting among vegetation. Rarely flies, preferring to swim into cover if disturbed. **Voice** Clear, sharp whistle *whit-wee* or *si-wee-wheet*, not unlike whistling duck, when swimming and in flight. [CD5:17b]

Marbled Duck *Marmaronetta angustirostris* 39–42 cm; WS 63–70 cm

Small, slender, ***pale*** grey-brown duck with ***dark eye patch***, rather large head and slim grey bill. Upperparts with large buffish spots. In flight, relatively long neck, wings and tail, pale secondaries, no speculum nor wingbars, white underwing. **HH** Regular Palearctic winter visitor in very small numbers (mainly Dec–Feb), Senegal (mainly Djoudj; also Niokolo-Koba). Bred N Senegal (one pair, 1979). Frequents freshwater and coastal wetlands. Relatively shy and often remains concealed in emergent vegetation. Red List category: Vulnerable. **Voice** Mostly silent. [CD1:56]

ad
Fulvous Whistling Duck
ad
White-faced
Whistling Duck
juv
African Pygmy Goose
ad ♂
ad
Black-bellied
Whistling Duck
ad ♀
ad ♂
ad
White-backed Duck
ad
Marbled Duck

***DABBLING DUCKS** feed in water by up-ending or skimming the surface with bill. Rise vertically from the water without any foot-pattering. Upperwing pattern and speculum colour are often important identification marks. In Palearctic species males adopt female-like 'eclipse' plumage soon after breeding, from May/June until October/November. Females and juveniles of African species usually very similar to males.*

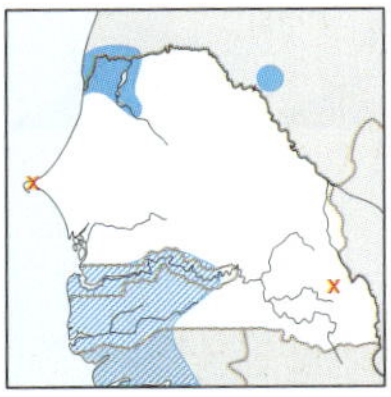

Northern Pintail *Anas acuta* — 51–62 cm; WS 79–87 cm

A. a. acuta. Large, long-necked, ***slim*** duck with pointed tail. In flight, characteristic silhouette with ***long neck*** and ***pointed tail***, broad white trailing edge to speculum and greyish underwing. **Adult male** ***Head and hindneck dark brown; neck-sides, breast and centre of underparts white;*** yellowish-buff at rear of flanks; ***contrasting black undertail-coverts***. Long, finely pointed tail. Speculum dark green glossed with bronze and bordered with pale rufous at front and white at rear. **Adult female** Rather plain, buff-brown head; slim grey bill. Speculum brownish bordered white. **HH** Locally common (N Senegal) to uncommon Palearctic visitor (mainly Oct–Mar) to freshwater and coastal wetlands. **Voice** Male has clear, mellow *krrup*, reminiscent of Common Teal; female a low quack and Eurasian Wigeon-like growl. [CD1:53]

Northern Shoveler *Anas clypeata* — 44–52 cm; WS 73–82 cm

Distinguished in all plumages by ***long, broad bill***, producing characteristic front-heavy appearance at rest and in flight. **Adult male** Head dark green, breast white, ***flanks chestnut***, rear end black and white. In flight, conspicuous ***blue-grey panel on forewing***, separated by white wingbar from dark green speculum. **Adult female** Best distinguished by bill shape. In flight, wing pattern similar but duller than male; underwing-coverts all white, contrasting with dark belly. **HH** Locally common to uncommon Palearctic visitor (mainly Oct–Apr) to freshwater and coastal wetlands. **Voice** Mostly silent. Male utters gruff, nasal *took-took*; female a variety of quacks. [CD1:55]

Eurasian Wigeon *Anas penelope* — 45–51 cm; WS 71–85 cm

Adult male Chestnut head with yellow forehead and crown; pinkish breast. In flight, ***striking white panel on upper forewing*** and ***white belly***. **Adult female** Rather uniform rusty-brown to brownish-grey with darker smudge around eye. In flight, appears long winged with short pointed tail and relatively small bulbous head; from below has ***conspicuous white belly*** and grey underwing. **HH** Scarce Palearctic winter visitor to freshwater and coastal wetlands, N Senegal; vagrant, Gambia (Jan 2010). **Voice** Male utters distinctive melodious, whistling *swheeoooh*; female a growling *krrr*, sometimes repeated. [CD1:49]

American Wigeon *Anas americana* — 45–56 cm

Resembles Eurasian Wigeon in size and structure. In flight, ***white axillaries contrast with greyish underwing*** (all plumages). **Adult male** Crown cream-white; ***dark metallic green band from eye to nape***. **Adult female** As Eurasian Wigeon but head, neck and upperparts greyer, more contrasting with rusty-orange breast and flanks. **HH** N American vagrant, N Senegal (Feb 1975; winter 2005/06). **Voice** Similar to Eurasian Wigeon but male's whistle more throaty and disyllabic *wheeoh-woh*. [CD5:20a]

Gadwall *Anas strepera* — 46–55 cm; WS 78–90 cm

A. s. strepera. Medium-sized dabbling duck with ***diagnostic small white speculum*** (conspicuous in flight; also often visible at rest), white belly and white underwing. **Adult male** Mainly greyish; black rear end; lead-grey bill. **Adult female** Mottled brownish with greyer head; bill blackish on culmen with well-defined dull orange sides. White speculum smaller than in male. Female Mallard superficially similar but plumage warmer brown, orange on bill less clear-cut, belly brownish, tail with more white. **HH** Rare Palearctic visitor, N Senegal (Jan–Mar); one claim from Gambia (Nov 2002). Freshwater wetlands. **Voice** Mainly silent. Female gives loud quack. [CD1:50]

Mallard *Anas platyrhynchos* — 50–60 cm; WS 81–95 cm

A. p. platyrhynchos. ***Large***, heavily built duck. In flight, has dark blue speculum prominently bordered with white at front and rear; underwing largely white. **Adult male** Green head; narrow white collar; purplish-brown breast; yellow bill. **Adult female** Pale brown marked darker; bill orange with irregular dark area on culmen. In flight, brown belly. Compare female Gadwall. **HH** Palearctic vagrant, N Senegal, to freshwater wetlands. **Voice** Male has low, nasal *vrreb*; female a loud quack. [CD1:52]

Northern Pintail
Northern Shoveler
Eurasian Wigeon
American Wigeon
Gadwall
Mallard

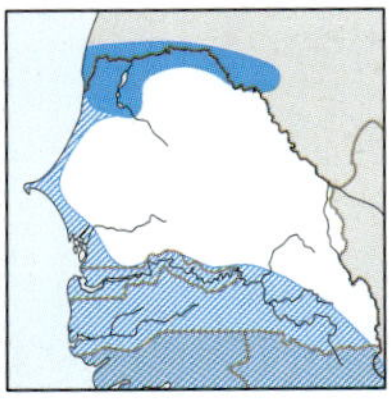

Garganey *Anas querquedula* 37–41 cm; WS 59–67 cm

Adult male ***Head dark purplish-brown with bold white stripe from above eye curving onto nape***, breast dark brown, flanks pale grey, scapulars pointed and blackish striped white. In flight, ***pale grey-blue panel*** on forewing and green speculum bordered with white at front and rear; dark breast contrasts with whitish belly. **Adult female** Mottled dull brown with ***pale supercilium bordered by dark crown and eye-stripe***, and ***whitish loral spot*** and throat usually separated by dusky stripe on ear-coverts (sometimes lacking). In flight, dull greyish forewing and white trailing edge (rear border to speculum broader than front bar; reverse of Common Teal). **HH** Locally very common to uncommon Palearctic visitor (mainly Sep–Mar) to freshwater and coastal wetlands. **Voice** Male has distinctive dry rattling call; female a harsh, nasal *kheh* and soft quack. [CD1:54]

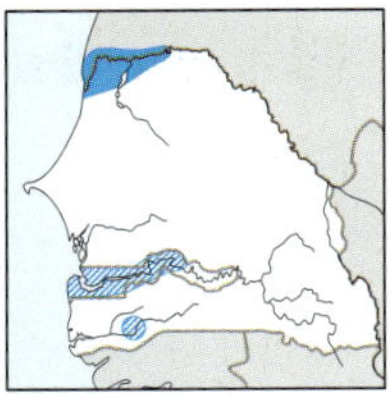

Common Teal *Anas crecca* 34–38 cm; WS 53–59 cm

A. c. crecca. Small dabbling duck, slightly smaller than Garganey. In flight, appears mainly dark and small with narrow pointed wings. Upperwing has short white wingbar bordering dark green speculum, and narrow white trailing edge. **Adult male** ***Head chestnut with dark green band from eye to nape narrowly bordered with yellow***, body greyish with white stripe on scapulars, sides of undertail-coverts creamy-yellow bordered with black. **Adult female** Head rather plain with darker crown and indistinct eye-stripe. Body greyish-brown speckled and mottled dark; whitish stripe at sides of undertail-coverts. Bill brownish-grey, often with dull orange base. **HH** Locally fairly common to uncommon Palearctic visitor (Oct–Mar) to freshwater wetlands. **Voice** Male has clear, liquid *kreek*, frequently uttered. Female gives sharp nasal quack and low growling *trrr*. [CD1:51]

Blue-winged Teal *Anas discors* 37–41 cm

Adult male Dull violet-blue head with ***large white crescent in front of eye*** and ***white patch at sides of black vent***. In flight, has ***bright blue panel on forewing*** separated by white wingbar from green speculum, and ***no white trailing edge*** to secondaries. **Adult female** Head pattern as female Garganey but more indistinct; ***prominent pale loral spot***. Legs yellowish. In flight, wing pattern as male but duller. **HH** N American vagrant, NW Senegal (Mar 1975, 1979, 1989 & 2005) and near Dakar (Lake Youl, Mar 2006). Freshwater and coastal wetlands. **Voice** Mostly silent.

DIVING DUCKS feed by diving. Swim under water and run along the surface to take off.

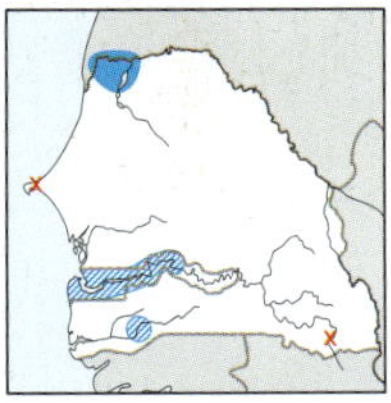

Common Pochard *Aythya ferina* 42–49 cm; WS 67–75 cm

Medium-sized with peaked crown, ***sloping forehead and rather long bill*** creating distinctive head profile. In flight, has ***grey band across flight feathers on upperwing***. **Adult male** Unmistakable combination of ***bright chestnut head, black breast and rear end, and pale grey body***. In eclipse, duller, more female-like, but head plain, breast darker and eye reddish. **Adult female** Head brownish with diffuse pale loral patch, eye-ring and line behind eye; breast brownish; body greyish-brown. In flight, has greyish underwing and belly. **HH** Uncommon to rare Palearctic visitor (late Oct–Feb) to freshwater and coastal wetlands. **Voice** Mainly silent. [CD1:58]

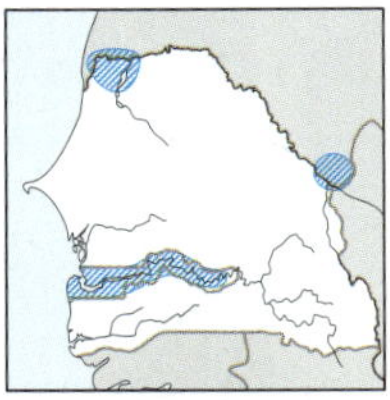

Ferruginous Duck *Aythya nyroca* 38–42 cm; WS 60–67 cm

Mainly dark chestnut with conspicuous ***pure white undertail-coverts***. Bill longish, forehead high, crown peaked. In flight, ***broad white band across flight feathers strongly contrasting with rest of upperwing***, underwing mainly white, well-defined white belly and undertail contrast with dark body. **Adult male** Diagnostic ***white eye***. **Adult female** Duller, with dark eye. **HH** Uncommon to rare Palearctic visitor (late Oct–Feb) to freshwater and coastal wetlands. Red List category: Near Threatened. **Voice** Mainly silent. Male utters soft *wheeoo*. [CD1:59]

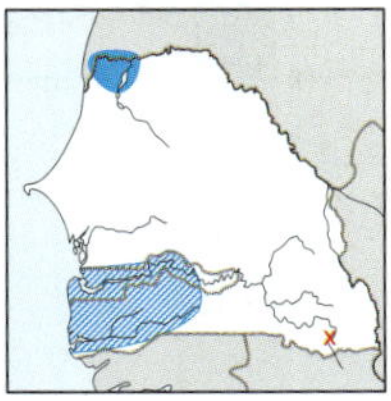

Tufted Duck *Aythya fuligula* 40–47 cm

Small with steep rounded forehead and flattish crown. In flight, has conspicuous white band across upperside of secondaries and inner primaries, and extensive white belly. **Adult male** ***Black with striking white flanks and belly and long, drooping crest. Eye yellow***. **Adult female** Dark brown with short tuft at rear of crown, flanks paler, yellow eye. Often has variable amount of white on lores or around base of bill, and whitish undertail-coverts. **HH** Scarce to locally uncommon Palearctic visitor (Nov–Mar), Senegal. Rare or vagrant, Gambia. In freshwater and coastal wetlands. **Voice** Mainly silent. Female has growling *krrr*. [CD1:60]

Garganey
Common Teal
Blue-winged Teal
Common Pochard
Ferruginous Duck
Tufted Duck

PLATE 19: OSPREY, FISH EAGLE AND PALM-NUT VULTURE

***ACCIPITRIDAE* (Plates 19–32)** *A large and diverse-looking assemblage of diurnal raptors. Its members characteristically have hooked bills and powerful talons. Exhibit a wide variety of wing shapes, tail shapes and flight actions, according to habitat and hunting techniques. Sexes more or less alike; females almost always larger. Normal flight consists of regular wingbeats alternated with gliding. Almost all species carnivorous. Conspicuous aerial displays used by many. Identification often problematic; correct assessment of jizz (combination of structure, proportions, wing attitudes, and flight actions) essential.*

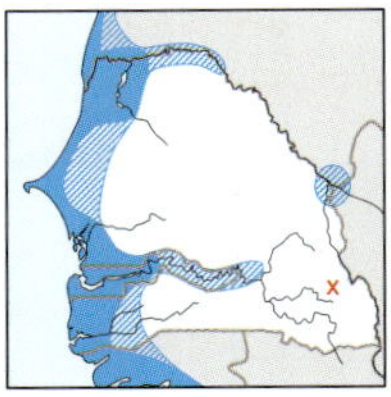

Osprey *Pandion haliaetus*

52–61 cm; WS 145–173 cm

P. h. haliaetus. Large, with mainly dark upperparts and white underparts. Wings long and narrow, with long 'hand' and only four 'fingers'. Tail relatively short and square. Flies with shallow, powerful wingbeats interspersed with long glides. Soars and glides on arched wings held angled at carpal joint. **Adult** Head white with short nuchal crest and broad dark stripe through eye. Some brown streaking on forehead. Pale brown streaks form indistinct band on upper breast (usually more prominent in female). In flight, from below, black carpal patches contrast with white underwing-coverts; greater coverts blackish, forming dark line across centre of wing. **Juvenile** Similar, but feathers of upperparts narrowly fringed buffish. Greater coverts white barred dark; flight feathers coarsely barred dark. Eye orange. Adult plumage acquired after c.18 months. **HH** Seasonally common Palearctic visitor (mainly Oct–Apr; immatures may remain during rains). Perches conspicuously on prominent vantage point at large rivers, lagoons, lakes, etc. Exclusively preys on fish caught by plunge-diving. **Voice** Usually silent in Africa. [CD1:69]

African Fish Eagle *Haliaeetus vocifer*

63–73 cm; WS 190–240 cm

Large eagle associated with water. Soars on broad flat wings. **Adult *White head, mantle and breast contrast with black of upperparts and chestnut of 'shoulders' and lower underparts***. Short white tail often hidden by wings when perched. In flight from below, white head and upper breast and white tail contrast with chestnut underwing-coverts and lower underparts and black flight feathers. Adult female averages 10% larger than male; white of breast does not extend as far down as in male and is squarer. **Juvenile** Drab mottled brown above; breast whitish variably streaked brown; rest of underparts heavily mottled brown and white, with darker belly. Tail whitish broadly tipped blackish. In flight from below, underwing-coverts contrast with blackish flight feathers; primaries with pale window; ***tail whitish with black terminal band*** (diagnostic). **Immature** Following first moult, paler than juvenile, with head and underparts mainly whitish streaked brown; often has dark superciliary streak. Terminal dark band on tail narrower. In 3rd–4th year, gradually acquires more adult features: head, mantle and breast white, initially with some black streaking; tail gradually whitens and loses terminal band; chestnut appears on shoulders, belly and underwing. Cere pinkish. Full adult plumage attained in at least 5 years. **HH** Fairly common to uncommon on large rivers, lagoons, lakes, reservoirs. Feeds mainly on fish, which is caught with talons. **Voice** Vocal. Highly distinctive, far-carrying, ringing *WEEE-AH kleeuw kleeuw kleew*, uttered by both sexes at rest and in flight, with head thrown back; male higher pitched than female. Often duets. [CD5:29]

Palm-nut Vulture *Gypohierax angolensis*

*c.*60 cm; WS 140–150 cm

Chunky, with broad, rounded wings and very short, rounded tail. **Adult** Readily identified by ***black-and-white plumage and bare reddish-pink face***. Bill heavy, yellowish-horn. In flight, mainly white with black secondaries and black tail broadly tipped white. Primaries tipped black. **Juvenile** Entirely drab brown with darkish bill, dull yellow face and cere. In flight, dark brown with blackish-brown flight and tail feathers. In immature, brown gradually replaced by white, producing mottled appearance. Adult plumage acquired in 3–4 years. **HH** Fairly common to locally common. Singly, often near water. Relishes nuts of oil and raphia palms, but also feeds on various small animals, incl. fish, amphibians and crabs, occasionally mammals, birds and large insects; also scavenges. **Voice** Usually silent. [CD5:30]

ad
ad
Osprey
ad
African Fish Eagle
ad
juv
imm
juv
imm
ad
ad
Palm-nut Vulture
ad
juv
ad
juv

VULTURES (Plates 20–22) Medium-sized to huge with long, broad, strongly 'fingered' wings, and usually unfeathered head and neck. Larger species have long neck with ruff at base, and short tail. Carrion eaters. Given to soaring, often at great height. In flight head is tucked between shoulders and appears relatively small.

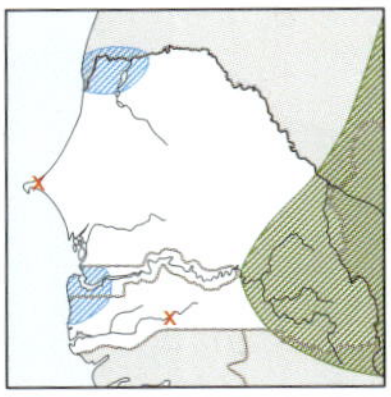

Egyptian Vulture *Neophron percnopterus* 55–75 cm; WS 155–175 cm

N. p. percnopterus. In all plumages, distinctive flight silhouette with ***wedge-shaped tail*** and small, narrow head. Soars with wings held flat; when gliding, wings slightly arched. **Adult *Mainly white with bare yellow face, long, narrow bill and long feathers at back of head***. In flight from below, black flight feathers contrast with white underwing-coverts. **Juvenile** Dark brown; face and legs greyish. Gradually acquires white feathers. Adult plumage attained in c.5 years. **HH** Scarce and local visitor (and resident?), Senegal; rare visitor, Gambia. Favours arid savanna. Red List category: Endangered. **Voice** Usually silent. [CD1:76]

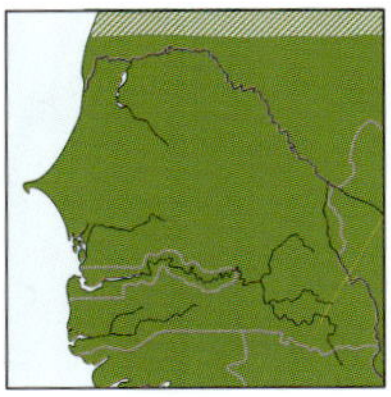

Hooded Vulture *Necrosyrtes monachus* 65–75 cm; WS 170–182 cm

Smallish, dark brown vulture with bare face and throat. In flight, uniformly dark with broad wings and short, rounded tail. **Adult** Bare face and neck pink, flushing red in excitement. Back of head and neck covered in creamy down. Small ruff at base of neck. Bill long, slender and black with pink base. **Juvenile** Similar, but feathers of upperparts tipped buff; bare face and throat whitish (can flush red as in adult); back of head and neck covered in brownish down. Juvenile Palm-nut Vulture (Plate 19) has heavy bill, feathered head and short, rounded tail. Juvenile Lappet-faced Vulture (Plate 21) has similar silhouette and can be difficult to distinguish at distance, if size cannot be gauged, but has larger head with relatively short bill. **HH** Still common throughout but declining. Gregarious and omnivorous scavenger, occurring in various habitats. Often near habitation, scavenging for offal and refuse; also in open country, with other vultures. Red List category: Endangered. **Voice** Usually silent. [CD5:31]

White-headed Vulture *Trigonoceps occipitalis* 78–85 cm; WS 202–230 cm

Distinctive vulture with characteristic peaked hindcrown. Upperparts black, face and neck white to pink. Breast black, rest of underparts white. ***Bill red*** tipped black, cere blue. **Adult male** In flight from below, ***white-tipped greater coverts form diagnostic narrow band separating coverts from flight feathers***. **Adult female** Similar but with ***inner secondaries white***, conspicuous both perched and in flight. **Juvenile** Mainly dark brown; bare parts duller. In flight, entirely dark with contrasting white band on underwing as adult. Gradually acquires adult plumage; sexual dimorphism apparent after first moult (2nd year). Full adult plumage attained in c.6 years. **HH** Scarce to locally uncommon resident in dry to lightly wooded savanna. Red List category: Vulnerable. **Voice** Usually silent.

African White-backed Vulture *Gyps africanus* 80–98 cm; WS 212–218 cm

Very large, mainly brownish vulture with, in all plumages, blackish bill. Soars with wings held in shallow V. **Adult** Mainly brown with ***white back and rump*** (visible when wings are opened and in flight), and bare blackish face. Plumage pales with age; very old birds buffish-white. Skin of neck blackish, sparsely covered with whitish down. White ruff at base of neck. In flight from below, ***black flight feathers contrast with whitish underwing-coverts***. **Juvenile** Darker brown with narrow whitish streaking; ***no white on back***. In flight from below, mainly dark brown with long, narrow white bar near leading edge of wing; short white streaking on underwing-coverts visible at close range. Not always safely separable from juvenile Rüppell's Griffon Vulture. Gradually becomes plainer and paler; white on back appears in 4th year. Adult plumage acquired in 6–7 years. **HH** Fairly common to locally common in lightly wooded savanna; also near villages. Gregarious. Red List category: Near Threatened. **Voice** Usually silent. [CD5:32]

ad
juv
Egyptian Vulture
juv
ad
ad
juv
Hooded Vulture
ad
♀
♂
White-headed Vulture
juv
ad
ad
African White-backed Vulture
juv
juv

Rüppell's Griffon Vulture *Gyps rueppellii* 85–107 cm; WS 220–250 cm

G. r. rueppellii. **Adult** Very large, with dark grey-brown feathers broadly tipped buffish or whitish, producing scaly appearance. Bare skin on head and neck greyish, sparsely covered with whitish down. White ruff at base of neck. Flight and tail feathers blackish. Bill heavy and ***yellowish-horn*** with grey cere. Eye pale yellow. In flight from below, very dark with long, narrow white bar on leading edge of wing and two long and more indistinct white lines formed by white-tipped coverts. **Juvenile** Dull tawny-brown; rather plain, not scaly-looking, above and broadly streaked paler below. Head and neck covered in brownish down; skin initially pinkish, soon darkening. Bill blackish. Eye dark. In flight from below, very young birds with single long, narrow white bar near leading edge of wing not safely separable from juvenile White-backed Vulture. In immature, white tips to underwing-coverts form additional white lines, permitting separation from young White-backed in flight. Adult plumage acquired in 6–7 years. **HH** Fairly common dry-season breeder; less common, S Senegal. In dry, open savannas. Red List category: Near Threatened. **Voice** Usually silent. [CD5:33]

Eurasian Griffon Vulture *Gyps fulvus* 95–110 cm; WS 230–280 cm

G. f. fulvus. **Adult** Huge, ***mainly sandy-brown*** vulture with whitish ruff and yellowish bill and eye. Greater upperwing-coverts have broad buff tips. In flight from below, blackish flight feathers contrast with pale brownish underwing-coverts, which have 1–2 paler lines. Soars with wings held in shallow V. African White-backed Vulture is darker brown and smaller; adult has white patch on back. **Juvenile** Slightly darker and streaky with pale brownish ruff and dark bill and eye. Greater upperwing-coverts lack broad buff tips. In flight, underwing-coverts paler and plainer than adult. Adult plumage acquired in 6–7 years; differences in age groups slight. **HH** Scarce to uncommon Palearctic visitor (late Oct–early May). **Voice** Usually silent. [CD1:77]

Lappet-faced Vulture *Torgos tracheliotus* 98–115 cm; WS 255–290 cm

T. t. tracheliotus. Huge, mainly dark vulture with bare head and neck and massive bill. Soars with wings held flat or slightly arched. **Adult** Mainly blackish with white streaks on underparts and white 'trousers'. Head and neck deep pink with folded skin (lappets). Bill yellowish-horn; cere blue-grey. In flight, broad wings with saw-toothed trailing edge and slightly bulging secondaries; from below, distinct white bar along leading edge of wing and white thighs contrast with mainly black plumage. **Juvenile** Duller and more uniformly darkish. In flight, appears wholly dark. Juvenile/immature White-backed Vulture and Rüppell's Griffon Vulture have more square-tipped tails (slightly more wedge-shaped in Lappet-faced). In immature, bare skin on head becomes more reddish, white gradually appears on underparts, white bar on underwing starts appearing in 2nd year, 'trousers' half white in 4th year. Adult plumage acquired in 6–7 years. **HH** Uncommon, Senegal (a few breeding records); scarce to locally uncommon, Gambia. In dry country. Red List category: Vulnerable. **Voice** Usually silent. [CD1:79]

ad
juv
juv
Rüppell's Griffon Vulture
ad
ad
juv
juv
Eurasian Griffon Vulture
juv
ad
juv
Lappet-faced Vulture

Eurasian Black Vulture *Aegypius monachus* 100–115 cm; WS 250–285 cm

Huge, mainly dark vulture, similar in size to Lappet-faced Vulture. Soars with wings held flat or slightly arched. **Adult** Very dark brown (appearing blackish) with largely pale head. Base of bill and cere pale blue. In flight, appears all dark but wing-coverts darker than flight feathers, wings rather evenly broad with saw-toothed trailing edge; tail slightly wedge-shaped. Differs from Lappet-faced Vulture in having more rectangular wings, uniformly dark lower underparts with contrasting pale feet, and somewhat longer and more wedge-shaped tail. **Juvenile** Plumage and head blackish; base of bill and cere pinkish. In immature, plumage gradually becomes browner after c.2 years, head turns paler. Adult plumage acquired in 6–7 years. **HH** Palearctic vagrant, Senegal (an immature, Feb 2007). Red List category: Near Threatened. **Voice** Usually silent. [CD1:78]

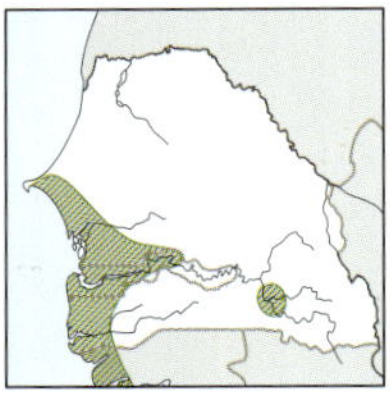

Bat Hawk *Macheiramphus alcinus* 45 cm; WS 110 cm

M. a. anderssoni. Characteristic shape with ***long, broad and pointed wings*** and ***rather long, square tail*** reminiscent of large dark falcon. **Adult** Mainly blackish-brown with dark stripe on centre of white throat and short occipital crest. Occasionally some white below; throat sometimes dark. **Juvenile** Underparts with variable amount of white (often mainly white with irregular dark brown breast-band). **HH** Rare. Frequents forest edge and wooded savanna; often near rivers. Crepuscular. Catches prey (mainly small bats, also birds) in feet and swallows it whole in flight. **Voice** Rather hoarse, high-pitched *kwheet-kwheet-wheet-wheet-*... and *kwik-kwikik-kwik-kwikik-*... [CD5:26]

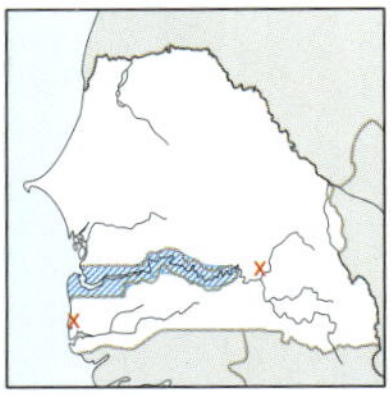

Secretarybird *Sagittarius serpentarius* 125–150 cm; WS 212 cm

Unmistakable. A very large, unique, strange-looking and mainly terrestrial bird of prey with very long, strong legs and short toes adapted for walking and killing prey by impact. ***Mainly grey with bare reddish face and long plume-like erectile feathers on nape tipped black***. Rump and belly black. Tail narrowly tipped white with black subterminal band. ***Long legs yellow with black-feathered tibia*** ('trousers'). In flight, black flight feathers contrast with pale grey wing-coverts and body; very long central rectrices project beyond long legs, producing distinctive silhouette. **Juvenile** similar, but facial skin yellow and central tail feathers shorter. Underwing-coverts tinged brownish. **HH** Rare; no recent records. Usually singly, in open savanna. Strides slowly across grassland, searching for prey (reptiles, small rodents, insects). Also soars high, like vulture. Non-breeding dry-season visitor. Red List category: Vulnerable. **Voice** Usually silent. [CD5:66]

Eurasian Black Vulture

Bat Hawk

Secretarybird

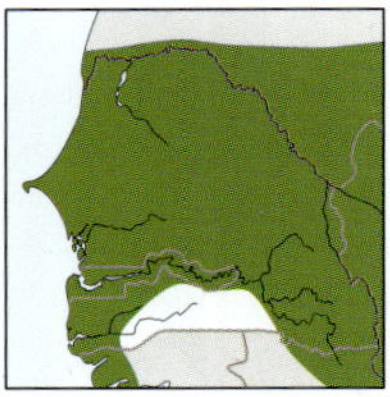

Black-shouldered Kite *Elanus caeruleus* 31–35 cm; WS 75–85 cm

E. c. caeruleus. Smallish, ***very pale*** raptor with relatively broad head and long wings. ***In flight all-white below except contrasting black primaries***. **Adult** Blue-grey above with small dark eye patch and black 'shoulders', white below. When perched, wings project beyond shortish, square tail. **Juvenile/immature** Similar, but washed brown above, feathers edged white, breast washed rufous. **HH** Fairly common to locally and seasonally common in various open habitats. Uses prominent perches. Hovers frequently and persistently, often with trailing feet. Feeds mainly on small rodents. **Voice** Mostly silent. [CD1:71]

African Swallow-tailed Kite *Chelictinia riocourii* 35–38 cm; WS 90 cm

Adult Small, ***very slender and graceful*** with long pointed wings and ***deeply forked tail***. Pale grey above, pure white below with ***narrow black patch on coverts***, along edge of underwing. Tail projects well beyond wingtips when perched. **Juvenile** Tail short, feathers of upperparts edged rufous, breast washed pale rufous. **HH** Uncommon to locally and seasonally common non-breeding intra-African migrant in dry, open country. Gregarious, often in small flocks; occasionally singly or in pairs. Hovers, soars and hangs motionless against wind with spread tail. Catches prey (mainly insects; also lizards and, rarely, small rodents) by dropping from hover; insects also taken in flight. Attracted to bush fires. Roosts in groups (largest near Kaolack, Senegal:16,000, Jan 2007). **Voice** Usually silent. [CD5:27]

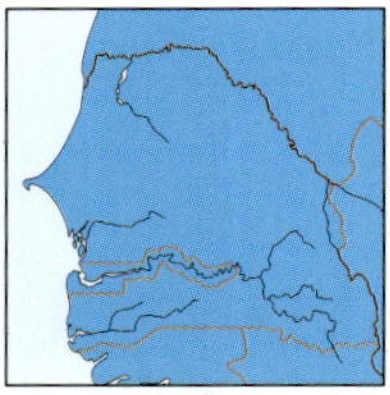

Black Kite *Milvus migrans* 50–60 cm; WS 130–155 cm

M. m. migrans. **Adult** As Yellow-billed Kite, but with ***pale greyish head*** narrowly streaked black and ***black bill*** (contrasting with yellow cere); wing more contrasting, with diffuse, broad, pale diagonal bar across coverts above and pale area at base of primaries below. Eye colour variable, from pale brown (at c.2–3 years old) to pale yellowish (at c.6–7 years). **Juvenile** Very similar to juvenile Yellow-billed Kite but more streaked on head and underparts. **HH** Fairly common to locally common Palearctic passage migrant and visitor (Sep–May) in various habitats. **Voice** As Yellow-billed Kite. [CD1:72; 5:28]

Yellow-billed Kite *Milvus migrans parasitus* 50–60 cm; WS 130–155 cm

M. m. parasitus. Commonest African raptor. Medium-sized with long, ***slightly forked tail*** (which may appear square or only slightly notched when spread). Flight graceful and buoyant with slow wingbeats and long, narrow wings angled back at carpal joint. **Adult** Entirely chocolate-brown (sometimes appearing quite rufous) with ***yellow bill*** and dark eye. Female/immature Western Marsh Harrier have square tails and typically soar and glide with wings in shallow V and less tail twisting. **Juvenile** Rather paler below, with dark bill. Feathers of upperparts have pale tips and underparts some pale streaking. Distinguished from adult Black Kite by lack of pale greyish on head. Beware confusing rufous-looking birds with Red Kite. **HH** Common intra-African migrant; late dry-season breeder (Mar–May). Records throughout the year but situation obscured due to confusion with Black Kite. Ubiquitous and gregarious. Omnivorous; takes any animal matter, incl. offal; often searches for road kills. Attracted to bush fires. Soars and glides with wings slightly arched and tail constantly twisting. Roosts in groups when not breeding. **Voice** Distinctive. Plaintive, tremulous *keeeey-aarrrr*. [CD1:72; 5:28] **Note** If split from Black Kite, scientific name becomes *M. aegyptius parasitus*.

Red Kite *Milvus milvus* 55–72 cm; WS 140–180 cm

M. m. milvus. Distinctive. Medium-sized, elegant raptor with long, narrow wings and long, ***deeply forked tail***. Flight action as Black Kite, but even more graceful. **Adult *Mainly rufous*** narrowly streaked black; pale greyish head. Wings have conspicuous pale diagonal band across coverts above and strongly contrasting, large whitish 'window' on primaries below. ***Tail rusty-red*** above, paler below. **Juvenile** Similar to adult, but underparts paler; upperwing-coverts narrowly tipped white; tail browner. **HH** Palearctic vagrant, Gambia (Dec 1987, Feb & May 1996). Red List category: Near Threatened. **Voice** Usually silent. [CD1:73]

ad
Black-shouldered Kite
juv
ad
juv
ad
African Swallow-tailed Kite
juv
ad
ad
Black Kite
ad
juv
ad
Yellow-billed Kite
ad
Red Kite

HARRIERS *Medium-sized and slender with long wings and tails. Sexually dimorphic; females mainly brownish, some posing identification problems. Adult plumage attained during 2nd calendar year. Flight buoyant with wings characteristically held in shallow V. Hunt low, quartering the ground and dropping on prey (frogs, small rodents, birds, insects). Frequent open country, cultivated areas and marshes. Roost on the ground. Usually silent.*

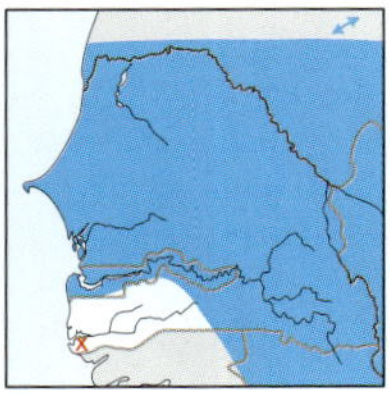

Pallid Harrier *Circus macrourus* 40–50 cm; WS 95–120 cm

Adult male ***Pale grey*** above with ***black wedge at wingtip***, mainly white below. **Adult female** Dark brown above with ***white uppertail-coverts***. Distinguished from adult female Montagu's Harrier by darker upperwing with almost ***uniformly dark secondaries*** and ***less neatly marked underwing***, with paler primaries contrasting with darker secondaries and darkish greater secondary coverts. Also has ***narrow pale collar*** (usually absent in Montagu's) and less white around eye. **Juvenile** Plain rufous below; dark cheek patch reaching base of bill, bordered by dark eye-stripe and ***narrow pale collar*** (usually absent in juvenile Montagu's); white around eye less extensive than in Montagu's; neck-sides uniform dark brown. Underwing lacks dark tips to inner primaries. **HH** Uncommon or scarce to locally fairly common Palearctic visitor (mainly Nov–Mar) in open habitats. Prey includes insects, esp. grasshoppers, small mammals and small birds; also frogs. Red List category: Near Threatened. **Voice** [CD1:82]

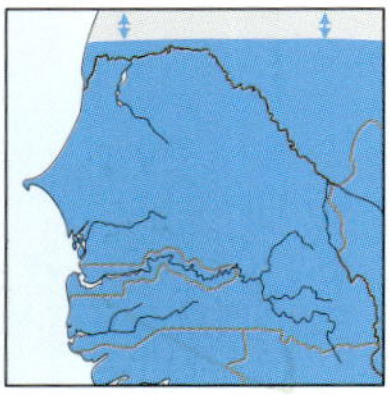

Montagu's Harrier *Circus pygargus* 38–50 cm; WS 96–120 cm

Most elegant harrier with narrow pointed wings, long tail and very agile, tern-like flight. **Adult male** Mainly blue-grey; belly white variably streaked chestnut. In flight, wings with black tips and one black bar above, two below. Rare dark morph is blackish with greyish tail. **Adult female** Facial pattern less distinct than Pallid Harrier, white, claw-shaped area around eye larger. Upperwing paler than Pallid, often with ***dark bar on base of secondaries*** (never shown by Pallid); underwing more uniform, with ***three well-defined dark bands and broad pale band along dark trailing edge***. Rare dark morph is blackish-brown with pale greyish on underwing and banded tail. **Juvenile** Usually deeper rufous below than Pallid, cheek patch smaller, ***no prominent pale collar***. Underwing has dark tips to primaries. **HH** Uncommon to locally common Palearctic visitor (mainly Oct–Apr) in open habitats. Feeds mainly on insects, esp. grasshoppers. Roosts on ground, also during hottest part of day, sometimes in groups (largest night roost near Darou Khoudoss, Senegal: 1300, Feb 2008). Wintering grounds in Sahel between c.11°N and 16°N. **Voice** [CD1:83]

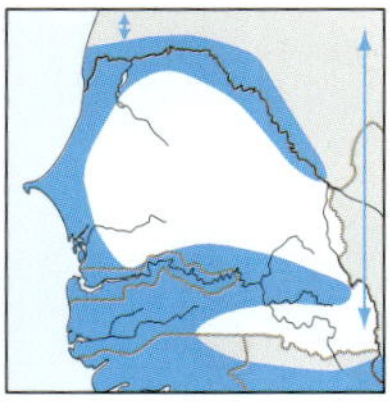

Western Marsh Harrier *Circus aeruginosus* 42–56 cm; WS 110–140 cm

C. a. aeruginosus. **Adult male** Mainly brown, with buff-brown head and breast streaked darker. In flight, striking ***tricoloured wings*** with large black tips and plain, blue-grey tail. Rare dark morph is blackish-brown with whitish band on underwing. **Adult female** Mainly dark brown with creamy crown, nape and throat. Leading edge of wing usually creamy, breast often has pale patch. **Juvenile** Similar to adult female but darker brown and lacks pale breast patch. Creamy areas usually slightly deeper in colour and restricted to crown and throat, but sometimes also on leading edge of wing; rarely entirely lacking. Adult male plumage acquired in c.3 years. **HH** Fairly common to locally common Palearctic visitor (mainly Oct–Mar). Singly or in pairs in or near marshy areas; also hunts over other open habitats, esp. grassland. Takes frogs, birds and small mammals; also reptiles and insects. **Voice** [CD1:84]

♀
♂
Pallid Harrier
juv
juv
♀
♂
♂
dark morph
♂
juv
juv
Montagu's Harrier
♀
♀
♀
dark morph
♂
♂
dark morph
♀
♂
Western Marsh Harrier
♂

***SNAKE EAGLES** Medium-sized eagles, with relatively large heads and large yellow eyes (producing somewhat owl-like appearance when perched), long, broad wings, bare tarsi and short toes. Given to soaring and perching for long periods. Drop on their prey (mainly snakes and other reptiles) and kill it on ground with their powerful feet. Occur in open country and woodland.*

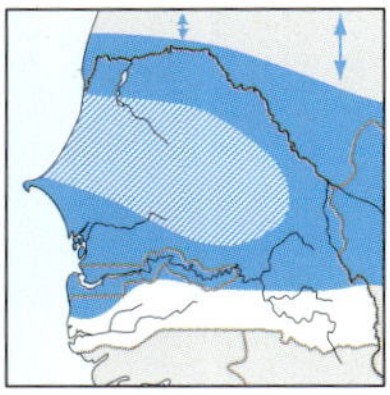

Short-toed Snake Eagle *Circaetus gallicus* 62–69 cm; WS 162–185 cm

Adult male Head and upperparts typically grey-brown. Below largely white, with upper breast mottled brown or brown streaked whitish, and rest of underparts irregularly blotched and barred brown. Upperwings have variable contrast between greyish-brown coverts and darker brown flight feathers. Underwings white with irregular, narrow dark lines on coverts and three narrow dark bands and dusky trailing edge to flight feathers. Tail has 3–4 brown bars with subterminal bar broadest. Some have brown parts darker, creating dark hood, and more distinct markings below. **Adult female** Similar, but usually with throat and upper breast all dark without pale streaking. **Juvenile** Head and breast more rufous-brown, blotches on underparts and underwing pattern usually paler and more indistinct. By 3rd (or 4th) calendar-year head, breast and upper mantle plumage replaced by pale feathering with dark shaft streaks. In 4th–6th calendar-year gradually acquires adult plumage. Full adult plumage typically attained in 8th or 9th calendar-year. **HH** Uncommon to locally fairly common Palearctic visitor (mainly Oct–Mar). In open savanna woodland. Some immatures remain in wintering areas during northern breeding season. Frequently hovers or hangs motionless in wind. **Voice** Usually silent in Africa. [CD1:80]

Beaudouin's Snake Eagle *Circaetus beaudouini* 62–69 cm; WS *c.* 170 cm

Adult male Similar to Short-toed Snake Eagle but head, upperparts and upper breast typically ***darker grey-brown*** and lower breast and belly white with variable amount of ***narrow*** dark bars (sometimes restricted to breast-sides and flanks). Throat has variable amount of white. Underwing usually plain white with dark trailing edge and two narrow parallel blackish bars across flight feathers (even the palest Short-toed has underwing-coverts with dark markings). Upperwing more uniformly dark. Head appears large and owl-like on perched bird, but not so in flight. **Adult female** Throat and upper breast usually all dark. **Juvenile** Dark brown above, more rufous-brown below. Underwing-coverts brownish; remiges with indistinct dusky bars. Tail has indistinct dusky bands. Differs from juvenile Brown Snake Eagle mainly in narrow dark bars on underside of flight feathers, extensive dark tips to primaries and lack of white bands in tail. Pale buzzards and Osprey have dark carpal patches in flight. **HH** Rare to scare or locally frequent in open savanna woodland. Frequently hovers or hangs motionless in wind. Red List category: Vulnerable. **Voice** Clear *kee-u*, uttered in series. Usually silent. [CD5:34]

Brown Snake Eagle *Circaetus cinereus* 66–75 cm, WS 160–185 cm

Adult ***Entirely dark brown***. In flight, ***unmarked whitish remiges*** contrast with dark brown body and underwing-coverts. Tail dark brown with three narrow pale bars. **Juvenile** Similar to adult, or with white streaks on head or white feathers on lower underparts. Immature acquires variable amount of white mottling on underparts. **HH** Uncommon to locally fairly common in savanna woodland. Northward movement with rains recorded. Perches very upright and for long periods in prominent vantage point. Soars, but rarely hovers or hunts in flight. **Voice** Usually silent. Guttural *khok-khok-khok-khaw* in soaring display. [CD5:36]

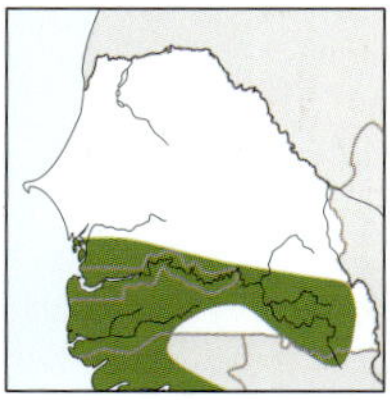

Western Banded Snake Eagle *Circaetus cinerascens* 50–58 cm; WS 120–134 cm

Stocky, shorter tailed than other snake eagles. **Adult male** Grey-brown; indistinct whitish barring on lower underparts. ***Tail with single broad white bar across centre*** (usually concealed when perched, but conspicuous in flight). Underwing white with four dark parallel bars on flight feathers. Cere and base of bill pinkish or orange-yellow. **Adult female** Usually darker; some all brown as illustrated. **Juvenile** Much paler. Top of head whitish streaked dark; upperparts pale brown, feathers edged buff. Underparts buffish-white, slightly darker on breast and with some brown on belly and thighs. Underwing and tail pattern as adult but duller. Immature is variably mottled and streaked brown on underparts, later apparently acquiring brown plumage with varying amounts of barring on lower underparts. **HH** Uncommon in savanna woodland; more frequent during rains. Perches unobtrusively for long periods on large tree, often in riverine woodland. **Voice** Vocal. Loud *kho-kho-kho-kho-*... descending in pitch, in aerial display or from perch. Also a plaintive, nasal *ko-ah ko-ah*... and *ko-ah ko-koaaah* from perch. [CD5:37]

juv
♀
♂
pale ad/juv
♀
Short-toed Snake Eagle
juv
♀
juv
♀
♂
Beaudouin's Snake Eagle
♂
imm
ad
ad
Brown Snake Eagle
juv
ad
juv
brown ♀/imm
ad
Western Banded Snake Eagle

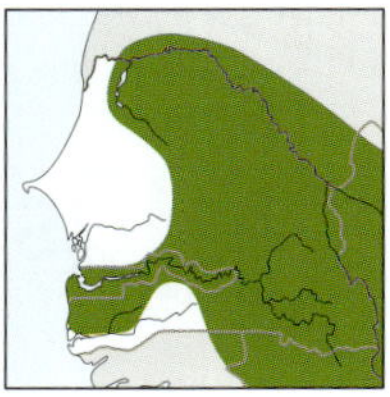

Bateleur *Terathopius ecaudatus* 55–70 cm; WS 170–187 cm

Stocky, large-headed, almost tail-less eagle. In flight, ***unique silhouette with long, broad-based wings curving to a point*** and ***very short tail***. Soars with wings held in marked V, frequently canting. **Adult male** ***Black with chestnut mantle to uppertail-coverts***, grey lesser and median upperwing-coverts, and ***chestnut tail***. Loose feathers of head give cowled appearance. Bare facial skin, cere and feet bright red (sometimes fading to yellowish). In flight, white underwing-coverts contrast with black inner primaries and secondaries. Some have creamy or pale brown upperparts. **Adult female** Similar but with grey panel on secondaries and inner primaries. In flight, underwing largely white with black trailing edge. **Juvenile** Entirely brown. Bare facial skin and cere greenish-blue. Note large head, long wings, invisible tail, dark eye. In flight, silhouette similar to adult, but tail slightly longer. In immature, plumage becomes sooty-brown, tail shortens, face turns yellowish, feet pinkish. Recognisable adult plumage gradually acquired at 5–7 years; full adult plumage attained in 7–8 years. **HH** Generally uncommon and decreasing, but may locally still be fairly common. Usually singly or in pairs, occasionally in small groups, in savanna. Feeds on small animals, incl. termites, and carrion. Northern limit of range expands northwards during rains. Red List category: Near Threatened. **Voice** Far-carrying, barking *kow-aw* and soft *ko-ko-ko-ko-koaaagh*. Usually silent. [CD5:38]

Long-crested Eagle *Lophaetus occipitalis* 53–58 cm; WS 115 cm

Unmistakable. Medium-sized ***blackish eagle with long, loose crest*** and, ***in flight, large white patches ('windows') on primaries both above and below***. Flies with rapid shallow wingbeats interspersed by glides; soars with wings held flat. **Juvenile** Crest shorter (attains adult length after 3 months), eye grey (yellow after 9 months). **HH** Uncommon in south and east; rarely recorded in north. Frequents wooded savanna, cultivation and forest edge. Perches conspicuously and for long periods atop trees or poles. Captures prey (mainly rodents) on ground. **Voice** Vocal. High-pitched, nasal *keeee-aah* and a series of sharp *kikikikikikeeah* with variations, from perch or in soaring display flight. Also short nasal *kwoh*. [CD5:62]

juv
♀
♂
Bateleur
ad
juv
♀
♂
Long-crested Eagle
ad
ad

African Harrier Hawk *Polyboroides typus* 60–68 cm; WS *c.*160 cm

P. t. pectoralis. Medium-sized raptor with small head, broad wings and relatively long, slightly rounded tail. Flight with slow, buoyant wingbeats alternated with glides; frequently soars. **Adult** Mainly grey with bare, yellow to orange face and lower underparts finely barred black and white. ***Tail black with white band across centre*** and narrow white tip. ***In flight, has characteristic pattern of grey wings broadly tipped black, and black tail with distinctive white bar***; on underwing, narrow white line separates grey from broad black trailing edge and tips. **Juvenile/immature** Dark brown to tawny, variably mottled, streaked and barred. In flight, remiges and rectrices more or less distinctly barred; greater coverts tipped dark. Full adult plumage acquired in *c.*2–3 years. Young best identified by small head and jizz; beware confusing with other brown raptors (especially eagles). **HH** Generally uncommon to rare, N Senegal; fairly common, Gambia and S Senegal. Northern movements during rains recorded. Singly or in pairs in forest and wooded savanna. Long legs very flexible at 'knee', enabling insertion into tree holes and crevices. Clings to tree trunks with flapping wings while probing cavities. Feeds on variety of small animals; also oil-palm husks and eggs. Regularly raids nests, esp. of colonial species, such as weavers. Male performs undulating display flight. **Voice** Plaintive whistling *sueeeee* and high-pitched *hueeeup-hueeeup-hueeeup-*... [CD5:40]

African Cuckoo Hawk *Aviceda cuculoides* 40 cm; WS 91 cm

A. c. cuculoides. Plumage pattern and flight silhouette with long wings and longish square tail superficially reminiscent of large cuckoo. When perched, wings almost reach tip of tail. **Adult** Dark grey above and on breast, lower underparts white with broad chestnut bars. In flight, shows ***chestnut underwing-coverts***; flight feathers whitish with long, narrow, black parallel bars. **Juvenile** Dark brown above with whitish supercilium, white below variably marked with brown drop-like spots (occasionally only some coarse barring on flanks). Underwing-coverts buffish speckled dark brown; remiges barred like adult. **HH** Rare to scarce non-breeding visitor, S Senegal and Gambia. Favours dense woodland. Usually unobtrusive. Pairs perform undulating and tumbling display flights, with much calling. Hunts from perch, catching prey (insects, lizards) on ground or in fast, agile flight. **Voice** Usually silent. Loud, plaintive *peee-uuw*; also sharp *wheet-wheet*. [CD5:25]

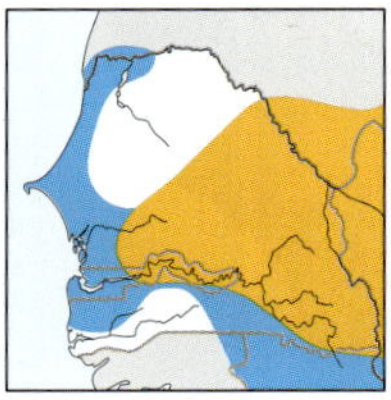

Grasshopper Buzzard *Butastur rufipennis* 41–44 cm; WS 102 cm

Slender with long wings, tail and legs and, ***in flight, conspicuous rufous, black-tipped 'hand'***. **Adult** Dark grey-brown above, rufous with narrow dark streaking below. Underwing has well-defined blackish border to pale rufous flight feathers. **Juvenile** More rufous than adult; feathers of upperparts edged rufous. **HH** Common intra-African migrant, breeding locally inland in semi-arid belt during rains, moving south and west in dry season. In Gambia, common Apr–Jul, uncommon to locally scarce Nov–Jan; single breeding record, Kiang West, Jul 1996. In open country. Perches conspicuously and for long periods atop small trees and bushes. Often near bush fires, catching disturbed insects (mainly grasshoppers). Flight low and buoyant, rather 'harrier-like'; also soars on flat wings. **Voice** Loud *ki-ki-ki-ki-kee* when breeding. Otherwise silent. [CD5:53]

ad
juv/imm
African Harrier Hawk
juv/imm
juv/imm
ad
juv
ad
African Cuckoo Hawk
ad
juv
ad
ad
Grasshopper Buzzard

Lizard Buzzard *Kaupifalco monogrammicus* 35–37 cm; WS 79 cm

K. m. monogrammicus. Medium-sized, stocky grey raptor with ***diagnostic black stripe on centre of white throat*** and ***single white band on black tail. Uppertail-coverts white***. Lower underparts white narrowly barred grey. Underwing white narrowly barred grey; appears all white at distance. Rarely, tail has two white bands. ***Cere and legs pinkish-red*.** **Juvenile** similar but mantle and wing-coverts edged buff, underparts washed buffish, bare parts paler. Adult plumage attained within one year. **HH** Fairly common resident in various types of woodland and cultivation. Perches for long periods in trees or on wires and poles, from which it catches prey (lizards, small snakes, insects, also small mammals and birds) with quick dash. **Voice** Far-carrying, characteristic, melodious *KLEEUUu-klukluklukluklu*. [CD5:54]

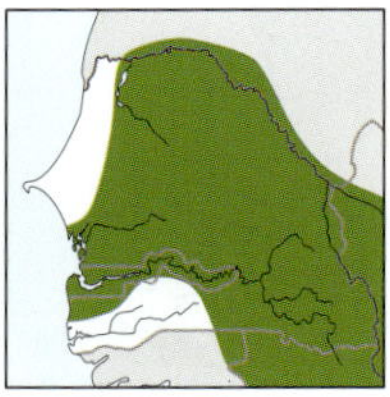

Gabar Goshawk *Micronisus gabar* 28–36 cm; WS 60 cm

Smallish *Accipiter*-like raptor with ***prominent white uppertail-coverts***. **Adult** Plain grey above and on breast, rest of underparts narrowly barred grey. Cere and legs pinkish-red. Underwing narrowly barred; four black bars on tail. Dark Chanting Goshawk is much larger. Melanistic morph is all black except for white barring on upperside of primaries and faint white bars on tail; no white uppertail-coverts. **Juvenile/immature** Pale, streaky head; underparts whitish streaked rufous-brown on breast and barred rufous-brown on belly and flanks; cere and legs yellowish to orange. Underwing-coverts narrowly barred rufous-brown. **HH** Fairly common to uncommon resident in wooded savanna and thornbush. Some movements recorded, south in dry season, north in rains. Hunts from perch or on the wing, feeding mainly on small birds. Often robs nests. **Voice** Rapid, high-pitched *kik-kik-kik-...* or *kwk-kwik-kwik-...*; also more melodious *kwee-kwee-kwee-...* [CD5:42]

Dark Chanting Goshawk *Melierax metabates* 38–48 cm; WS 95–110 cm

M. m. metabates. Medium-sized, grey raptor with long red legs, long broad wings and longish, rounded tail. **Adult** Plain grey above and on breast, rest of underparts narrowly barred grey. ***Uppertail-coverts white finely barred grey***. In flight, dark wingtips above and below. Cere and legs pinkish-red. **Juvenile/immature** Brown above, variably marked brown and white below. Cere and legs dull yellowish to orangey. In flight, note shape of wings and tail; uppertail-coverts white narrowly barred grey-brown. **HH** Fairly common to locally common resident in open woodland and thornbush.. Some irregular southward movements during dry season. Perches prominently and upright on poles or atop trees and bushes. Feeds on variety of small animals, incl. lizards, snakes, mammals and insects; also birds. **Voice** Usually silent except when breeding. Melodious call, a loud, accelerating, piping *wheeeow-whew-whew-whew* or *kleeu-kleeu-klu-klu-klu* by both sexes, from perch or in flight. [CD1:85]

ACCIPITER *HAWKS* ***(Plates 28–29)*** *Small to medium-sized, with rather short, rounded wings and long tails. Tarsi bare and long. Large size difference between sexes; females much larger. Swift and agile. Hunt by dashing on their prey (mainly birds and small mammals), following a stealthy approach behind cover.*

Ovambo Sparrowhawk *Accipiter ovampensis* 33–40 cm; WS 67 cm

Medium-sized *Accipiter*. **Adult** Grey above; ***entire underparts***, incl. throat, ***narrowly barred grey***; Tail grey above with four broad blackish bars, narrow white tip and ***rather faint white shaft streaks in grey bars***. Underwing barred blackish on remiges. Eye dark red. Cere and legs orange-red to yellow. Compare Gabar Goshawk. Rare melanistic form is all black with paler tail bars; remiges barred dark below. **Juvenile/immature** Two forms. Head and underparts either rufous or whitish variably (usually lightly) streaked dark brown; some barring on flanks. Both forms have brown upperparts with buffish edges to feathers (producing scaly effect when fresh), ***whitish supercilium and dark patch on ear-coverts***. **HH** Intra-African migrant, frequenting woodland. Scarce (Gambia, Jun–Jan) to rare (S Senegal, Jun–Jul). Status inadequately known. Fast, dashing flight; also soars. Feeds mainly on birds. **Voice** High-pitched *keep-keep-keep-...* Usually silent except when breeding. [CD5:49]

ad
Lizard Buzzard
ad
ad
melanistic morph
juv
ad
juv
ad
Gabar Goshawk
ad
juv
juv
juv
ad
Dark Chanting Goshawk
ad
ad
melanistic
morph
juv
rufous
morph
juv
whitish
morph
ad
ad
Ovambo Sparrowhawk

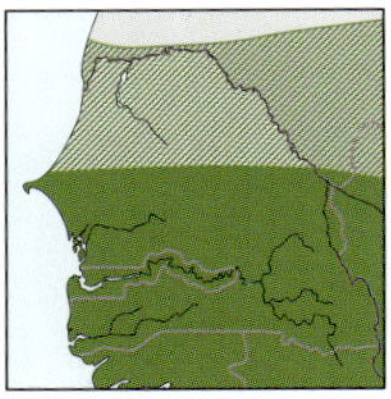

Shikra *Accipiter badius* 28–30 cm; WS 58–60 cm

A. b. sphenurus. Small, pale *Accipiter*. **Adult** Blue-grey above, white below with fine rufous barring. ***In flight uniformly grey above*** (no white on rump or tail), ***very pale below***; wingtips dusky. Eye orange to red. Cere and legs yellow. **Juvenile/immature** Brown above, dark throat-stripe, blotched and barred below, tail barred. In flight from below, indistinctly barred on wings and tail. Eye yellow. Adult plumage acquired in two years. **HH** Fairly common to locally common throughout in savanna, woodland and other open habitats. Mainly wet-season visitor in extreme north. Soars frequently. Less unobtrusive and more vocal than congeners. Usually hunts from perch, dropping onto prey (mostly lizards) on ground; also takes insects and birds. **Voice** Vocal. Fast, sharp *ki-ki-ki-ki-…* Also *kiwik-kiwik-kiwik-kiwik-…* (male), and plaintive *keeu-keeu-keeu-…* (female). [CD5:46]

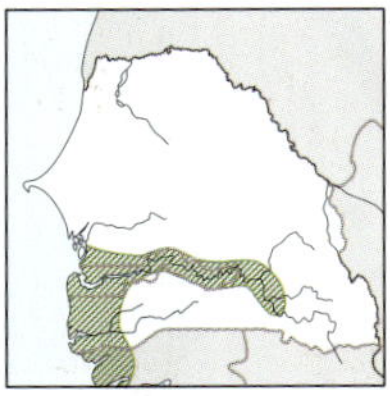

Red-thighed Sparrowhawk *Accipiter erythropus* 23–28 cm; WS 40 cm

A. e. erythropus. Very small (size of wood dove). **Adult** Blackish above with ***white crescent on uppertail-coverts*** (conspicuous in flight) ***and three small, broken white bars on blackish tail***. Grey-white tinged pinkish below; underwings and undertail barred blackish. Eye orange to red with bright red orbital ring. Female is noticeably larger with browner upperparts. **Juvenile** Brown above, variably barred brown below. **HH** Rare forest resident. Unobtrusive; principally active in early morning or late afternoon. Feeds mainly on small birds, lizards and large insects. **Voice** Usually silent. High-pitched *kik-kik-kik-kik-…* or *ki-ki-ki-ki-kiw-kiw.* [CD5:47]

African Goshawk *Accipiter tachiro* 36–48 cm; WS 70 cm

A. t. macroscelides (Red-chested Goshawk). Medium-sized forest *Accipiter* with three white spots in centre of blackish tail. **Adult** Slate-grey above, narrowly barred orange-chestnut below, flanks and thighs orange-chestnut. Underside of flight feathers barred black and white, underwing-coverts barred orange. **Juvenile** Dark brown above with whitish supercilium, whitish below with dark throat-stripe, spotted breast and barred flanks and thighs. **HH** Locally uncommon to fairly common resident, coastal Gambia and SW Senegal (Casamance); occasionally recorded further inland. In forest and outliers in savanna. Unobtrusive, usually keeping to dense cover, except when soaring and calling over territory. Feeds mainly on small birds and mammals; also insects. **Voice** Series of sharp, abrupt *kwit!* notes, in (usually high) display flight over territory or from perch, esp. at dawn. Resembles call of Velvet-mantled Drongo. Juvenile has plaintive, nasal *we-aaaaauw.* [CD5:43,44] **Note** If split from African Goshawk, scientific name becomes *A. toussenelii macroscelides*.

Eurasian Sparrowhawk *Accipiter nisus* 28–41 cm; WS 58–80 cm

A. n. nisus. Small to medium-sized. Tail long, narrow-based and square with sharp corners, narrow white tip, broad subterminal band and 4–5 narrower dark bars. **Adult male** Dark slate-grey above, narrowly barred rufous below; rufous ear-coverts. **Adult female** Much larger and browner with white supercilium and brownish barring on underparts. **Juvenile** Browner than female; barring more irregular. **HH** Palearctic vagrant, Gambia (Feb). In open country and woodland. **Voice** Usually silent on migration. [CD1:86]

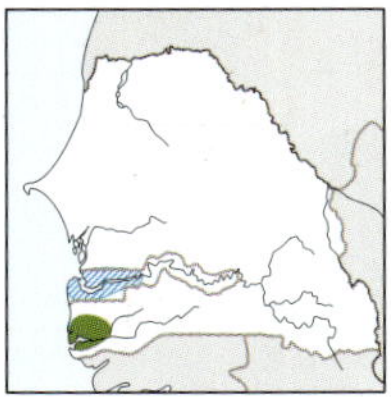

Black Sparrowhawk *Accipiter melanoleucus* 46–55 cm; WS 102 cm

A. m. temminckii. Largest *Accipiter*. **Adult** ***Head and upperparts black contrasting with pure white underparts; mottled black patches on breast-sides, flanks and thighs***. Tail long, dark brown above, whitish below, barred blackish on both surfaces (terminal band broadest). Underwing white with narrow parallel blackish bars on flight feathers. Rare melanistic morph all black with white throat. **Juvenile/immature** Two forms. Upperparts dark brown with rusty edges to feathers. Underparts and underwing either rufous or white, streaked brown (most heavily on breast). Underwing barred blackish on remiges. Attains adult plumage after c.30 months. **HH** Rare non-breeding visitor, coastal Gambia (Mar–May, Nov) and SW Senegal (Casamance, Dec–Feb, Jun) in various types of forest. Feeds mainly on birds, esp. doves and pigeons, caught inside or outside forest. **Voice** Usually silent; vocal when breeding. Call *kyip* or *klee-ep* (male) and deeper *chep* (female); duet an alternating series. [CD5:51]

juv
ad
ad
ad
juv
Shikra
juv
ad
ad
Red-thighed Sparrowhawk
ad
♂
♀
juv
juv
ad
ad
African Goshawk
♀
♂
Eurasian Sparrowhawk
melanistic morph
ad
juv rufous-breasted form
juv white-breasted form
ad
Black Sparrowhawk
ad
juv rufous-breasted form

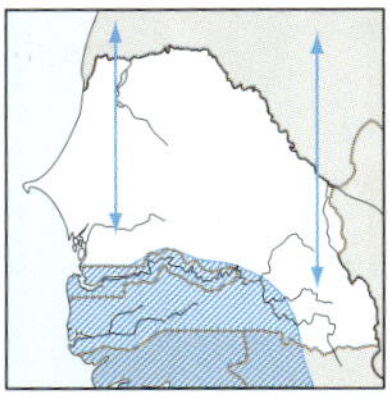

European Honey Buzzard *Pernis apivorus* 51–60 cm; WS 113–150 cm

Medium-sized raptor with highly variable plumage (from blackish through various shades of brown to whitish below), but with characteristic ***small-looking head and slender neck, protruding well in flight***. Tail rather long, with rounded corners. Bill small. Soars on flat wings, ***slightly bowed when gliding***. **Adult male** Blue-grey head, brown upperparts, white underparts with bold, irregular dark bars. In all plumages, ***diagnostic tail pattern with dark terminal band and two narrower bars near base***. Underwing has dark carpal patches, ***prominent broad black trailing edge*** and, typically, parallel bands across base of pale grey flight feathers. Primaries with sharply demarcated, small black tips. Eye yellow. **Adult female** Usually browner on head and upperparts than male. Tail bands narrower, two bars near base wider spaced. Flight feathers usually with more, and more evenly spread, bars. **Juvenile** Flight and tail feathers with 4–5 more evenly spread bars. Uppertail-coverts often showing whitish U. Eye dark. **HH** Rare to scarce Palearctic migrant (mainly Nov–May; some immatures remain during rainy season). Favours wooded habitats, but can occur anywhere on migration. Principally insectivorous. **Voice** Usually silent in Africa. [CD1:70]

BUTEO *BUZZARDS* *Medium-sized, with broad-winged, eagle-like shape, but generally smaller and with only moderately 'fingered' wings. Tarsi bare. Plumages very variable, sometimes creating identification problems. Mostly seen soaring, wings held in shallow V, or perching on vantage points. Prey mainly upon mammals and reptiles caught on ground. Occur in woodland and open country.*

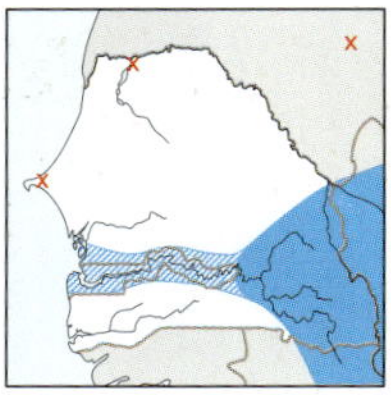

Red-necked Buzzard *Buteo auguralis* 40–50 cm; WS 95 cm

Adult Dark brown above with chestnut head-sides and nape. Below, throat and breast dark brown; rest of underparts white variably blotched dark brown. Relatively short rounded tail ***bright rufous*** with narrow black subterminal band. Upperwing has pale panel at base of primaries. Underwing white with black tips to flight feathers and black carpal markings continuing as irregular line along coverts. **Juvenile** Brown above, tail brown barred black, ***underparts entirely white*** blotched black on breast and flanks. **HH** Scarce to locally uncommon intra-African migrant, Senegal (mainly during wet season); rare, Gambia. In woodland and cultivated areas. **Voice** Loud, drawn-out, mewing *peeee-ah*. [CD5:56]

Common Buzzard *Buteo buteo* 45–58 cm; WS 110–132 cm

B. b. vulpinus (Steppe Buzzard). Rather compact raptor with broad, rounded wings and tail. In flight from below, flight feathers whitish indistinctly barred grey and tipped blackish (forming dark trailing edge), dark carpal patch. Plumage variable. Underparts and tail typically pale rufous-brown. In foxy morph these are uniform dark chestnut; tail with subterminal band. Rare dark morph mainly blackish. ***B. b. buteo*** very dark to pale brown (less rufous), usually with diffuse pale breast-band; underwing pattern less contrasting, barring on flight and tail feathers broader and less distinct. **HH** Rare Palearctic migrant, Nov–Feb (*B. b. vulpinus*). Occasional presence of *B. b. buteo* suspected. **Voice** Usually silent in Africa. [CD1:88]

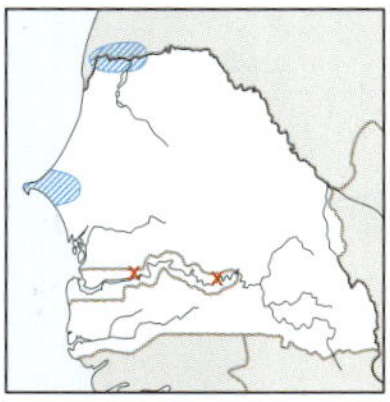

Long-legged Buzzard *Buteo rufinus* 50–65 cm; WS 115–163 cm

B. r. rufinus & cirtensis. Resembles Common Buzzard but slightly larger, looking heavier when perched, with larger bill and longer legs, and has dark rufous-brown belly and 'trousers' contrasting with pale buff, almost unmarked, head and upper breast. Tail plain orange-rufous, often paling at base. In flight, appears more eagle-like, with longer wings and slightly longer tail and has large black carpal patch and trailing edge to underwing. Plumage variable, but typically mainly buffish and orange-rufous. Dark morph mainly blackish; below, greyish flight feathers narrowly barred darker and with dark trailing edge; tail barred with broad subterminal band. Races indistinguishable in field; N African *cirtensis* slightly smaller. **Juvenile** Similar but wing lacks distinct dark trailing edge and tail more buff-grey with faint barring (rarely with narrow dark terminal band). **HH** Rare Palearctic migrant (Oct–Mar), Senegal (both races); vagrant, Gambia. **Voice** Usually silent. [CD1:90]

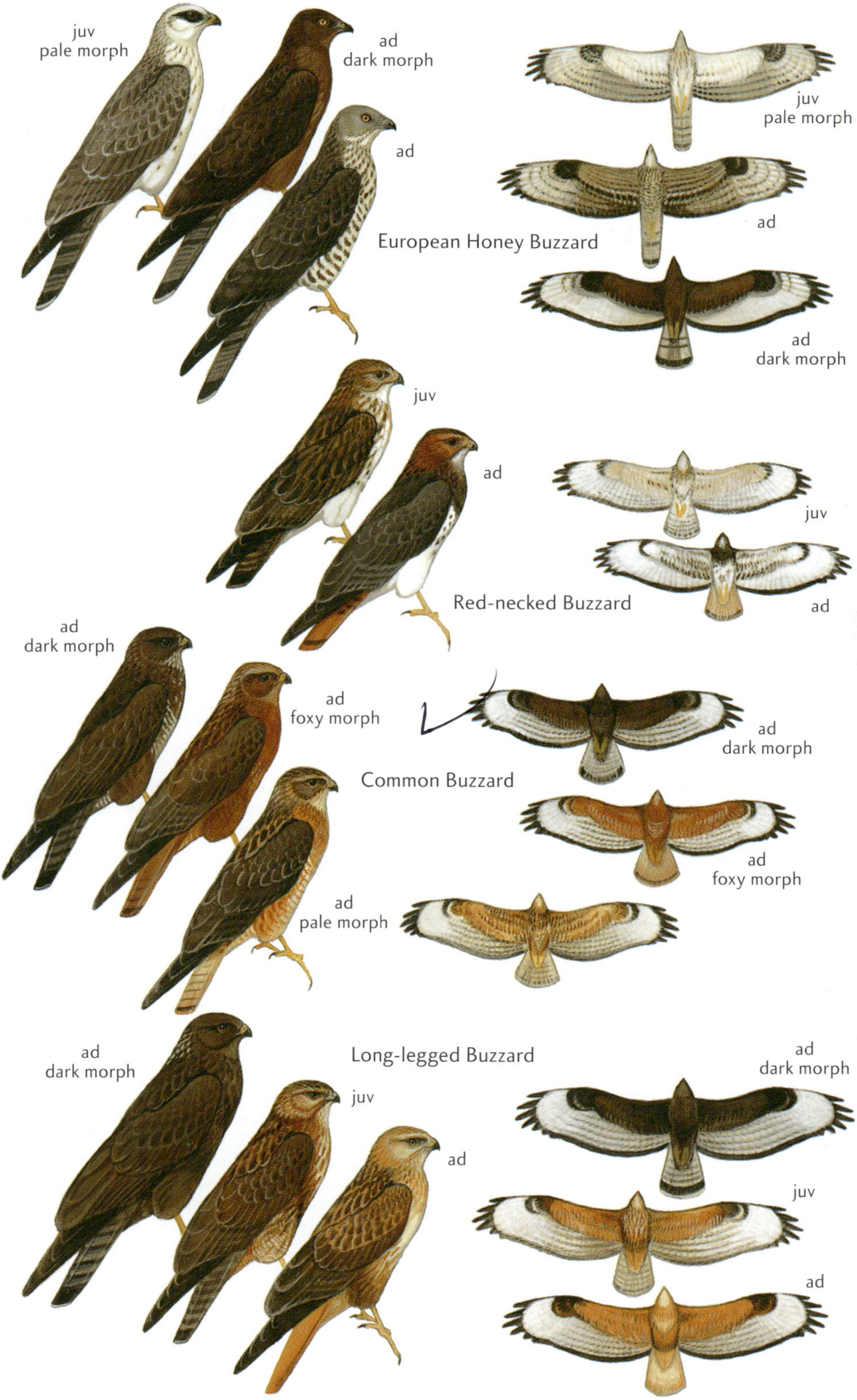
juv
pale morph
ad
dark morph
ad
juv
pale morph
ad
European Honey Buzzard
ad
dark morph
juv
ad
juv
ad
Red-necked Buzzard
ad
dark morph
ad
foxy morph
ad
dark morph
Common Buzzard
ad
foxy morph
ad
pale morph
ad
dark morph
Long-legged Buzzard
ad
dark morph
juv
ad
juv
ad

PLATE 31: SMALL AND MEDIUM-SIZED EAGLES

EAGLES (Plates 31–32) *Medium-sized to very large raptors with broad 'fingered' wings. Tarsi feathered. Given to soaring. Prey on large and small mammals, birds, and reptiles; some eat carrion and insects. In general fiercer and more active than buzzards. Occur in open country, woodland and forest.*

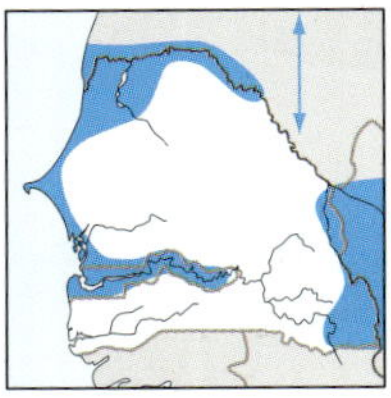

Booted Eagle *Hieraaetus pennatus* 45–55 cm; WS 110–132 cm

Small, buzzard-sized eagle with, in all plumages, ***small white spot at base of neck*** on leading edge of wing ('landing lights'), broad buff diagonal band on upperwing and narrow whitish crescent at base of square tail. Legs heavily feathered (hence name). Two morphs. Pale morph brown above, ***creamy-white below with contrasting blackish flight feathers***. Tail greyish and unbarred. Dark morph entirely dark brown. Black Kite may appear similar to dark morph from above, but is more slender with shallowly forked tail and has flapping and gliding flight. Wahlberg's Eagle is longer winged, less stocky and usually soars with closed tail. **HH** Uncommon to scarce Palearctic visitor (mainly Nov–Mar). In wooded savanna. **Voice** Usually silent on migration. [CD1:96]

Bonelli's Eagle *Aquila fasciata* 55–72 cm; WS 145–180 cm

A. f. fasciata. Medium-sized eagle with relatively small, protruding head and longish tail. **Adult** Dark brown above, white, lightly streaked dark below. In flight, small white patch on mantle; no pale 'windows' on primaries above; white body contrasts with dark underwings; longish tail with black terminal band. Adult African Hawk Eagle has conspicuous white 'windows' on primaries above and whiter secondaries below. **Juvenile** Rusty-brown below. In flight, resembles juvenile African Hawk Eagle, but lacks obvious pale areas on primaries above and has dusky secondaries below. Distinctive silhouette in flight should distinguish it from most other raptors. Gradually acquires adult plumage over 3 years, becoming darker brown; underparts streaked dark; broad dark bands on greater coverts and tail starting to show. **HH** Rare Palearctic winter visitor, N Senegal. Soars less than most other raptors, with wings held flat or slightly arched and tail often closed. Glides on slightly arched wings with carpal joints pressed forward and almost straight rear edge. **Voice** Usually silent. [CD1:95]

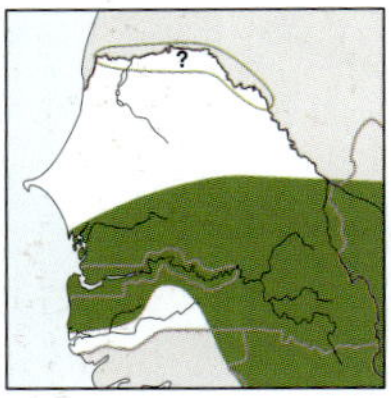

African Hawk Eagle *Aquila spilogaster* 60–74 cm; WS 142 cm

Medium-sized eagle with long wings narrowing towards body and longish tail. **Adult** Blackish above, white streaked black below, thighs unstreaked. Female larger than male and more heavily streaked below. In flight, ***pale greyish 'windows' on primaries*** above, underwing with ***black-mottled underwing-coverts*** and ***black trailing edge***, ***tail pale grey with broad black terminal band*** and several indistinct narrow dark bars. Ayres's Hawk Eagle is noticeably smaller, with more heavily marked underparts, heavily barred underwings and no 'windows' on primaries. **Juvenile** Dark brown above, rufous below. Underwing-coverts rufous bordered by dark line; narrowly barred, grey tail. Gradually acquires adult plumage in 3–4 years. Juvenile Black Sparrowhawk of rufous-breasted form has similarly coloured underparts, but is smaller, has bare legs and different behaviour. Juvenile Ayres's Hawk Eagle is smaller, with paler rufous underparts and much bolder barring on flight feathers below. **HH** Uncommon to locally fairly common in savanna. Occurrence in N Senegal unclear due to confusion with Bonelli's Eagle. **Voice** Melodious *klu-klu-klu-kluee* or *kluee-kluee* rather reminiscent of Wahlberg's Eagle but shorter. [CD5:60]

Ayres's Hawk Eagle *Hieraaetus ayresii* 45–60 cm; WS 124 cm

Small, stocky eagle with broad, rounded wings and fairly long tail. **Adult** Blackish-brown above, white below boldly streaked and blotched black, incl. thighs. Short, erectile occipital crest. Plumage variable; variations include: white forehead, short white supercilium, sparsely marked underparts. In flight, ***white 'landing lights'*** (small white spot at base of neck on leading edge of wing). Below: heavily mottled black-and-white wing-coverts, heavily barred flight and tail feathers, black terminal tail band. Female larger and usually more heavily marked below. **Juvenile** Pale rufous to whitish head and underparts. Wing-coverts pale rufous, flight feathers and tail white barred blackish. **HH** Rare to scarce visitor, Gambia. Normally in forest and its outliers. **Voice** Usually silent. Melodious *whueeep-whip-whip-whip-whip-whueep* in aerial display; also *whip-whip-whip-*… [CD5:61]

ad
dark morph
ad
pale morph
ad
Booted Eagle
ad
dark morph
juv
ad
ad
juv
Bonelli's Eagle
ad
juv
ad
African Hawk Eagle
juv
ad
white-headed
morph
juv
ad
Ayres's Hawk Eagle
juv
ad

PLATE 32: MEDIUM-SIZED AND LARGE EAGLES

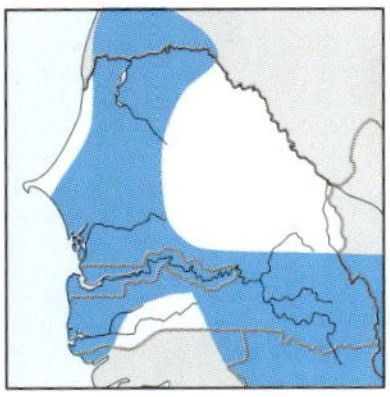

Wahlberg's Eagle *Hieraaetus wahlbergi* — 55–61 cm; WS 141 cm

Smallish, rather slender and variably coloured eagle with slight occipital crest (not always visible), baggy 'trousers' and dark eye. Characteristic silhouette when soaring, with head protruding, ***long, parallel-edged wings held flat or slightly downcurved and long, square tail usually held closed***. **Adult** Commonest morphs dark to warm brown. In flight, appears all dark, with slightly paler flight feathers contrasting faintly with underwing-coverts. Uncommon pale morph has head and underparts white to buffy. In flight, dark flight and tail feathers contrast with white underparts and underwing-coverts. Intermediate plumage variations may occur. **Juvenile** As adult. Compare dark morph Booted Eagle (Plate 31) and Brown Snake Eagle (Plate 25). Black Kite is only slightly smaller but has forked tail. **HH** Scarce to uncommon non-breeding visitor in savanna, Senegal (records in all months); claimed to be fairly common, Gambia (unsubstantiated breeding records) but status unclear due to confusion with similar-looking raptors. Perches inconspicuously in tree or soars for long periods. Takes mainly birds, rodents and lizards. **Voice** Usually silent. Distinctive, clear and loud *kleeeee-ay*. Also fast *kyip-kyip-kyip-kyip-*… [CD5:58]

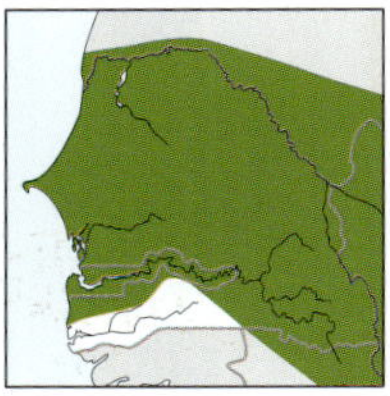

Tawny Eagle *Aquila rapax* — 62–75 cm; WS 165–185 cm

A. r. belisarius. Often rather ragged-looking eagle. Wings fairly long; tail broad and rounded. Flight and tail feathers dark greyish with faint dark bars; 'fingers' contrastingly darker. Soars with wings held flat or 'hand' slightly depressed and tail fanned; when gliding, 'hand' slightly more arched. **Adult** Variable. Dark brown to tawny and even pale buffish, with dark-centred body feathers and upperwing-coverts giving streaked appearance. Flight and tail feathers dark with very indistinct or no dark bars. Above, creamy-white patch from lower back to uppertail-coverts; pale patch on inner primaries very faint, if present. In pale birds three inner primaries usually contrastingly pale, producing 'window' in dark area of underwing. **Juvenile** Typically paler, rufous-tawny bleaching to creamy-white (a few darker). Wings have narrow white trailing edge; tail tipped white. Pale tips to greater coverts form narrow line on centre of wing, both above and, more irregularly and less distinct, below. Pale 'window' on underside of three inner primaries always present. **Immature** Variable. White trailing edge lost. Eye becomes yellow. Adult plumage acquired in *c.*4–5 years. **HH** Fairly common to scarce in wooded habitats throughout. Moves south in dry season. Hunts from perch or on the wing. Feeds on any animal matter, also carrion; robs other large birds of their prey. **Voice** Usually silent. Barking *kyow* and guttural *kwork* in display. [CD1:91]

Crowned Eagle *Stephanoaetus coronatus* — 80–99 cm; WS 163–209 cm

Very large, powerful forest eagle with broad rounded wings and long tail. **Adult male** Dark brown above and on throat, boldly barred and blotched black, rufous and buff below. Tail white below with three broad blackish bars. Short, erectile occipital crest. Underwings have diagnostic ***rufous coverts*** bordered by bold black-and-white greater coverts, white flight feathers with broad black trailing edge and three narrower parallel black bars. **Adult female** Noticeably larger and usually more heavily marked below; flight feathers have only two narrow black bars. **Juvenile** Head and underparts mainly white. Underwing-coverts pale rufous-buff; 3–4 bars on remiges. Gradually acquires adult plumage in *c.*4 years. **HH** Rare resident, SW Senegal (Casamance); vagrant, Gambia. In forest and outliers in savanna. Preys on monkeys and other mammals to size of duiker. **Voice** Far-carrying, melodious *kewee-kewee-kewee-*… (male) and deeper *kowi-kowi-kowi-*… (female), in undulating display flight, often at great height, esp. in hot midday hours. [CD5:64]

Martial Eagle *Polemaetus bellicosus* — 78–86 cm; WS 195–260 cm

Very large, powerful savanna eagle with broad, flat-crowned head, long broad wings and ***short tail***. **Adult male** Dark grey-brown above and on breast; white below speckled dark brown. Tail greyish below narrowly barred blackish. Short, erectile occipital crest. In flight, dark wings and breast contrast with white lower underparts. Flight feathers narrowly and indistinctly barred. **Adult female** Larger and usually more heavily spotted below. **Juvenile** Top of head and upperparts grey with pale edges to feathers, underparts plain white. In flight mainly white below with narrowly and indistinctly barred flight feathers and tail. Acquires adult plumage in *c.*6–7 years. **HH** Widespread resident in small and declining numbers, Senegal; rare, Gambia. Soars at great height with wings held flat and primary tips slightly upturned. Occasionally hovers. Hunts mostly on the wing, also from prominent perch, preying on mammals, large birds and reptiles; also eats carrion. Red List category: Near Threatened. **Voice** Usually silent. Rapid *klee-klee-klee-klooee-klooee*… in display flight and from perch. [CD5:65]

ad
pale morph
ad
dark morph
ad
pale morph
ad
dark morph
Wahlberg's Eagle
juv
ad
pale morph
ad
dark morph
juv
ad
Tawny Eagle
ad
dark morph
juv
juv
ad
ad
juv
Crowned Eagle
juv
juv
ad
ad
Martial Eagle

FALCONS (Plates 33–35) *Small to medium-sized diurnal birds of prey with long, pointed wings and narrow, longish tail. Sexual dimorphism marked or at least present to some degree; females usually larger. Juveniles most like adult females but often darker and more streaked; adult plumage usually acquired during 2nd calendar year. Aerial hunters; hunt by stooping or hovering. Prey include birds, small rodents, reptiles and insects. Normal flight fast and strong, consisting of fast wingbeats alternating with glides. Occur in open or relatively open habitats.*

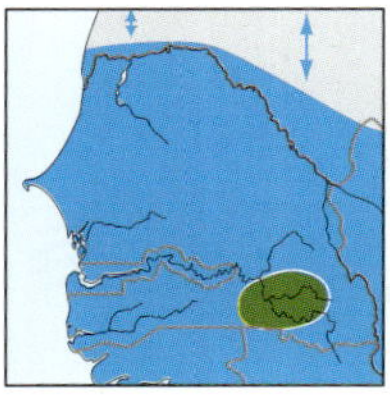

Common Kestrel *Falco tinnunculus* 30–38 cm; WS 65–80 cm

Adult male *tinnunculus* Head grey with ***dusky moustachial stripe***, upperparts rufous-brown spotted black, tail grey with broad black subterminal band and narrow white tip. Underparts buff densely spotted black. In flight, dark flight feathers contrast with chestnut upperwing-coverts and upperparts; pale underwing densely barred and spotted dark. **Adult male *rufescens*** overall darker and more heavily marked; head darker slate; underparts more rufous; tail has some barring. **Adult female** Head brownish streaked black, upperparts and tail brown barred black. Underwing more heavily barred than male. **Juvenile** As female but more boldly streaked below. **HH** Fairly common, throughout. *F. t. tinnunculus* is Palearctic visitor (Nov–Apr); status and distribution of resident *rufescens* unclear. Singly or in pairs in all open habitats, esp. cultivation and savannas; also rocky outcrops and towns. Frequently hovers with tail fanned. Also frequently soars on slightly blunt-looking wings (recalling small *Accipiter*). **Voice** Sharp, piercing *kee-kee-kee-kee-*...; also shrill, vibrant *krree-e-e-e-krree-e-e-e-e*... [CD2:1b]

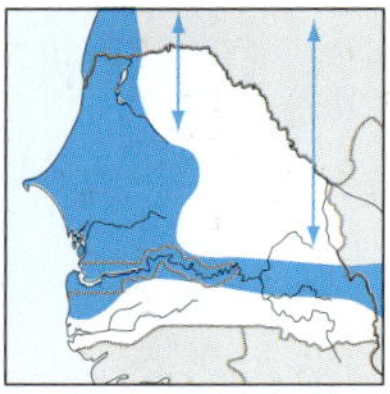

Lesser Kestrel *Falco naumanni* 25–33 cm; WS 58–74 cm

Resembles Common Kestrel but slightly smaller and central tail feathers usually slightly protruding. **Adult male** Head pale blue-grey, upperparts plain rufous-brown with blue-grey greater coverts, underparts lightly marked. In flight, ***tricoloured upperwings exhibit diagnostic blue-grey panel between dark outer wing and rufous coverts***. Underwing mainly white with dark tips and often diffuse trailing edge. **Adult female** As Common Kestrel of nominate race but paler, with less-marked head and underparts. Underwing whiter, dusky tip more distinct. ***Claws pale***. **HH** Locally common to frequent Palearctic winter visitor. In open habitats. Feeds principally on insects, which are caught in the air or on ground. Roosts in groups (largest near Kaolack, Senegal: 28,600, Jan 2007). **Voice** Usually silent away from breeding grounds. [CD2:1a]

Fox Kestrel *Falco alopex* 35–42 cm; WS *c.*90 cm

Very rufous, long-winged and long-tailed falcon. **Adult** Entirely foxy-red narrowly streaked black; tail narrowly barred black. In flight, mainly blackish flight feathers contrast with rufous upperwing-coverts. Underwing tipped black, flight feathers whitish indistinctly barred black, underwing-coverts pale rufous-buff. **Juvenile** Streaking more prominent; barring on tail broader. **HH** Vagrant, S Senegal and Gambia. On inselbergs in savanna. Flight buoyant; rarely hovers. Feeds mainly on insects; also lizards and small mammals. **Voice** Sharp *kee-kee-kee-kee-*... near breeding sites. [CD5:67]

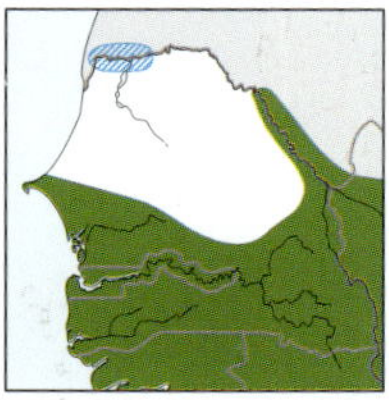

Grey Kestrel *Falco ardosiaceus* 30–39 cm; WS *c.*70 cm

Unmistakable. ***Wholly slate-grey*** with bright yellow cere, orbital ring and legs; primaries blackish. **Juvenile** very slightly tinged brown. **HH** Common resident; scarce, NW Senegal. Singly in savanna, farmland, large clearings and recently burnt areas. Perches on conspicuous vantage point. Never hovers. Preys largely on insects and reptiles, also small mammals and birds. **Voice** Shrill, rasping, vibrant twitter. Usually silent. [CD5:68]

Merlin *Falco columbarius* 24–33 cm; WS 50–69 cm

F. c. aesalon. ***Smallest falcon of the region***. Compact, with relatively short, pointed wings and moderately long tail. When perched, wingtips do not reach tip of tail. **Adult male** Head pattern rather indistinct, with darkish crown, faint moustachial stripe and diffuse pale nape patch. ***Upperparts blue-grey***; tail with broad black subterminal band. Underparts off-white to pale rusty, variably streaked dark. **Adult female** Brown upperparts fringed rufous or buff; tail barred. Underparts buffish heavily streaked dark brownish. Faint moustachial stripe. In flight, coarsely barred underwing; barred tail. **HH** Palearctic vagrant, N Senegal (Nov 1982). Favours open habitats. Preys mainly on small birds, caught after low flight. **Voice** Usually silent. [CD2:3a]

tinnunculus
♂
♀
rufescens
♀
♂
♀
♂
Lesser Kestrel
Common Kestrel
♂
♂
♀
tinnunculus
♀
♂
Fox Kestrel
ad
ad
Grey Kestrel
ad
ad
♂
♀
Merlin
♀
♂

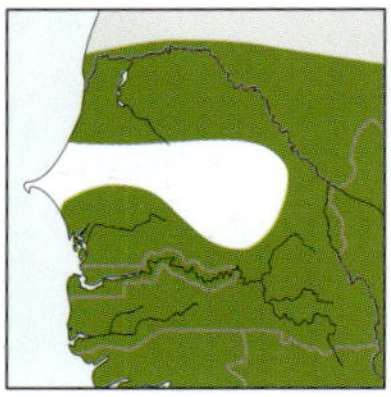

Red-necked Falcon *Falco chicquera* 29–36 cm; WS 65–80 cm

F. c. ruficollis. **Adult** *Top of head and nape bright chestnut contrasting with white cheeks*; small dark moustachial stripe. Upperparts blue-grey densely barred black. Underparts white narrowly and densely barred black except on throat; upper breast has pale rufous-buff band. In flight, dark above with blackish primaries; below, dense barring appears greyish from distance; black subterminal tail band. **Juvenile** Duller, with pale rufous-buff underparts and underwing-coverts. **HH** Uncommon to locally fairly common. Typically associated with palm trees, esp. *Borassus*. Feeds mainly on birds caught in the air. Esp. active at dusk. **Voice** Shrill, harsh, scolding *kheep-kheep-kheep-...* and soft *k-krrree-up*. [CD5:70]

African Hobby *Falco cuvierii* 28–31 cm; WS *c.*70 cm

Distinctive, swift and slender falcon with long, pointed wings and rufous underparts. **Adult** Blackish-slate above, ***orange-rufous below, palest on throat, with some indistinct, fine black streaks;*** short moustachial stripe. (Beware: cheeks occasionally whitish causing confusion with Eurasian Hobby). In flight from below, mainly orange-rufous; flight feathers narrowly barred dusky. Tail barred below. **Juvenile** Duller, with underparts heavily streaked black. Feathers of crown, nape and upperparts narrowly fringed rufous (often forming variable rusty nuchal patch). **HH** Uncommon to fairly common south of 15°N, in wooded savanna, forest–savanna mosaic and derived savanna. Feeds mainly on insects caught and eaten on the wing; also takes small birds. Esp. active at dusk. **Voice** Shrill *kee-kee-kee-...* and short, sharp *kik*. [CD5:71]

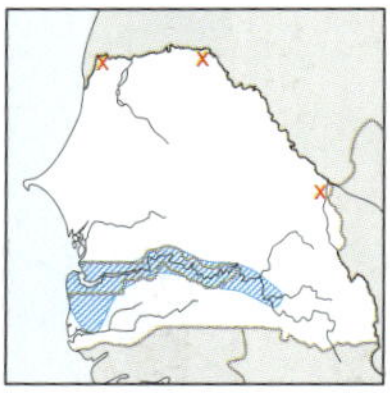

Eurasian Hobby *Falco subbuteo* 29–36 cm; WS 70–92 cm

F. s. subbuteo. Elegant falcon with long, pointed wings. Flight action and silhouette may recall swift *Apus*. **Adult** Dark top of head with ***well-defined, narrow black moustachial stripe contrasting with white cheeks and throat***. Upperparts dark slate-grey. Underparts whitish boldly streaked blackish; ***vent and thighs bright rufous***. In flight, uniformly dark above, ***underwing densely barred greyish*** (appearing dark from distance). **Juvenile** Browner, with pale forehead and no rufous on vent and thighs. **HH** Scarce to rare Palearctic visitor/passage migrant. In open habitats. Feeds mainly on insects and small birds, caught in the air. **Voice** Usually silent away from breeding grounds. [CD2:3b]

Red-footed Falcon *Falco vespertinus* 28–31 cm; WS 60–78 cm

Adult male ***Dark blue-grey with rufous vent and thighs***. Cere, orbital ring and feet reddish. In flight, ***silvery grey flight feathers contrast with rest of plumage***. **Adult female** ***Crown, nape and underparts pale creamy buff to deep orangey***, small black mask around eye, upperparts slate-grey densely barred dark. In flight, barred flight and tail feathers with dark trailing edge contrast with pale orangey underwing-coverts and body. Contrasting pale head consipicuous from afar. **Juvenile** Top of head brown with pale forehead and broad pale neck collar; small black mask around eye and short moustachial stripe contrast with white cheeks. Upperparts greyish-brown, underparts buffish variably streaked dark brown. In flight, dark flight feathers contrast with rest of upperparts; underwing with black-barred flight feathers, dark trailing edge, and variably dark-barred underwing-coverts. **HH** Palearctic vagrant. Feeds principally on insects, caught and eaten on the wing. Regularly hovers. Red List category: Near Threatened. **Voice** Usually silent away from breeding grounds. [CD2:2a]

Eleonora's Falcon *Falco eleonorae* 36–42 cm; WS 84–105 cm

Medium-large, very slender falcon with long wings and tail. All plumages possess diagnostic underwing pattern with ***darker underwing-coverts contrasting with paler flight feathers***. **Adult** Two morphs, with intermediates. Pale morph has blackish top of head and well-defined moustachial stripe, contrasting strongly with white cheeks and throat. Upperparts blackish. ***Underparts orange-buff to rufous-orange narrowly streaked blackish***. Dark morph is entirely brownish-black. **Juvenile** Browner than pale morph; underparts buffish. **HH** Palearctic vagrant (satellite record: an immature, N Senegal, Jul 2006); possibly more regular migrant. Feeds on insects and small birds, usually caught in effortless, fluid flight. **Voice** Usually silent away from breeding grounds. [CD2:2b]

juv
ad
juv
ad
Red-necked Falcon
juv
ad
African Hobby
juv
ad
juv
ad
ad
Eurasian Hobby
juv
♀
♂
juv
♀
♂
Red-footed Falcon
ad
dark morph
juv
ad
pale morph
juv
ad
dark morph
ad
pale morph
Eleonora's Falcon

PLATE 35: FALCONS III

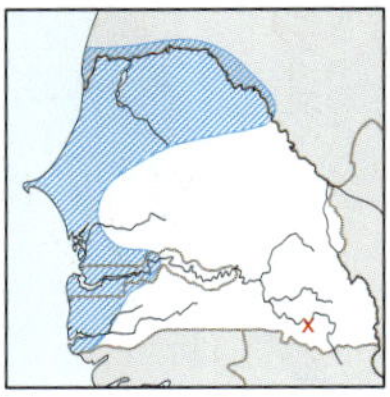

Peregrine Falcon *Falco peregrinus* 33–50 cm; WS 80–115 cm

Stocky, powerful falcon with broad-based, pointed wings and relatively short tail. **Adult *calidus*** (Palearctic migrant) ***Blackish top of head and broad moustachial stripe contrast strongly with white cheeks and throat***. Slate-grey above with paler blue-grey rump, white below finely and densely spotted and barred black. Underwing narrowly and evenly barred blackish (appearing grey from distance). **Adult *minor*** (resident) smaller and darker, with broader moustachial stripe; underparts more heavily marked. **Juvenile** Dark brown above with rufous-fringed feathers, heavily streaked below. Head pattern as adult, but with pale forehead, (often) pale supercilium, usually narrower moustachial streak, and whitish to buff cheeks. Underwing narrowly and uniformly barred. In juvenile ***minor*** streaking on underparts broader. Juvenile Lanner is paler above than juvenile Peregrine and more boldly streaked below; underwing-coverts darker, contrasting with paler flight feathers. **HH** Uncommon to rare Palearctic migrant (mainly Nov–Mar); perhaps resident, SE Senegal. Migrants in variety of open habitats; residents frequent inselbergs and cliffs. Feeds mainly on birds, caught on the wing. **Voice** Loud, shrill *khyeh-khyeh-khyeh-...* Usually silent except near breeding site. [CD2: 5a].

Lanner Falcon *Falco biarmicus* 38–52 cm; WS 90–115 cm

Large falcon with long, relatively blunt-tipped wings. **Adult *abyssinicus Crown and nape rufous*** bordered by black stripe from eye to nape; narrow blackish moustachial stripe contrasts with whitish cheeks. Upperparts blue-grey or grey-brown; underparts whitish washed pinkish or creamy-buff from breast downwards and variably, finely and sparsely spotted brown, mostly on flanks and thighs. Underwing pale, flight feathers narrowly barred dusky. N African ***erlangeri*** paler. **Juvenile** Crown and nape paler, upperparts dark brown, often including crown, throat whitish, rest of underparts heavily streaked blackish-brown. Underwing-coverts dark with some pale mottling, contrasting with pale flight feathers. Peregrine Falcon has shorter, more pointed wings and shorter tail; also darker, with blackish crown and broader moustachial streak; adult Peregrine barred below. **HH** Uncommon to locally fairly common in open habitats. Migrant *F. b. erlangeri* recorded N Senegal. Preys mainly on medium-sized birds, caught on ground and in flight. Often glides and soars with upcurved wingtips. **Voice** Raucous *kreh-kreh-kreh-...* and shrill *kirrree-kirrree-*. Usually silent except near breeding site. [CD2:4a]

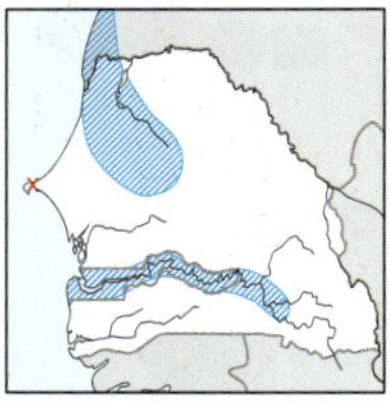

Barbary Falcon *Falco pelegrinoides* 32–45 cm; WS 76–100 cm

F. p. pelegrinoides. Resembles pale Peregrine Falcon in flight action and silhouette. **Adult** Head pattern as Peregrine, but variable amount of ***rufous or cinnamon on forehead, sides of crown and nape***, moustachial stripe narrower, white cheek patch larger. Underparts pale pinkish-buff narrowly and indistinctly barred brown. **Juvenile** As juvenile Peregrine but head with buff or rufous (reflecting adult plumage); underparts sandy-buff with finer streaking. In flight, note uniformly marked, dark-tipped underwing with sandy-buff ground colour. **HH** Rare Palearctic migrant. In open habitats. **Voice** Sharp *kek-kek-kek-...* Usually silent away from breeding grounds. [CD2:5b]

Saker Falcon *Falco cherrug* 43–60 cm; WS 102–135 cm

F. c. cherrug. Large, powerful falcon, resembling Lanner Falcon but usually heavier. **Adult *Head conspicuously pale***, whitish-buff with ill-defined eye-stripe and narrow, variably distinct moustachial stripe and crown streaking. Upperparts brown fringed rufous. Tail feathers barred; barring on central pair indistinct or absent. Underparts buff-white variably streaked dark brown, heavier on flanks; ***'trousers' usually dark. Underwing-coverts typically dark, contrasting with paler flight feathers***. **Juvenile** Head pattern usually more pronounced with more densely streaked crown and more prominent eye-stripe and moustache, underparts more boldly streaked, underwing-coverts darker and even more contrasting. Juvenile Lanner difficult to separate, but typically darker above; underwing less contrasting; head more strongly patterned, 'trousers' never dark. Some juvenile Peregrines of ssp. *calidus* similar to some juvenile Saker, but distinguished by lack of rufous tinge to upperparts, less heavily streaked underparts, lack of contrasting dark underwing-coverts and structural differences. **HH** Palearctic vagrant, N Senegal. In open habitats. Preys mainly on medium-sized birds and mammals, caught on ground after low flight. Red List category: Vulnerable. **Voice** Usually silent away from breeding grounds. [CD2:4b]

juv
ad
calidus
ad
juv
ad
ad
minor
Peregrine Falcon
juv
juv
ad
ad
Lanner Falcon
ad
ad
ad
juv
ad
juv
Barbary Falcon
ad
ad
juv
Saker Falcon
ad

PLATE 36: FRANCOLINS AND GUINEAFOWL

***FRANCOLINS** Thickset terrestrial birds with small head, short, robust bill, short wings and tail, and strong feet. Usually in pairs or small groups. Reluctantly take flight; prefer to run. Identification features include bare part coloration, vocalisations, habitat and range. Feed on variety of plant and animal matter, including seeds, green leaves, fruits, roots, insects and their larvae, millipedes, and small molluscs.*

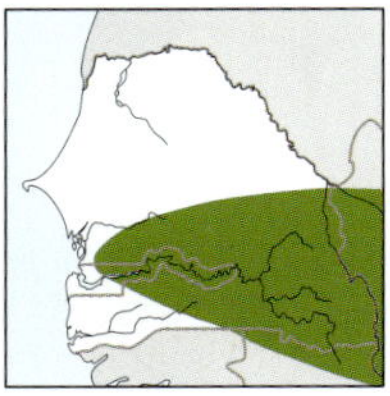

White-throated Francolin *Francolinus albogularis* — *c.*23 cm

F. a. albogularis. Small francolin with ***white throat and chestnut wings*** (conspicuous in flight). **Adult male** Crown greyish and chestnut bordered by creamy-white supercilium, upperparts largely brown-grey variably marked with black and rusty, underparts buff with some chestnut streaks on flanks. **Adult female** Variably and finely barred dark on breast and flanks. **HH** Uncommon, Senegal; rare, Gambia. In various types of wooded grassland. Shy and skulking. Hard to flush; prefers to escape danger by slipping away into vegetation. **Voice** Far-carrying, very fast, high-pitched *kulikulikulikulikuli...* and shrill, rasping *kREEK-krik kREEk-krikikikikew*, accelerating at end and fading away. [CD5:82]

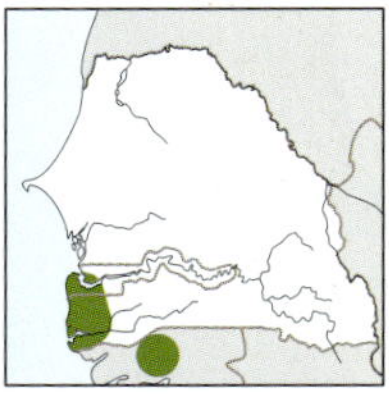

Ahanta Francolin *Francolinus ahantensis* — *c.*33 cm

Dark francolin with ***orange-red bill and legs***. Dark brown above, streaked white on hindneck and mantle. Throat white, rest of underparts grey-brown narrowly streaked white. **HH** Uncommon and local. At forest edge and in second growth, cultivation and thickets. **Voice** Repeated, loud, grating, rather high-pitched *kee-kee-keRRREE kee-kee-keRRREE...* usually in duet. Occasionally calls at night. [CD5:90]

Double-spurred Francolin *Francolinus bicalcaratus* — 30–35 cm

F. b. bicalcaratus. Large, mainly brownish francolin with ***conspicuous white supercilium*** and greenish-yellow bill and legs. Female smaller, spurs reduced. **HH** Common, throughout. In pairs and small coveys in various types of grasslands, farmbush, cultivation and scrub at forest edge. Principally active in early morning and late afternoon. Often seen on tracks. **Voice** Series of loud, grating syllables, *rrrraak kerRRAK kerRRAK...*; flight call *krrror krrror krrror...* [CD2:7b]

***GUINEAFOWL** Afrotropical family of medium-large terrestrial birds with small, bare head, relatively long neck and thickset body. Bare skin of head and neck often brightly coloured. Wings and tail short; feet strong. Sexes similar; males usually larger. Sedentary. Reluctantly take flight; prefer to run. Gregarious when not breeding. Often vocal. Roost in trees. Feed on a variety of vegetable and animal matter, including seeds, fruits, green leaves, insects, millipedes, spiders, small molluscs and frogs.*

Helmeted Guineafowl *Numida meleagris* — 53–63 cm

N. m. galeatus. **Adult *Large*** guineafowl with bare pale bluish head and upper neck, ***bony casque*** ('helmet') on crown and red gape wattle. Casque reddish at base, paling to horn at top. ***Plumage dark grey densely speckled white*** except on lower neck and breast. Flight feathers dark grey finely barred white. **Juvenile** Head tawny-buff with two blackish lateral crown-stripes and no casque, breast buffish scaled dusky. Rest of plumage dull grey-brown speckled buffish-white; upperparts with irregular rusty-buff barring. Immature as adult, but with smaller casque, shorter wattles and feathered neck. **HH** Common, Senegal; locally common, Gambia. In flocks of 10–40 (occasionally more) in wide variety of open habitats and savanna. Forages on ground in open areas. Regularly visits waterholes. Runs fast, with half-raised wings (secondaries arched over back). **Voice** Commonest call a series of hard, raucous notes interspersed with a grating rattle *chek-chek-chek krrrrrr chek-chek-chek-...* Contact note a soft, metallic *chink*. Female also utters a rather high-pitched, piping whistle *huu-i huu-i huui-...*; male responds with single *chek* or *cheenk*. [CD2:6a]

♂
♀
White-throated Francolin
ad
Ahanta Francolin
ad
Double-spurred Francolin
ad
Helmeted Guineafowl
juv

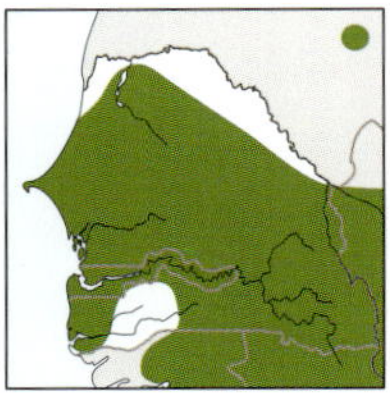

Stone Partridge *Ptilopachus petrosus* *c.*25 cm

Small dark brown ***chicken-like bird with relatively long, broad tail usually held cocked***. Fine feather markings only visible at close range. Legs dark red. Bare skin around eye dull red. **HH** Common. In small groups in various grassland habitats, often with rocks. Runs when disturbed; flies reluctantly. **Voice** Highly distinctive. An abruptly started, rising series of clear, fluty notes with cheerful, pleasant quality, *weet-weet-weet-weet-...*, often in duet or chorus and mostly at dawn and dusk. [CD5:79]

QUAILS *Very small and dumpy. Sexually dimorphic. Secretive. Usually seen when flushed. Reluctant to fly, but migratory and capable of sustained flight.*

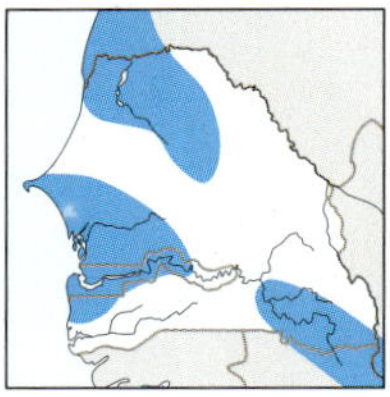

Common Quail *Coturnix coturnix* 16–19 cm; WS 32–35 cm

C. c. coturnix. **Adult male** Strong head pattern with ***blackish anchor on throat***; ***breast-sides and flanks streaked buff, black and chestnut***. **Adult female** Duller; no anchor on throat. In flight, plain wings; streaked upperparts. **HH** Uncommon to locally fairly common Palearctic migrant (Oct–Mar), Senegal; scarce, Gambia. On migration in all types of habitats, but usually favours open grassland. Very skulking and difficult to flush; prefers to run swiftly through grass, or to crouch and remain still. When pressed, flies low on bowed, whirring wings, usually swiftly returning to cover. **Voice** Far-carrying, rhythmic *KWIk-ik-wik*. A soft, rolling *wrreee* when flushed. [CD2:6b]

Harlequin Quail *Coturnix delegorguei* 15–18 cm

C. d. delegorguei. **Adult male** Small, ***dark quail with distinctive black and white head pattern and black-and-chestnut underparts***. Top of head blackish-brown bordered by long whitish supercilium extending onto neck-sides; throat white with black stripe on centre extending laterally and forming 'anchor'. **Adult female** Resembles female Common Quail but darker, with pale tawny underparts and chestnut-buff undertail-coverts. **HH** Afrotropical vagrant, Senegal. In grasslands, rank herbage and cultivation. Nomadic. **Voice** Emphatic, rhythmic *kwit kwit-kwit-kwit kwit-kwit-kwit kwit-kwit*, reminiscent of Common Quail, but slightly slower, more deliberate, marginally higher pitched and more monotonous. A squeaky *skreeee* when flushed. [CD5:78]

BUTTONQUAILS *Very small, quail-like, terrestrial birds with short, rounded body and wings, and cryptic plumage. Lack hind toe. Females larger and brighter coloured than males; sex roles reversed. Occur in savanna, grassland and scrub, occasionally forest edge. Secretive and difficult to flush; usually do not fly far.*

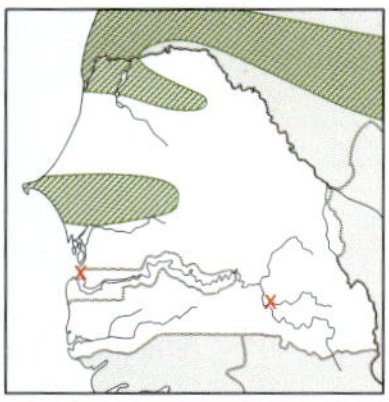

Quail-plover *Ortyxelos meiffrenii* 11–13 cm

Very small, short-tailed grassland bird. Mainly sandy-rufous above and whitish below, with ***distinctive wing pattern and flight action***. Greater coverts white, contrasting with blackish, white-tipped remiges and forming conspicuous diagonal band on upperwing. ***Flight fluttering and erratic, reminiscent of bush lark or butterfly***. Common Buttonquail and quails have different wing pattern, low whirring flight and different jizz. Quails also have darker upperparts and no, or much less, white on underparts. **HH** Scarce and local, Senegal; vagrant, Gambia. In arid and semi-arid grassland and thorn scrub. Secretive, creeping rapidly through grass with hesitant, to-and-fro, chameleon-like gait, but also running courser-like in the open. Sits tight until almost trodden upon; rises silently, flying off in jerky, undulating flight and usually landing only a short distance away. Dry-season breeding visitor. **Voice** Described as a very soft low whistle; silent when flushed.

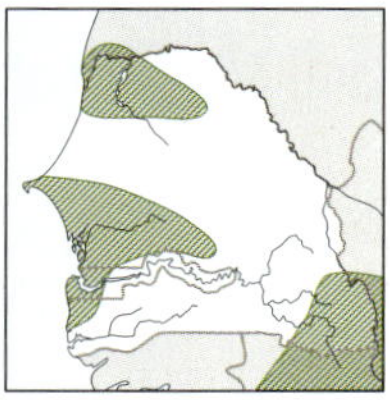

Common Buttonquail *Turnix sylvaticus* 14–16 cm

T. s. lepurana. Head and upperparts cryptically and variably coloured, being marked with shades of brown, rufous and buff; head-sides pale buff finely spotted blackish; median and greater wing-coverts largely pale buff with bold blackish spots. ***Breast orangey, spotted black on sides***; rest of underparts whitish. In flight, ***brown flight feathers contrast with pale coverts***. **HH** Scarce, in various types of grassland and cultivation. Prefers escaping danger by running rather than flying. When flushed, flies low and for only short distance before dropping into cover. Rarely flushed a second time. Secretive, but sometimes permits close approach by quiet observer. Most active in early morning and late afternoon, when often venturing into the open along tracks. **Voice** Strange, low, far-carrying *hooooooo...* like distant lowing cow or foghorn. Difficult to locate. Uttered by female mostly at dawn and dusk, also at night and during day. [CD2:8a]

ad
Stone Partridge
♂
♂
♀
Common Quail
♀
♂
♀
♂
Harlequin Quail
ad
ad
Quail-plover
Common Buttonquail

White-spotted Flufftail *Sarothrura pulchra* 15 cm

S. p. pulchra. Small and secretive. **Adult male** Head, breast and ***tail reddish-chestnut***, rest of plumage ***black densely spotted white***. **Adult female** More brownish and ***entirely barred reddish-buff; tail reddish-chestnut barred blackish***. **Juvenile** As respective adult but duller. **HH** Uncommon and local in south-west. In swampy forest, rank vegetation and shrubby growth along forest trails, edges and abandoned plantations, not always near water; also gallery forest. Skulking and shy, creeping swiftly through dense vegetation without disturbing it and rarely venturing into the open. **Voice** Vocal. Rapid series of short, resonant notes *too-too-too-too-too-...* or *hu-hu-hu-hu-hu.* Duets; second bird joining with similar song or with low, growling *krw krw krw...* When excited, a very fast, hard and shrill *kwipipipipipip...* [CD6:2b]

African Crake *Crex egregia* 20–23 cm

Adult Face to breast grey, upperparts dark brown with dark-centred feathers, lower underparts barred black and white. Flight feathers blackish. **Juvenile** Browner below, barring less distinct. **HH** Uncommon and local (mainly in rains) in wet and dry grassy habitats, edges of cultivation, etc. Active all day, but mostly at dawn and dusk. Skulking, but not shy. Flick short tail while walking. When flushed, flies reluctantly for short distance with dangling legs. **Voice** Single hard *kluk, kruk, krw* or *kip*, occasionally in fast series. [CD6:5a]

African Water Rail *Rallus caerulescens* 28 cm

Adult Medium-sized rail with distinctive ***long red bill*** and red legs; head-sides and underparts grey. **Juvenile** Browner. **HH** Rare, Gambia (first recorded May 1994). Skulking and usually keeping to cover of reedbeds and dense vegetation in swamps or beside ponds and rivers. Most active at dusk. **Voice** Vocal. A fast series of shrill notes, starting with rapid trill, speeding up, then gradually slowing and descending in intensity *trrrrri-kew-kew-kew-kew-kew-...* Others may join with low clucking and grunting notes. [CD6:5b]

Spotted Crake *Porzana porzana* 22–24 cm

Compact, freckled-looking crake with diagnostic ***plain buff undertail-coverts***. Bill yellow with orange-red base. **Adult male** Upperparts olive-brown streaked black; underparts paler, tinged grey and covered with fine white flecks. **Adult female** Usually browner and more speckled. **HH** Generally rare Palearctic visitor (Nov–Apr), Senegal (but common in Senegal delta). Vagrant, Gambia (Marakissa, Jan 2005). Probably overlooked. In marshes and similar habitats. Usually skulking but not particularly shy; also forages in the open. Flies reluctantly. Mainly active at dusk. **Voice** Rhythmic series of far-carrying, explosive *hwit!* notes ('whiplash'), also heard in winter quarters, mainly at dusk and at night. Contact call of wintering birds a hard *whee-up*. [CD2:10b]

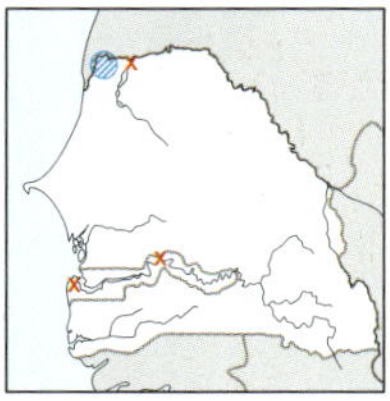

Baillon's Crake *Porzana pusilla* 17–19 cm

P. p. intermedia. Smallest crake. **Adult male** Darker and better marked than Little Crake, with red-brown upperparts more densely and conspicuously speckled white, short primary projection (***primaries barely projecting beyond tertials***, with rarely more than 3 primary tips visible), obvious black-and-white barring on flanks extending forward of legs. ***Bill wholly greyish-green***. **Adult female** Throat and underparts often paler, ear-coverts brownish. **HH** Palearctic visitor, N Senegal delta (Richard-Toll, Nov & Jan) and Gambia (Kartung, Feb 2001; Central River Division, Feb 2009). Breeding discovered 2009 (Senegal delta & Gambia). In marshes and similar habitats. Very skulking and rarely in the open, but not shy and often tame. Most active at dawn and dusk. **Voice** Probably silent in winter quarters. Song of male a dry rattle *tk tk tk tk rrrrrkkkkkkk*, mainly at night. [CD2:10a]

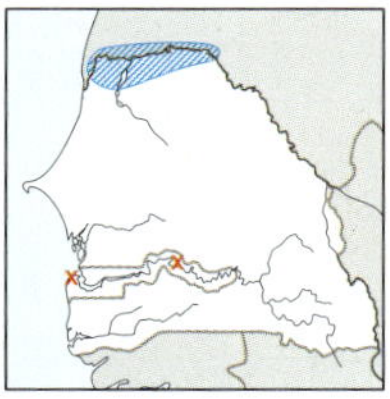

Little Crake *Porzana parva* 18–20 cm

Slightly larger and more slender than Baillon's Crake, with olive-brown upperparts having some white spots and streaks, inner webs of tertials fringed buff, forming creamy line, and ***long primary projection*** (almost as long as exposed tertials, with at least 5 primary tips visible). ***Bill lime-green with red base***. **Adult male** Face and underparts blue-grey. **Adult female** Whitish throat; buffish underparts. **HH** Rare Palearctic visitor, Senegal delta (Oct–Apr); vagrant, Gambia. Probably overlooked. In marshes and similar habitats. Usually secretive but not shy; sometimes in the open and allowing close approach. Only reluctantly flies. **Voice** Probably silent in winter quarters. Song of male *ik ikik ik*, a rhythmic series *kik-kik-kik-kik-kik-...*, or an accelerating series of notes descending in pitch *kwek, kwek, kwek kwek kwek-kwek-kwek...*, mainly at night. [CD2:9b]

juv
♀
White-spotted Flufftail
♂
ad
African Crake
juv
ad
African Water Rail
juv
♂
Spotted Crake
♀
♂
Little Crake
♂
Baillon's Crake
♀
♀

Black Crake *Amaurornis flavirostra* — 19–23 cm

Adult ***All-black*** crake with conspicuous ***yellow bill*** and ***red legs***. Eye and orbital ring red. **Juvenile** Duller; bill and legs dusky. **HH** Locally common. Frequents a variety of swampy places and edges of lakes, ponds, rivers and reedbeds. Less shy than other crakes; often in the open. **Voice** Vocal. A variety of clucking, purring and growling sounds, uttered in stereotyped duet. [CD6:7]

MOORHENS AND GALLINULES Less secretive than rails. Frequent open water or clamber about in dense waterside vegetation.

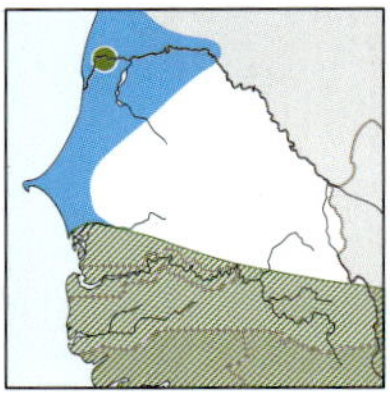

Common Moorhen *Gallinula chloropus* — 30–36 cm

Adult ***Mainly blackish*** with ragged white line along flanks, white-sided undertail, ***red, rounded frontal shield*** and ***red, yellow-tipped bill***. Legs and long toes greenish. ***G. c. chloropus*** has wing-coverts concolorous with upperparts and tinged olive; in slightly smaller Afrotropical ***meridionalis*** upperwing-coverts are slaty blue-grey without olive tinge. **Juvenile** Mainly greyish-brown, with whitish throat and belly, white flank line and white on undertail as adult, and greenish-brown bill. **HH** Locally and seasonally uncommon to fairly common Palearctic visitor (Oct–Apr); presence of Afrotropical *meridionalis* suspected. In variety of freshwater habitats with fringing vegetation. Rather shy. Tail cocked when walking. Swims readily with jerky movements of head and tail. **Voice** Sudden, single *yerrrrp* and short hard *kik*, sometimes in rapid series *kikikikikik-kik-kik-kik*… [CD2:11b]

Lesser Moorhen *Gallinula angulata* — 23 cm

Adult Resembles Common Moorhen, but has ***bright yellow bill with red culmen*** and is smaller and paler grey below. Frontal shield red and pointed. Female somewhat paler than male, with smaller frontal shield. **Juvenile** Dark brown above, head-sides, neck and breast brownish-buff, throat and centre of belly whitish, becoming pale greyish on flanks, bill dusky-yellow with darkish culmen. **HH** Locally and seasonally rare to uncommon in variety of wet habitats, incl. temporarily flooded places, ricefields, swamps and rank vegetation. Wet-season breeding visitor. **Voice** Short, sharp *khup* and *prp*, also in series. [CD6:9]

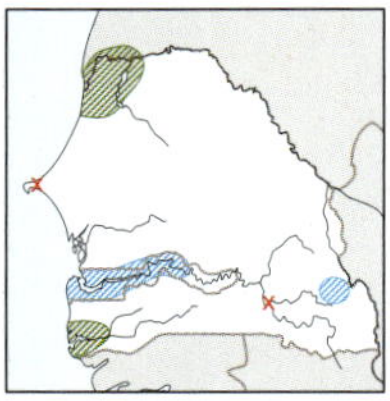

Allen's Gallinule *Porphyrio alleni* — 25 cm

Adult Rather small and elegant, glossy purplish-blue and dark green gallinule with white undertail, ***bluish frontal shield***, red bill and ***long red legs and feet***. At start of breeding, frontal shield turquoise-blue in male, apple-green in female; subsequently, shield becomes blue in both adults; following breeding, shield dark grey or dark blue. **Juvenile** Mainly warm buffish-brown with dark-centred upperpart feathers giving scaly appearance. Throat and belly white, undertail-coverts buff. **HH** Seasonally rare to uncommon in various freshwater wetlands. Usually skulking; mostly seen when it dashes from dense cover or clambers in tangled vegetation. Occasionally feeds in the open. Jerks tail when moving; walks on floating plants. **Voice** Various hard, dry calls, singly or in series, e.g. *kuk kuk kuk kk* and *kip-kip-kip-kip-kirrrr*. In flight, high-pitched *kli-kli-kli-*… [CD6:8]

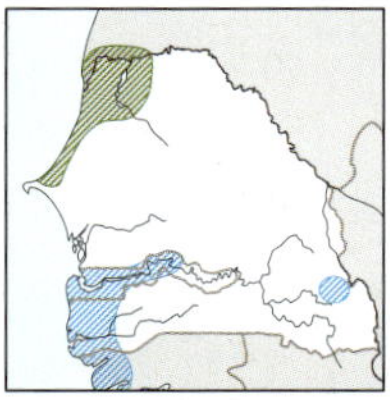

Purple Swamphen *Porphyrio porphyrio* — 45–50 cm

P. p. madagascariensis (African Swamphen). Unmistakable. **Adult** ***Very large***, glossy purple and blue-green rail with white undertail, ***massive bright red bill, red frontal shield*** and long red legs. Eye red. **Juvenile** Paler and greyer, with duskier bare parts. **HH** Locally and seasonally uncommon to fairly common. Solitary, in pairs or small groups in marshes, reedbeds, and shallow ponds and lakes with rank herbage. Rather shy, usually keeping to dense vegetation, but occasionally foraging in the open. **Voice** Various nasal, groaning, trumpeting, grunting, clucking, hooting and wailing calls. [CD2:11a]

Eurasian Coot *Fulica atra* — 36–42 cm

F. a. atra. Plump, thickset, ***all-black*** waterbird with ***white bill and frontal shield***. Legs and lobed toes grey-green. **HH** Rare to locally common Palearctic winter visitor to Sahel. Recorded from N Senegal (delta); vagrant, Gambia (Kaur, Feb 1994). Breeding recorded, N Senegal (two pairs, Jan 2001). Dives frequently. Takes off in long pattering run across water with fluttering wings. **Voice** Loud, sharp *kut* or *khuk*. [CD2:12a]

juv
ad
Black Crake
ad
juv
Common Moorhen
ad
juv
Lesser Moorhen
♀ br
ad
juv
Allen's Gallinule
ad
juv
Purple Swamphen
ad
Eurasian Coot

PLATE 40: OSTRICH, CRANES AND JACANAS

OSTRICH A huge, flightless bird, with long, almost bare neck, powerful legs and two, forward-directed, toes. The largest and heaviest living bird, now endemic to Africa (since extinction in Arabia). Principally herbivorous.

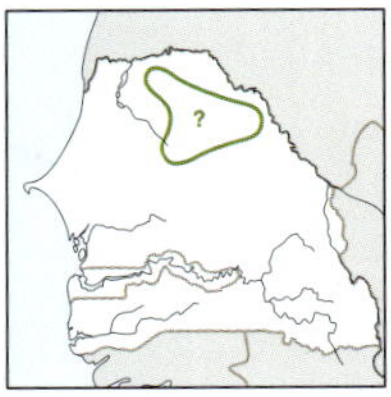

Ostrich *Struthio camelus* Height ♂ 210–275 cm, ♀ 175–190 cm

S. c. camelus. Unmistakable. **Adult male** Huge. Black body, white primaries and tail. In breeding season bill and bare skin (esp. on neck) become redder. **Adult female** Smaller; dull grey-brown plumage. **Juvenile** Similar to adult female. Adult height attained at *c.*12 months; full adult plumage of males acquired in *c.*2 years. Chick has buffish, spiky down, tipped black, and black spots on neck. Juvenile plumage emerges at *c.*3 months. **HH** Formerly in N Senegal (Ferlo); now probably exterminated. Usually in pairs or small groups, in open plains, thorn scrub, semi-desert. Often wary. **Voice** Largely silent. Displaying males utter a far-carrying, deep, booming sound. [CD5:1a]

CRANES Large terrestrial and wading birds with long neck and legs and relatively short, straight bill. Sexes similar; male usually larger. Flight slow and laboured with narrow neck outstretched and long legs projecting. Feed on plant and animal matter.

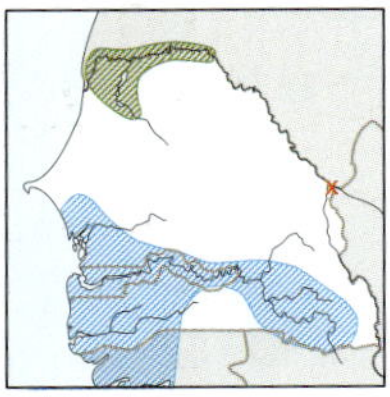

Black Crowned Crane *Balearica pavonina* 100 cm; WS 180–200 cm

B. p. pavonina. Unmistakable. **Adult *Very dark, slate-grey and black, with large white wing panel and straw-coloured crest***. Bare area on head-sides half white, half pink. In flight, conspicuous white upper- and underwing-coverts contrast with dark remiges and body; head and legs held slightly below horizontal, giving hump-backed appearance. **Immature** Washed rusty, crest shorter, no bare area on head-sides. Adult plumage attained after *c.*1 year, full development of facial pattern may require another year. **HH** Uncommon to locally fairly common but declining non-breeding visitor (has bred on Mauritanian side of Senegal River). In pairs or small groups. Frequents moist and dry open habitats, incl. wet plains, ricefields, seasonally flooded areas, and margins of lakes and rivers; also agricultural fields and dry savanna. Red List category: Vulnerable. **Voice** Far-carrying, mellow trumpeting *honk* or *ka-wonk*, usually in flight. [CD6:11]

Common Crane *Grus grus* 100–120 cm; WS 180–220 cm

G. g. grus. **Adult *Large***; mainly grey with bushy feathers hanging over tail. ***White stripe from eye*** across neck-sides. In flight, all grey with contrasting black remiges. **Immature** Browner overall. **HH** Palearctic vagrant, NW Senegal (Djoudj, Mar 2010). In open habitats. **Voice** Far-carrying, nasal trumpeting, usually in flight. [CD2:13a]

JACANAS Long-legged waterbirds with extremely long toes and claws, enabling them to walk on floating vegetation.

African Jacana *Actophilornis africanus* **30 cm**

Combination of ***bright chestnut upperparts*** and ***long greyish legs with extraordinary long toes*** diagnostic. **Adult** Crown and hindneck black, bill and large frontal shield bright blue. Head-sides, throat and neck white bordered by golden-yellow upper breast; rest of underparts chestnut. **Juvenile** White supercilium contrasting with dark brown crown, hindneck and eye-stripe. Underparts white with breast-sides washed yellow, flanks chestnut. **HH** Common at pools, marshes and backwaters with floating and emergent vegetation, also temporarily flooded areas (even small) and ricefields. Usually in small, loose groups, occasionally in pairs or large gatherings. Not shy. Sex roles reversed. **Voice** Various nasal, strident and grating sounds, incl. husky drawn-out whining *kyowrrr*, shorter, repeated *kreep-kreep-kreep*… and high-pitched *weep-weep-weep*… [CD6:17]

Lesser Jacana *Microparra capensis* 16 cm

Small, inconspicuous, crake-like jacana. **Adult** Much smaller than African Jacana, with ***rufous crown and nape***, white supercilium, dark eye-stripe and no frontal shield. Underparts white; neck-sides tinged pale golden-yellow. Tail short, rufous-chestnut. ***In flight, white trailing edge contrasts with black flight feathers***. Pale brown panel in centre of upperwing; underwing largely black. Short, rounded wings and trailing feet may recall crake. **Juvenile** similar; rump blackish. **HH** Afrotropical vagrant, SE Senegal (Jul 2009). Frequents various freshwater wetlands with floating vegetation (water lilies) or emergent grass; also temporally flooded areas. Flies readily, briefly raising wings upon landing. **Voice** Soft, rapid *whoop-whoop-whoop-whoop-whoop*, chattering *kikikikikikikiki*, softer *ku-ku-ku*, plaintive *shreee shree shree shree* or *see sree srrr*, and soft *chrr-chrr-chrr*. [CD6:18]

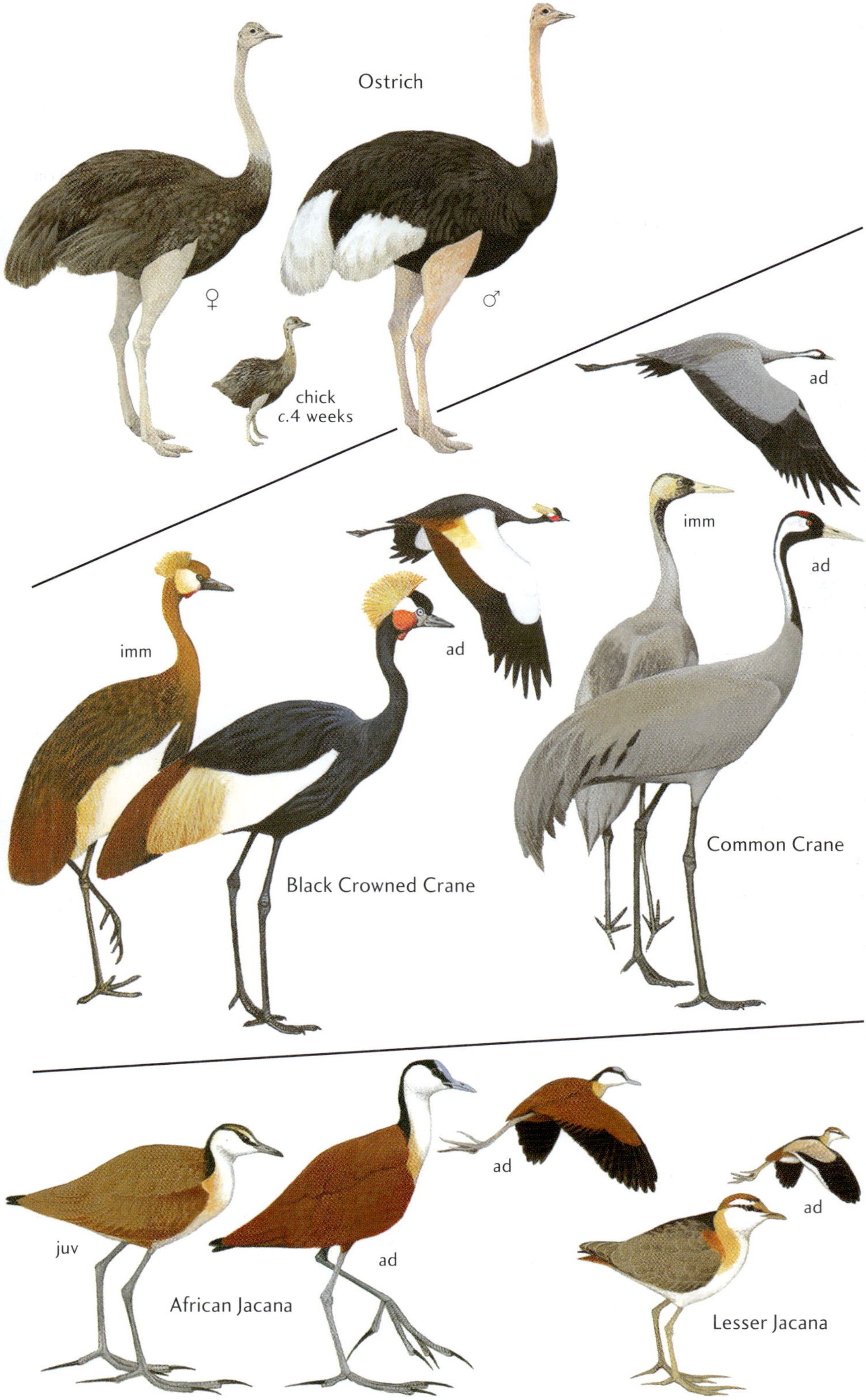
Ostrich
♀
♂
chick
c.4 weeks
ad
imm
ad
imm
ad
Black Crowned Crane
Common Crane
ad
juv
ad
African Jacana
ad
Lesser Jacana

PLATE 41: BUSTARDS

BUSTARDS *Medium-sized to very large terrestrial birds with long neck and legs, short bill and cryptically coloured upperparts. In large species, males much larger than females. Juveniles usually much like female. Occur in a variety of open-country habitats. Walk slowly and deliberately and only reluctantly take flight. Flight powerful with neck outstretched. Largely silent outside breeding season. Subject to seasonal movements, migrating north in response to rains. Breed mainly in rainy season. All species declining because of hunting and trapping.*

Savile's Bustard *Lophotis savilei* — 41 cm

Smallest bustard of the area, with relatively short neck and legs. **Adult male** Nape with tuft of rufous feathers (rarely seen; erected in display); head-sides creamy-white; neck deep buff. Throat black, foreneck grey, rest of underparts black. In flight, blackish remiges separated from tawny upperwing-coverts by whitish line. **Adult female** White throat, deep buff neck, broad white breast-band. **HH** Locally fairly common (Senegal) to scarce (Gambia). Generally considered sedentary, but seasonal movements suspected. Singly or in pairs in arid and semi-arid habitats. Secretive; flies reluctantly and for short distance only, preferring to avoid detection by remaining immobile in cover. **Voice** Clear, whistled *tuit-thit*. Male advertisement a whistled note followed by a rapid series of short whistles *tuit! tutututututututu*, or accelerating series of clear whistles *thut thut-thut-thut-thutututututut*; also series of frog-like notes in same rhythm. [CD6:14]

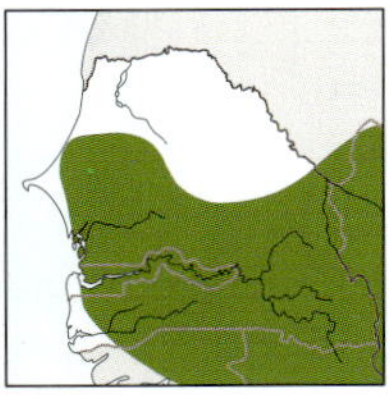

Black-bellied Bustard *Lissotis melanogaster* — 60 cm

L. m. melanogaster. Medium-sized bustard, with long, thin neck and long legs. **Adult male** Black stripe from eye ending in short nuchal crest.***Throat mottled blackish merging into black line bordered with white on foreneck; rest of underparts black***. Upperparts buffish-brown with dark blotches and vermiculations. In flight, upperwing has largest area of white of any bustard; underwing mainly black with white patch on primaries. **Adult female** Mainly brownish-buff with dark blotches and vermiculations. In flight, much less white in wing; flight feathers mainly blackish. **HH** Uncommon. Mainly sedentary, but northward movement with rains suspected. Singly or in pairs in open savanna; also farmland and derived savanna. Male display includes flight with deliberate wingbeats, alighting after glide with raised wings. **Voice** Generally silent. Male advertisement a hoarse, nasal sound followed by low, drawn-out growl leading into cheerful, popping note *vwok, rrorr-WIK!*. [CD6:16]

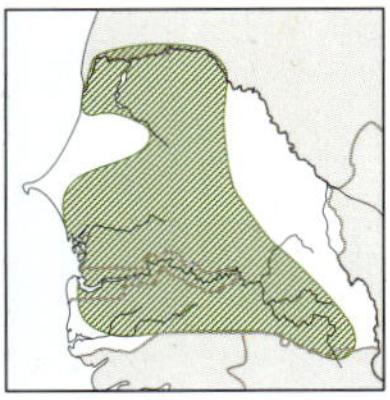

White-bellied Bustard *Eupodotis senegalensis* — 50–60 cm

E. s. senegalensis. ***Fairly small*** and slender, ***white-bellied*** bustard. **Adult male** Forehead and forecrown black becoming ***greyish-blue*** on hindcrown, bordered black on nape; head-sides and throat whitish except black patch on throat which extends onto sides of upper neck; neck greyish-blue; breast tawny. In flight, black flight feathers with some whitish on inner webs of primaries contrast with tawny-buff upperwing-coverts; underwing whitish with black trailing edge. **Adult female** Throat white; neck mainly tawny-buff. **HH** Uncommon, Senegal; rare visitor, Gambia. In open savanna, grassland, bush bordering cultivation and thorn scrub. Most active at dawn and dusk. Wary. **Voice** Vocal. Loud, sonorous, rather nasal, *kuk-kwarrak* often repeated; variation: *kuk-wrik-e-haah!* [CD6:15]

Arabian Bustard *Ardeotis arabs* — 80–100 cm; WS 205–250 cm

A. a. stieberi. **Adult male** ***Large***, with greyish, thick-looking neck and breast finely and densely barred blackish (appearing all grey at distance). Crown buffish broadly bordered black, with short black nuchal crest; supercilium creamy-white. Upperparts mainly dull earth-brown and tawny, underparts whitish. In flight, mainly dark above; outer primaries and coverts mainly black. **Adult female** Smaller. **HH** Uncommon, Senegal (no modern breeding records); last record Gambia in 1928. On arid grassy plains and in *Acacia* woodland. **Voice** Generally silent. [CD2:15b]

Denham's Bustard *Neotis denhami* — 80–100 cm; WS 170–250 cm

N. d. denhami. **Adult male** Large, with ***rufous hindneck and bold white markings on blackish wings*** (also conspicuous in flight). Head-sides greyish, foreneck and upper breast blue-grey becoming whitish on belly. **Adult female** Smaller; head-sides, throat and foreneck tinged buffish. **HH** Uncommon and local non-breeding visitor, Senegal; rare, Gambia. In wooded savanna and farmland; attracted to freshly burnt grassland. Usually wary. Male performs 'balloon' display, puffing out neck and breast feathers. Red List category: Near Threatened. **Voice** Mostly silent. Occasionally utters a guttural barking and booming sound.

ad
♂
♀
♂
Savile's Bustard
♂
♀
Black-bellied Bustard
♂
♀
White-bellied Bustard
ad
Denham's Bustard
♀
ad
Arabian Bustard
♂

PLATE 42: OYSTERCATCHERS, STILT AND AVOCET

OYSTERCATCHERS Conspicuous and stocky shorebirds with pied or all-black plumage, long orange-red bill and pinkish legs. Sexes similar, female slightly larger. Mainly coastal. Vocal.

Eurasian Oystercatcher *Haematopus ostralegus* 40–45 cm; WS 75–85 cm

H. o. ostralegus. Large, bulky, black-and-white wader with ***long, stout orange-red bill*** and ***sturdy, pinkish legs***. In flight, has broad white wingbar and white wedge from lower back to uppertail-coverts. In non-breeding plumage, has white white half-collar on throat and neck-sides. **HH** Uncommon Palearctic visitor (mainly Oct–Apr). Frequents coastal creeks, sandy beaches, estuaries, and saltpans. **Voice** Loud and vigorous, high-pitched *KLEEP!* or *K-PEEP!* Also sharp *kip kip kip…* [CD2:16b]

African Black Oystercatcher *Haematopus moquini* 42–45 cm

As Eurasian Oystercatcher but plumage entirely black. **HH** Sedentary S African species; seasonal short-distance movements recorded. Red List category: Near Threatened. Entirely black oystercatchers reported from Senegal (near Dakar, Feb 1970, one; Basse Casamance, Dec 1975, two); their identity remains uncertain, as African Black Oystercatcher not known north of 12°S in Angola and Canarian Black Oystercatcher *H. meadewaldoi* considered extinct (last collected in 1913; reportedly disappeared in 1940s).

STILTS AND AVOCETS Elegant, long-legged shorebirds with pied plumage and slender bills.

Black-winged Stilt *Himantopus himantopus* 35–40 cm; WS 67-83 cm

H. h. himantopus. Graceful black-and-white wader with ***long neck, black, needle-like bill*** and ***extraordinary long, pinkish legs***. In flight, legs extend far beyond tail; wings pointed, all black above and below; white wedge from back to tail. **Adult** Head white or with variable amount of black. **Juvenile** Duller; upperparts brownish with narrow buff fringes. **HH** Common Palearctic and Afrotropical visitor to various aquatic habitats, incl. coastal lagoons, shallow lakes, pools, inland reservoirs, ricefields and swamps. Seasonal local movements reported but poorly understood. Breeding attempt, Senegal (Saloum delta, Jul 1977). Feeds on land and in shallow water, picking food from mud, surface of water or floating vegetation; also probes and occasionally immerses head in water. **Voice** Vocal in breeding season, otherwise rather silent. Calls varied, incl. sharp, fast *kyik kyik kyik…* grating *kreet kreet…*, more drawn-out *krrrrrt* and high-pitched *kip-kip-kip-…* [CD2:17a]

Pied Avocet *Recurvirostra avosetta* 42–46 cm; WS 67–77 cm

Elegant, white-and-black shorebird with ***slender, upturned black bill*** (diagnostic) and grey-blue legs. In flight, distinctive black-and-white pattern. Female has slightly shorter and more upturned bill. Immature has black areas replaced by brownish. **HH** Locally fairly common to common Palearctic visitor. Usually in small groups, on mudflats, estuaries, saltpans and lake margins (mainly Sep–Mar; some present all year). Typically feeds by sweeping bill from side to side through water or soft mud; also pecks items from mud. Frequently swims and up-ends like a duck. **Voice** Loud, clear *klup klup…* [CD2:17b]

ad br
ad non-br
Eurasian Oystercatcher
ad non-br
ad
African Black Oystercatcher
ad
juv
ad
ad dark-headed variant
Black-winged Stilt
ad
Pied Avocet

THICK-KNEES *Cryptically patterned, with large head and large yellow eyes. Crepuscular and nocturnal, often encountered on roads at night. Calls melodious and far carrying. Note wing pattern.*

Senegal Thick-knee *Burhinus senegalensis* 32–39 cm; WS 75–80 cm

Closed wing has broad, pale greyish panel bordered above by single horizontal black bar. Upperpart feathers with dark shafts. In flight, prominent white patches in black primaries, pale wing panel. **HH** Widespread and locally common on river banks, sandbanks and lake shores, and in mangroves. Singly, in pairs or small groups. Local movements in response to changes in water levels reported. **Voice** A series of clear, piping notes, accelerating and increasing in volume, with last few notes fading away *pi pi pi-pi-pi-pi-pi-PII-PII-PII-pii-pii-pii-pii-piu*. [CD6:20]

Water Thick-knee *Burhinus vermiculatus* 38–41 cm

B. v. buettikoferi. ***Closed wing has broad, pale greyish panel bordered above by narrow horizontal white bar*** highlighted by black above. Upperpart feathers with dark shafts and finely vermiculated (only visible at close range). In flight, white patches in black primaries, pale wing panel; feet project slightly beyond tail. **HH** Rare, S Senegal. Singly, in pairs or small groups, frequenting riverbanks, lake shores, estuaries, mangroves, lagoons and beaches. Rests by day in light woodland or scrub near water. Local movements in response to changing water levels likely. **Voice** A series of plaintive, piping, whistled notes, first accelerating and rising in pitch and volume, then dying away with slower, drawn-out, plaintive notes, *pi-pi-pi-pi-pee-pee-PEE-PEE-PEE-PEE-PEE-PEE-PEE-peeu-peeeu-peeeu-peeeu-peeeu*. Also very rapid *pipipipipipipipipipi*. [CD6:21]

Stone-curlew *Burhinus oedicnemus* 38–45 cm; WS 76–88 cm

Adult male ***oedicnemus*** ***Closed wing has horizontal white bar bordered above and below by black, with broad greyish panel below.*** Bill yellow tipped black. In flight, white patches on outer and innermost primaries, pale wing panel bordered above by narrow black-bordered white bar. N African ***saharae*** slightly paler, more rufescent, but differences clinal, and much individual variation. Senegal Thick-knee has relatively longer bill with more extensive black tip, and single black horizontal bar bordering more distinct, pale grey wing panel. **Adult female** Very similar, but white covert bar less strongly bordered with black (noticeable when pair seen together). **HH** Scarce Palearctic visitor (Sep–Mar), Senegal. **Voice** Mostly silent in W Africa. [CD2:18a]

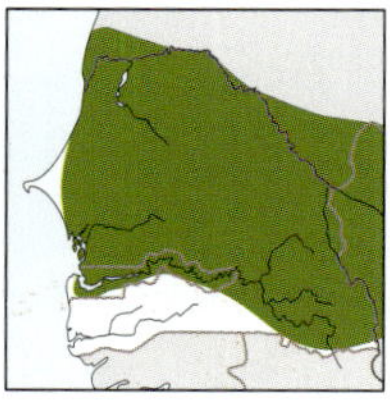

Spotted Thick-knee *Burhinus capensis* *c.*43 cm

B. c. maculosus. Largest and tallest African thick-knee, readily separated from others by lack of any bars or panel on wing. **Adult** Upperparts, incl. wing-coverts, warm tawny-brown densely spotted blackish. Head pattern less distinct than other thick-knees, with white areas smaller. In flight, small white primary patches. **Juvenile** Slightly duller; upperparts streaked. **HH** Widespread and fairly common, Senegal; locally fairly common to rare, Gambia. In arid areas and dry, open woodland. Mainly singly or in pairs. Rests by day on stony ground in the open or under shrub. Dry-season visitor. **Voice** A series of loud piping notes, rising in pitch and volume, then dying away, *pi-pi-pi-pi-plee-plee-PLEEW-PLEEW-WHEEW-WHEEW-wheew-wheew-...* and an accelerating, slightly rising series of similar notes *piu-piu-piu-piupiupiupiu...*; lower pitched than calls of other two thick-knees. Also harsh notes and rapid *pi-pi-pi-...* in alarm. [CD6:22]

Senegal Thick-knee
ad
ad
ad
Water Thick-knee
ad
Stone-curlew
juv
ad
ad
Spotted Thick-knee

PLATE 44: EGYPTIAN PLOVER, COURSERS AND PRATINCOLES

EGYPTIAN PLOVER Uniquely patterned, chunky, riverine wader of uncertain affinities, sometimes placed in its own monotypic family (Pluvianidae).

Egyptian Plover *Pluvianus aegyptius* — 20 cm; WS 42 cm

Adult Distinctive pattern of black, white, blue-grey and creamy-buff. Legs blue-grey. In flight, striking wing pattern with diagonal black band on white. **Juvenile** Some rusty on crown, mantle and wing-coverts. **HH** Locally fairly common. Singly, in pairs or small groups on sandbars in large rivers, where breeding. In various other aquatic habitats when not breeding. Undertakes irregular local movements in response to changes in water levels. **Voice** Harsh *chreek-chreek-chreek*. [CD6:23]

COURSERS Terrestrial, ground-feeding species with long legs and cryptically coloured plumage, occurring in dry habitats. Sexes similar. Prefer to run when disturbed.

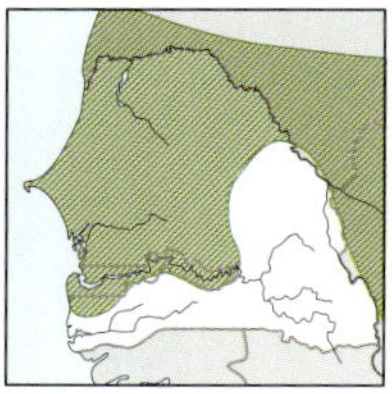

Temminck's Courser *Cursorius temminckii* — 19–21 cm

Small courser with ***blackish patch on centre of chestnut belly***. **Adult *Crown rufous*** bordered by white supercilium and broader black stripe back from eye, forming V on nape. In flight, black outer wing and secondaries contrast with grey-brown upperparts; underwing black. **Juvenile** Duller, head plainer, upperparts faintly scaled, belly patch more diffuse. **HH** Uncommon to locally fairly common in open and burnt grassland, dry farmland, semi-arid savanna and similar open habitats. Nomadic, with seasonal movements in response to rainfall and burning of grasslands. **Voice** Mostly silent. Nasal, metallic *het-het-het-herr…* like toy trumpet or squeaking hinge, in high, circular display flight. [CD6:24]

Cream-coloured Courser *Cursorius cursor* — 21–24 cm; WS 51–57 cm

C. c. cursor. **Adult *Pale sandy-cream***. White supercilium and black stripe from behind eye form V on nape and delimit ***grey hindcrown***. In flight, has black outer wing and black underwing; white tips to secondaries form narrow trailing edge. **Juvenile** Duller, head plainer, upperparts faintly scaled. **HH** Uncommon Palearctic visitor (Aug–May) and possible resident, N Senegal; rare, Gambia (Dec–Feb). Singly, in pairs or small groups in arid plains, stony and sandy desert and semi-desert; occasionally on bare farmland and saltpans. Forages by running fast, then stopping briefly to pick insects from ground or look around. Bobs head. Flight jerky. **Voice** Mostly silent. A short, sharp *kwit* or *krit* usually given in flight. [CD2:18b]

Bronze-winged Courser *Rhinoptilus chalcopterus* — 25–29 cm

Large, ***long-legged*** courser with ***relatively big, distinctively patterned head, black breast-band and pinkish-red legs***. In flight, black flight feathers contrast with brown coverts, white uppertail-coverts and black tail band; underwing-coverts buffish-white. Violet primary tips usually not visible in field. **HH** Uncommon to scarce (but easily overlooked). Singly or in pairs in wooded savanna. Partial to recently burnt ground. Active at dusk and at night. Rests in shade by day. Nomadic. **Voice** Short series of 3–4 ringing syllables (difficult to transcribe) *hu thu-WHEH-hep*, often rhythmically repeated. Also a harsh sound in flight. [CD6:25]

PRATINCOLES Principally aerial-feeding, gregarious and more water-dependent species with short legs, long, pointed wings, forked tail and graceful flight, reminiscent of terns. Sexes similar.

Collared Pratincole *Glareola pratincola* — 24–28 cm; WS 60–70 cm

G. p. pratincola & fuelleborni. Relatively large pratincole with, in flight, ***chestnut underwing-coverts and axillaries***, white rump, blackish flight feathers, narrow white trailing edge to inner wing (which may wear off), ***and strongly forked tail***. **Adult breeding** Creamy-yellow throat bordered by black line. **Adult non-breeding** Duller, throat outline indistinct, breast mottled. **Juvenile** Scaly upperparts; throat pattern indistinct; breast blotched. **HH** Fairly common to uncommon and local resident, intra-African migrant (*fuelleborni*) and Palearctic visitor (nominate); races indistinguishable in the field. Wet-season colonial breeder, N Senegal. Frequents various dry and wet, open habitats. Catches insects on the wing. Most active at dawn and dusk. Movements unclear. **Voice** Sharp *kik* and *kirrik*; also piercing, rattling *krrrrrrret*. [CD2:19a]

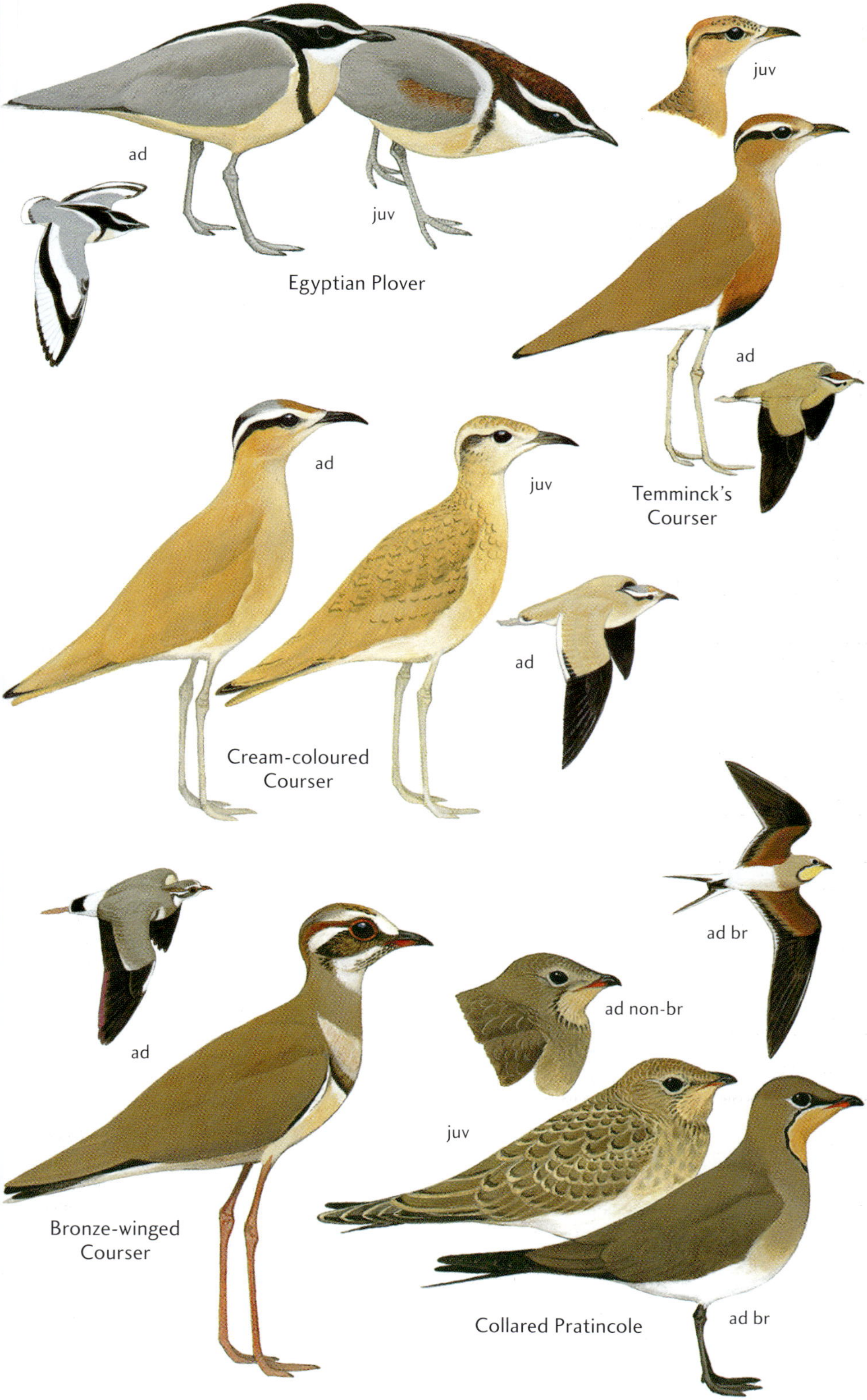
ad
juv
Egyptian Plover
juv
ad
Temminck's Courser
ad
juv
ad
Cream-coloured Courser
ad br
ad
ad non-br
juv
Bronze-winged Courser
Collared Pratincole
ad br

CHARADRIUS *PLOVERS* *Small waders with large, rounded head, relatively short bill, pointed wings and long legs. Underparts largely white; head and breast markings are important field marks. Feeding action typically consists of short, swift runs interrupted by abrupt stops followed by a pause or a dip.*

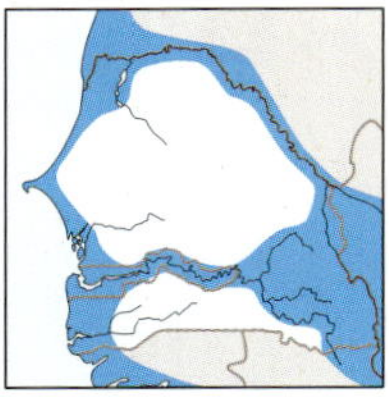

Little Ringed Plover *Charadrius dubius* 16–18 cm; WS 32–35 cm

C. d. curonicus. Resembles corresponding plumages of Common Ringed Plover, but slightly smaller and slimmer, with narrower bill and more tapering rear end; call different. In flight, ***no wingbar***. **Adult breeding** Black bill, distinct ***yellow orbital ring***, dull-coloured legs. **Adult non-breeding** Duller; head plainer than Common Ringed. **HH** Uncommon to locally common Palearctic visitor (mainly Oct–Mar). In various open habitats, incl. edges of lakes, ponds and rivers, farmland and airfields; less frequently on coast. **Voice** Piping, rather plaintive *PEE-uu*. [CD2:20b]

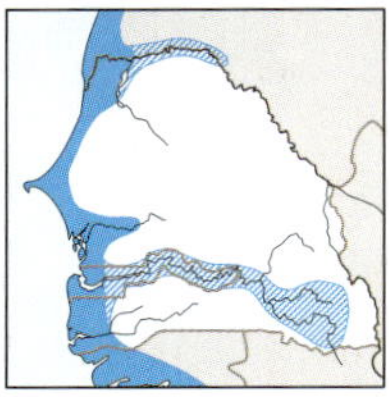

Common Ringed Plover *Charadrius hiaticula* 18–20 cm; WS 35–41 cm

C. h. hiaticula & tundrae. Resembles Little Ringed Plover, but more compact, with ***stubby bill*** and ***orange legs***; call different. In flight, ***prominent white wingbar***. **Adult breeding** Black-tipped, orange bill, orange legs. **Adult non-breeding** Duller, with head markings and breast-band brownish; forehead and supercilium remain white; bill dark. **HH** Common Palearctic visitor (mainly Oct–Apr; records in all months). On mudflats and inland shores; also other open habitats, incl. farmland and areas with short grass. **Voice** Mellow *too-EE* rising at end. [CD2:21]

Kentish Plover *Charadrius alexandrinus* 15–17 cm; WS 34 cm

C. a. alexandrinus. Small, pale plover with ***narrow dark patch at breast-sides*** and blackish legs. Rather rounded with relatively large head and short rear end. In flight, white wingbar and broad white sides to rump and tail. Compare White-fronted Plover. **Male breeding** White forehead and supercilium, black bar across forecrown, ***variable amount of rufous on crown*** and black mask. **Male non-breeding/female** Duller, head markings and breast patches grey-brown, little or no rufous on crown. **HH** Fairly common to locally common Palearctic visitor. Rare resident, Senegal (breeding recorded: south of Dakar, 1970 & 1979; Casamance, 1981). On coastal and inland shores. **Voice** Short *kip*. Alarm a hard *pirrrrr*. [CD2:22]

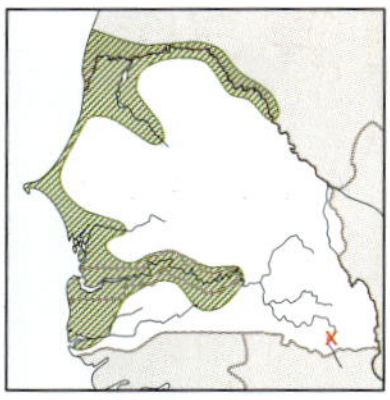

Kittlitz's Plover *Charadrius pecuarius* 13–16 cm; WS 33 cm

C. p. pecuarius. Small, rather long-legged plover with ***warm buff breast***. In flight, white patch on dark outer wing extending as narrow white line on inner wing. **Adult breeding** ***Distinctive head pattern*** with white forehead, broad white supercilium meeting on nape, and black band across forecrown extending through eye to neck-sides. **Adult non-breeding** Duller; supercilium orange-buff. **Juvenile** similar but upperparts scaly. **HH** Uncommon to locally common intra-African migrant (breeding recorded, N & S Senegal). In various open habitats, incl. areas with short grass, edges of reservoirs and lakes, mudflats, beaches and saltpans; not always near water. **Voice** Mostly silent. Calls include *tuweet* or *pipeep*, hard *trip* and dry rattling *trrrrr*. [CD6:28]

White-fronted Plover *Charadrius marginatus* *c.*18 cm

C. m. mechowi. Small plover with ***large white forehead*** and supercilium. In flight, white wingbar and broad white sides to rump and tail. **Adult male breeding** Black bar across forecrown, rusty grey-brown crown and hindneck, black eye-stripe, ***variable creamy-buff to orange wash on breast***. Female has less black on forecrown. **Adult non-breeding** Duller, with small, diffuse lateral breast patches. **Juvenile** similar but upperparts scaled sandy-buff. Resembles Kentish Plover but breast patches less distinct; tail longer, extending beyond folded wings. **HH** Scarce to locally uncommon resident. Mainly sedentary, but seasonal movements recorded. On sandy beaches and lagoons. **Voice** Low *wit* or *twirit*. Alarm a dry *trrrr*. [CD6:31]

Forbes's Plover *Charadrius forbesi* *c.*20 cm

Rather long-tailed plover with ***two blackish breast-bands*** and ***red orbital ring***. In flight, appears dark, without white wingbar. **Adult** White band from eye across nape. Upper breast-band merges with brown of throat. **Juvenile** Duller; band across nape buffish. **HH** Rare intra-African migrant. Frequents grassland, farmland, recently burnt ground, ricefields and edges of muddy pools, lakes and rivers; also rocky outcrops and gravel plains. **Voice** Plaintive *pee-oo*, sometimes repeated, and sharp *pee-pee-pee-*... [CD6:30]

ad br
ad non-br
Little Ringed Plover
ad non-br
ad br
Common Ringed Plover
ad non-br
♂ br
Kentish Plover
♂/♀ non-br
ad non-br
ad non-br
ad br
Kittlitz's Plover
ad br
♂/♀ non-br
White-fronted Plover
♂ br
juv
ad
ad
Forbes's Plover

LAPWINGS *Mostly well marked and easily identified. Several have wattles at base of bill or carpal spurs. Distinctive black-and-white wing pattern in flight. Flight rather slow and heavy. Sexes similar. Juveniles are duller versions of adults. Occur singly, in pairs or small groups.*

African Wattled Lapwing *Vanellus senegallus* — *c.*34 cm

V. s. senegallus. Large, mainly pale brown with ***conspicuous yellow wattles***, yellow bill and ***long yellow legs***. Forehead and forecrown white bordered blackish, neck streaked, throat black. In flight, broad, diagonal white band on inner wing contrasts with black flight feathers. **HH** Common throughout, in damp grassland, grassy edges of lakes and rivers, marshes, open and lightly wooded grassland, usually (but not always) near water; also mangrove fringes. **Voice** Shrill *kwip-kwip-kwip-...* and more nasal *ke-weep ke-weep...* [CD6:33]

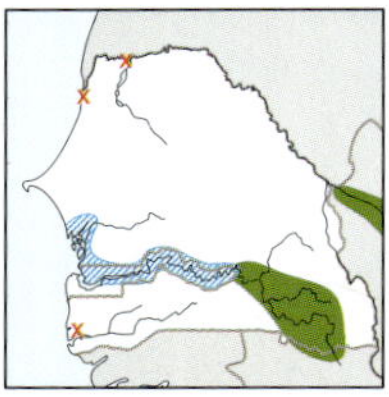

White-headed Lapwing *Vanellus albiceps* — 28–32 cm

Medium-sized waterside lapwing with ***white band on wing***, ***large yellow wattles*** hanging from base of yellow, black-tipped bill, and long greenish-yellow legs. Head and neck grey with white band from forehead to nape. In flight, ***mainly white wings*** with black outer three primaries and squarish black area on coverts. **HH** Uncommon, Senegal; scarce wet-season visitor, Gambia. On sand banks, mud or rocks in rivers; also forages in grassy areas near water. Seasonal movements related to water levels. **Voice** Sharp, high-pitched *kip, kip, kip,...* and *kwip, kwip, kwip,...* [CD6:34]

Black-headed Lapwing *Vanellus tectus* — *c.*25 cm; WS 59 cm

V. t. tectus. Rather small ***dry-country lapwing*** with distinctive black-and-white head pattern, wispy black crest, black-tipped red bill, and red legs. In flight, large white area on forewing extends as diagonal band on inner wing. **HH** Common throughout, in dry grassland with scattered bushes and patches of bare ground. Mostly active in early morning, in evening and at night, spending most of day resting in shade. Mainly dry-season visitor from Sahel. **Voice** Short, piercing *kir* and shrill *kwairr*. [CD6:35]

Spur-winged Lapwing *Vanellus spinosus* — 25–28 cm; WS 64 cm

Distinctively patterned, waterside lapwing. Top of head black; head-sides and neck white; underparts mainly black. In flight, diagonal white band from carpal joint to inner secondaries contrasts with black of flight feathers and brown coverts. **HH** Common throughout, in various wetlands, incl. marshes, rivers, lakes, reservoirs, estuaries, coastal saltpans and irrigated farmland. **Voice** Sharp, metallic *kit*, often rapidly repeated. Territorial call a harsh *ti-ti-terr-it*. [CD6:37]

Lesser Black-winged Lapwing *Vanellus lugubris* — 22–26 cm

Small, slim and rather plain-looking with ***white forehead***. In flight, ***white secondaries form broad, contrasting trailing edge to wing*** (diagnostic). **Adult** Head and neck brownish-grey; upperparts grey-brown. Breast brownish-grey becoming blackish on lower border; rest of underparts white. Legs dark brownish. **Juvenile** Head pattern duller; wing feathers and scapulars fringed buff. **HH** Rare intra-African migrant, Senegal (one breeding record: Saloum, 1980); rare/vagrant, Gambia (one, Dec 1997–Mar 1998; two, Nov 2009). In coastal savannas and lightly wooded grassland. **Voice** Melodious *thi-HUwit*, shorter *thu-WIT* or *thi-whoo*, and longer *tihi-hooee*. [CD6:39]

Northern Lapwing *Vanellus vanellus* — 28–31 cm; WS 67–72 cm

Rather thickset, relatively short-legged lapwing with wispy black crest. In flight, ***strikingly broad, rounded wings***; wings all dark above, tipped white on outer 3–4 primaries. Characteristic, flappy flight action. **Adult non-breeding** Upperparts dark glossy green (appearing black at distance), feathers tipped buff. Tail white with broad black subterminal band. Underparts white with broad black breast-band; undertail-coverts cinnamon. **HH** Rare Palearctic visitor, N Senegal (mainly Nov–Feb). Vagrant, Gambia. In various open habitats, incl. beaches, mudflats and ricefields. **Voice** Hoarse, shrill *cheew-ep* or *pee-witt*. [CD2:29]

ad
African Wattled Lapwing
ad
White-headed Lapwing
ad
Black-headed Lapwing
ad
Spur-winged Lapwing
ad
Lesser Black-winged Lapwing
juv
ad
ad
Northern Lapwing

Grey Plover *Pluvialis squatarola* 27–30 cm; WS 56–63 cm

Rather large, chunky, coastal wader with relatively large head, large dark eye, and ***stout blunt bill***. In flight, bold white wingbar, square white rump and uppertail-coverts, and whitish underwing with ***contrasting black axillaries*** ('armpits'). **Adult non-breeding** Mainly dull greyish; upperparts mottled darker; lower underparts white; diffuse supercilium. **Adult breeding** Head-sides to belly black bordered with white; upperparts spangled black and white. **Immature** As adult non-breeding but plumage tinged pale yellowish-buff; neatly chequered above, finely streaked below. **HH** Common Palearctic visitor (mainly Oct–Apr; records in all months). Singly or in loose groups; principally on tidal mudflats. Rare inland. **Voice** Far-carrying, clear, rather plaintive, drawn-out *tlueeee*. [CD2:27]

GODWITS Large shorebirds with long legs, necks and bills. Note bill shape, and wing and tail pattern in flight.

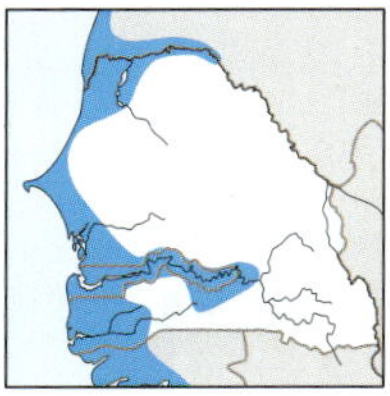

Black-tailed Godwit *Limosa limosa* 37–44 cm; WS 63–74 cm

L. l. limosa. Large, slender wader with ***very long, straight bill***, long neck, and long dark legs. In flight, distinctive pattern of ***broad white wingbar*** and ***square white uppertail contrasting with broad black terminal band***. Underwing white with dusky trailing edge. Legs project well beyond tail. **Adult non-breeding** Plain grey above. Bill pinkish with black tip. Longer necked and legged than Bar-tailed Godwit. **Adult breeding** Rufous-orange head and neck; breast and flanks variably barred. Upperparts dark brown mixed with rufous and grey. **Juvenile** Neck and breast variably washed pale rusty-orange; rusty-orange fringes to feathers above. **HH** Common Palearctic visitor (mainly Sep–Apr; records in all months). Singly or in groups on coastal lagoons, tidal mudflats and freshwater edges. Forages by probing deep in mud; often wades up to belly, even submerging head. Red List category: Near Threatened. **Voice** Short *kip* or *kip-kip-kip* and sharp *weeka-weeka*. [CD2:45]

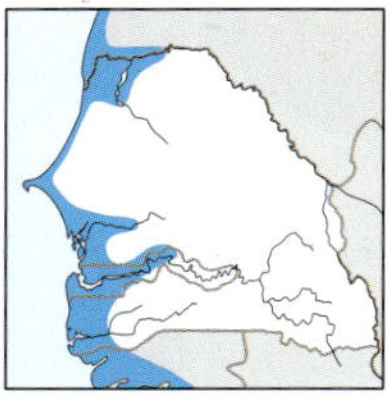

Bar-tailed Godwit *Limosa lapponica* 33–41 cm; WS 62–72 cm

L. l. taymyrensis. Resembles Black-tailed Godwit but more stocky, with somewhat shorter, ***slightly upturned bill*** and shorter legs. In flight, ***plain brown upperwing***, white wedge onto back and ***barred tail***. Underwing white. Feet project only slightly beyond tail. **Adult non-breeding** Pale brownish-grey above, appearing streaked. Long white supercilium. **Adult breeding** Underparts dark rufous. Female much paler than male. **Juvenile** Head and breast washed buffish; upperpart feathers notched buff. **HH** Common Palearctic visitor (mainly Sep–Apr; records in all months). In small groups or singly on tidal mudflats and sandy shores; occasionally inland. **Voice** Generally silent. Flight call a low, nasal *kirruk* or *kvip*, and sharper, also nasal, *keweep*. [CD2:46]

CURLEWS Large to very large shorebirds with long legs and necks and long, decurved bills. Plumage mainly brown and without seasonal variation. Note size and shape of bill, head pattern, and call.

Whimbrel *Numenius phaeopus* 37–46 cm; WS 78–88 cm

N. p. phaeopus. Smaller and darker than Eurasian Curlew, with shorter bill (more abruptly decurved at tip), shorter legs and boldly striped head pattern (dark brown crown with pale stripe on centre; pale supercilium bordered below by dusky eye-stripe). In flight, white wedge on back and barred tail; upperwing rather dark and uniform. **HH** Common Palearctic passage migrant and visitor (mainly Aug–Apr; some year-round) to entire coast and along Gambia River. Singly or in small, loose groups on tidal mudflats, sandy shores, mangrove creeks, estuaries and lagoons; occasionally on dry ground inland. **Voice** Clear, loud, rapid *bi-bi-bi-bi-bi-bi-bi*. Occasionally *kurrlee* (like Eurasian Curlew) and hard, rasping *krrreep*. [CD2:47]

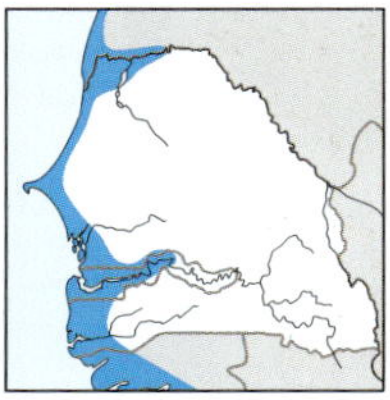

Eurasian Curlew *Numenius arquata* 48–57 cm; WS 89–106 cm

N. a. orientalis. Largest, streaky brown wader, with ***very long, evenly decurved bill***. In flight, white wedge on back and barred tail; dark outer wing contrasts with paler inner wing. Larger and paler than Whimbrel, bill longer, no bold stripes on head. **HH** Fairly common to scarce Palearctic passage migrant and visitor (mainly Aug–Apr; some year-round) to entire coast; uncommon to scarce along Gambia River. Singly or in small groups on tidal mudflats and estuaries. Red List category: Near Threatened. **Voice** Clear, liquid, bubbling *kur-lee* and *kwurrrr-lee*; also shorter *kwee-kwee*. [CD2:4

Grey Plover
juv
ad br
ad non-br
ad non-br
ad br
juv
Black-tailed
Godwit
ad
non-br
ad br
juv
Bar-tailed Godwit
ad non-br
ad
ad
Whimbrel
Eurasian Curlew

CALIDRIS *SANDPIPERS* *Feed by pecking busily. Their identification may be problematic and require careful observation. Important features include wing, rump and tail patterns, bill shape and length, and call. Juveniles (Aug–Oct) are neatly patterned with upperpart feathers having crisp pale fringes. First-winter plumage (Oct–Mar) is very similar to adult non-breeding.*

Red Knot *Calidris canutus* — 23–26 cm; WS 47–53 cm

C. c. canutus. Largest *Calidris*. Stocky, rounded, coastal wader with straight, relatively short and thick bill. Legs relatively short, greyish or greenish. In flight, narrow white wingbar, whitish-grey rump and ***grey tail***. **Adult non-breeding** Pale grey above with indistinct whitish supercilium and dusky eye-stripe; feathers narrowly fringed white. Underparts white with grey-streaked breast and flanks. **Adult breeding** Deep rufous face and underparts, rufous and black upperparts. **Juvenile** Scaly upperparts. **HH** Uncommon to locally frequent Palearctic visitor (mainly Oct–May). On tidal mudflats and estuaries. Vagrant inland (Sapu, Gambia). **Voice** Low, hoarse *nut, nut, …* [CD2:30]

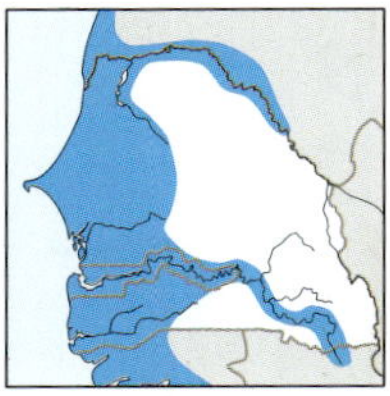

Little Stint *Calidris minuta* — 14–15 cm; WS 27–30 cm

Very small wader with ***relatively short, straight black bill*** and ***black legs***. In flight, narrow white wingbar and grey outer tail feathers. **Adult non-breeding** Grey above; feathers often with dark shaft streaks. Underparts white with grey breast-sides (sometimes forming breast-band). **Adult breeding** Head, breast and upperparts variably tinged rusty-orange with darkish-marked crown and creamy-buff V on mantle. **Juvenile** Crisp plumage, usually tinged rusty-orange above with whitish Vs on mantle and scapulars. **HH** Common Palearctic visitor (mainly Sep–May; records in all months). On coastal and inland mudflats and in wetlands, incl. lagoons, estuaries, edges of pools and rivers, ricefields, etc. **Voice** Sharp, high-pitched *kip* or *chit*. [CD2:32]

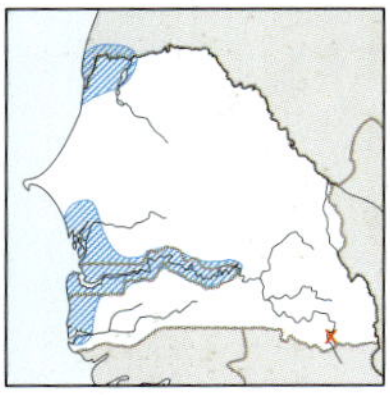

Temminck's Stint *Calidris temminckii* — 13–15 cm; WS 31 cm

Similar in size to Little Stint, but more elongated and much plainer, with (typically) complete grey-brown breast-band and shorter, brownish to olive-yellow legs. Recalls miniature Common Sandpiper. In flight, narrow white wingbar and white outer tail feathers. When flushed, flies high and erratically. **Adult non-breeding** Dull grey-brown above. **Adult breeding** Grey-brown above, some feathers dark centred and fringed rusty and greyish. **Juvenile** Buffish fringes above. **HH** Scarce to rare Palearctic visitor (mainly Oct–May). Singly, occasionally in small groups, at muddy edges of pools, lakes, rivers, marshes and ricefields. **Voice** High-pitched, dry trill *tirrr*. [CD2:33]

Sanderling *Calidris alba* — 18–21 cm; WS 39 cm

Active, rounded wader with relatively short, straight black bill and black legs. ***Lacks hind toe***. Runs very fast at edge of surf. In flight, bold white wingbar on blackish upperwing. **Adult non-breeding** ***Palest wader***. Very pale grey above, white below. **Adult breeding** Head and breast rufous marked black; upperparts black fringed rufous and grey. **Juvenile** Chequered black and white above. **HH** Common Palearctic visitor (mainly Oct–Mar; records in all months). Usually in small groups, on sandy beaches; also sandy river banks, lagoons, saltpans, occasionally inland. **Voice** Sharp *twik* or *krit*; sometimes accelerating into trill. [CD2:31]

Curlew Sandpiper *Calidris ferruginea* — 18–23 cm; WS 41 cm

Rather elegant with longish, evenly decurved black bill and longish black legs. In flight, clear white wingbar and ***broad white band on uppertail-coverts***. **Adult non-breeding** Grey above with prominent white supercilium. Underparts white with breast-sides washed grey. Compare Dunlin. **Adult breeding** Rusty-red head and underparts. **Juvenile** Scaly upperparts. **HH** Common Palearctic visitor (mainly Aug–Apr; records in all months). In groups on tidal mudflats, estuaries, lagoons and saltpans. Small numbers at inland waters. **Voice** Rather pleasing, rippling *chirrup*, less grating than Dunlin. [CD2:36]

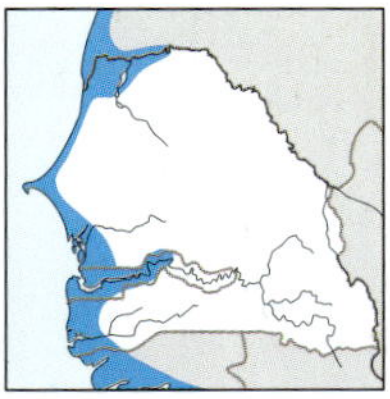

Dunlin *Calidris alpina* — 16–22 cm; WS 32–36 cm

C. a. schinzii & arctica. ***Rather dumpy*** with variably longish, black bill, slightly decurved at tip (shortest in *arctica*) and black legs. In flight, narrow white wingbar and dark-centred rump and uppertail-coverts. **Adult non-breeding** Less elegant than Curlew Sandpiper with less evenly decurved bill and shorter legs. Plain drab grey above with ***indistinct*** white supercilium. **Adult breeding** Black belly patch (diagnostic). **Juvenile** Scaly upperparts, streaked breast, variably spotted belly-sides. **HH** Uncommon to fairly common Palearctic visitor (mainly Sep–Apr). On tidal mudflats, saltpans, lagoons. **Voice** Shrill, rasping *krrreet* or *treerrp*. [CD2:38]

Red Knot
ad br
juv
ad non-br
Little Stint
ad br
juv
ad non-br
Temminck's Stint
juv
ad br
ad non-br
Sanderling
ad br
ad non-br
juv
Curlew
Sandpiper
ad br
juv
ad non-br
Dunlin
juv
ad br
ad non-br

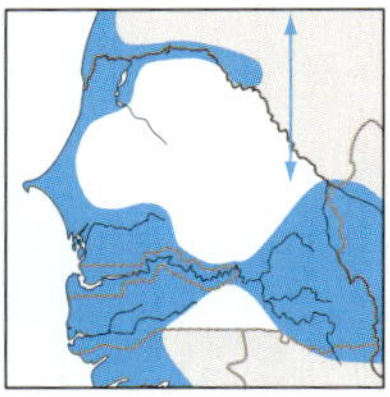

Ruff *Philomachus pugnax* ♂ 26–32 cm, WS 54–61 cm; ♀ 20–25 cm, WS 46–49 cm

Distinctly shaped wader with relatively small head, longish neck and slightly curved, medium-long bill. Plumage pattern, bare-part coloration and size variable; males considerably larger. ***Feathers of mantle appear loose and are often fluffed giving hunch-backed appearance***. In flight, narrow white wingbar and ***white oval patches on uppertail-sides***. Flying flocks often identifiable by marked size difference between sexes. **Adult non-breeding** Grey-brown and scaly above; legs orange-red to yellowish. Head and neck white in some males. Moulting males often have white on head and black blotching on breast. **Juvenile** Head, neck and breast warm buff; neatly patterned upperparts. **HH** Common Palearctic visitor (mainly Sep–Apr; some year-round). Frequents marshes, muddy edges of pools, lakes, reservoirs and lagoons, ricefields, flooded grassland and estuaries. **Voice** Usually silent. [CD2:40]

TRINGA *SANDPIPERS Medium-sized waders with relatively long bills and legs. Note bill shape, leg colour, upperpart pattern in flight, and call.*

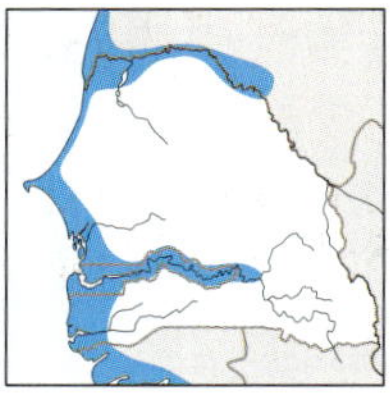

Spotted Redshank *Tringa erythropus* 29–33 cm; WS 52 cm

Elegant wader with ***long red legs*** and ***long fine bill*** with reddish base to lower mandible and slight but distinct downward kink at tip. In flight, white wedge on back and plain wings. **Adult non-breeding** Pale grey above, white supercilium highlighted by dusky eye-stripe; grey wash to breast. Legs and bill longer than in Common Redshank. **Adult breeding** Black; white spotting on upperparts. Legs dusky red to blackish. In moult has black-blotched plumage. **Juvenile** Closely streaked and barred grey below. **HH** Uncommon Palearctic visitor (mainly Oct–Apr; some year-round). Frequents estuaries, lagoons, mangroves, and various freshwater wetlands. Often wades deeply; frequently swims and up-ends. **Voice** Loud, abrupt *chu-wit!* [CD2:51]

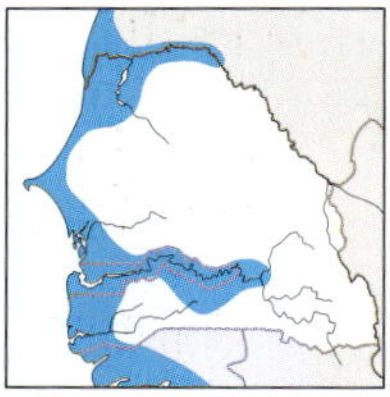

Common Redshank *Tringa totanus* 26–29 cm; WS 47–53 cm

T. t. totanus. Medium-sized wader with ***orange-red legs*** and ***red-based bill***. In flight, distinctive ***broad white trailing edge to wing*** and white wedge on back. **Adult non-breeding** Plain grey-brown above, white eye-ring, greyish breast. **Adult breeding** Densely streaked brown below. **Juvenile** Brown above with buff spots and notches. **HH** Common Palearctic visitor (mainly Aug–Apr; some year-round). Frequents tidal mudflats, estuaries, lagoons, mangroves and various freshwater habitats. **Voice** Mournful, drawn-out *tiuuu* and clear, ringing *tiu-lu tiu-lu-lu*, recalling Common Greenshank but higher pitched. [CD2:52]

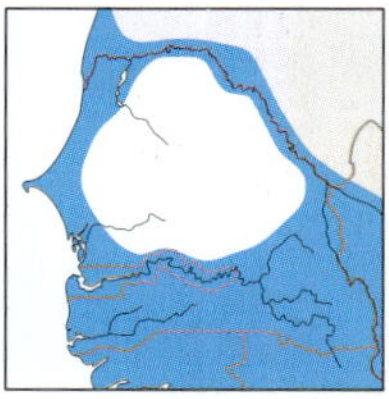

Common Greenshank *Tringa nebularia* 30–34 cm; WS 55–62 cm

Largest *Tringa*, with ***dull greenish legs*** and long, stout, ***slightly upturned, greyish bill***. In flight, as non-breeding Spotted Redshank but wings darker, more contrasting; rump and tail whiter. **Adult non-breeding** Pale grey above, white below, crown, hindneck and breast-sides finely streaked grey. **Adult breeding** Head, neck and breast heavily streaked. **Juvenile** Darker, with streaky head and breast, and white-fringed upperparts. **HH** Common Palearctic visitor (mainly Aug–Apr; some year-round). Frequents tidal mudflats, estuaries, lagoons, mangroves and various freshwater habitats. **Voice** Loud, ringing *tiu-tiu-tiu*. Occasionally a rasping *kruip-kruip-kruip* when flushed, and rapid *chip-chip-chip*. [CD2:54]

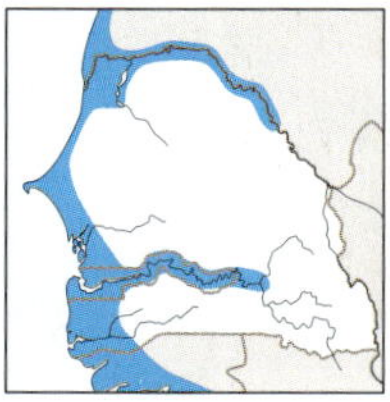

Marsh Sandpiper *Tringa stagnatilis* 22–25 cm; WS 42 cm

Recalls Common Greenshank, but distinctly smaller and more delicate, with ***straight, very fine, needle-like dark bill*** and relatively longer legs. In flight, plain dark upperwing and long white wedge on back; feet project well beyond tail. Often has dark carpal area (shoulder). Legs dull greenish. **Adult non-breeding** Plain pale grey above with ***white forehead and supercilium***, white below. **Adult breeding** Crown and neck finely streaked brown-grey, upperparts brown-grey blotched black, breast spotted, flanks with some V-shaped markings. Legs often tinged yellowish. **Juvenile** Finely streaked crown and hindneck, white-fringed upperparts. **HH** Uncommon Palearctic visitor (mainly Sep–Mar; some year-round). Frequents tidal mudflats, estuaries, lagoons, and muddy edges of various freshwater habitats. **Voice** Clear *keeuw*, singly or repeated up to 3–4 times, reminiscent of Common Greenshank, but less ringing. Alarm a sharp, rapidly repeated *chip*. [CD2:53]

juv
♂ non-br
ad br
♂ moulting
juv
♀ non-br
ad non-br
Ruff
♂ non-br
white-headed
individual
Spotted
Redshank
ad br
juv
ad non-br
Common Redshank
ad non-br
ad br
ad br
juv
juv
ad non-br
Common
Greenshank
ad non-br
Marsh
Sandpiper

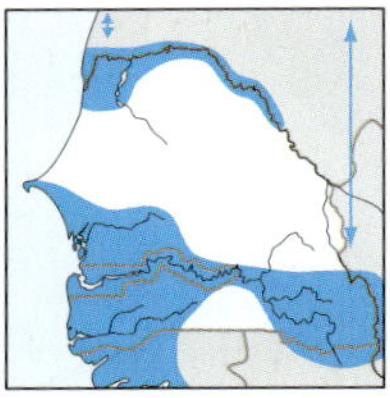

Green Sandpiper *Tringa ochropus*

20–24 cm; WS 39–44 cm

Medium-sized wader with contrasting, black-and-white-looking plumage. Resembles Wood Sandpiper but rounder, with darker, rather plain-looking upperparts, darker breast sharply contrasting with white belly, shorter white supercilium (only extending to eye) and more distinct white eye-ring; bill slightly longer, legs shorter, greyish-green. In flight, plain ***blackish upper- and underwings*** contrast strongly with pure white belly and rump, tail broadly barred blackish, feet barely project beyond tail. **Adult non-breeding** Rather plain, dark grey-brown above. **Adult breeding** Darker above, with more distinct streaking on head and breast and more contrasting speckling. **HH** Uncommon to common Palearctic visitor (mainly Oct–Mar; some year-round). Usually singly in various freshwater habitats, foraging inconspicuously at muddy edges of ditches, pools, lakes and rivers. Bobs rear as Common Sandpiper. **Voice** Clear, ringing *tlooeet weet-weet!* [CD2:55]

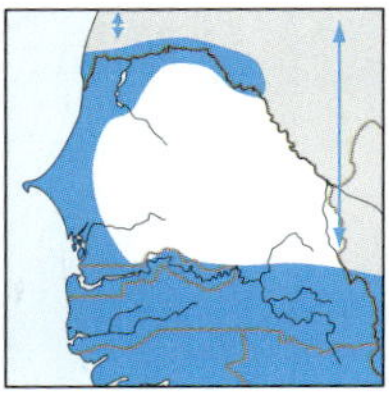

Wood Sandpiper *Tringa glareola*

19–21 cm; WS 35–39 cm

Smallish, slender wader with ***long, prominent supercilium*** and grey-brown, ***densely speckled upperparts***. Bill medium-long, straight, dark. Legs greenish to dull yellowish. In flight, plain dark upperwings, square white rump and pale grey underwing; feet clearly project beyond tail. Breeding plumage darker and more distinctly speckled above than non-breeding. **HH** Common Palearctic visitor (mainly Sep–Apr; some year-round). In small groups or singly, frequenting various freshwater habitats, incl. ricefields, muddy edges of ditches, pools and lakes, marshes, flooded grassland and sewage outlets. Bobs rear as Common Sandpiper. **Voice** Loud, ringing *chiff-iff-iff* or *chipipip*. [CD2:56]

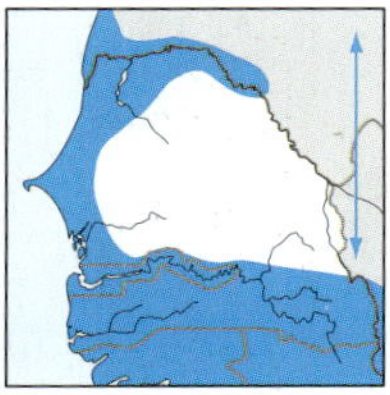

Common Sandpiper *Actitis hypoleucos*

18–21 cm; WS 32–35 cm

Smallish, rather compact wader readily identified by horizontal posture and ***constant nervous bobbing of rear body***. Flight action distinctive: ***low over water with rapid, fluttering wingbeats interspersed by short glides on stiff, bowed wings, revealing white wingbar***; rump and central tail dark, sides of tail white. Feet do not project beyond tail. **Adult non-breeding** Plain brown above with narrow white eye-ring and indistinct pale supercilium; white below with sharply demarcated brown breast; white extends onto breast-sides and forms wedge in front of closed wing. **Adult breeding** Dark markings above. **HH** Common Palearctic visitor (mainly Aug–Apr; some year-round). Usually singly, frequenting muddy edges of pools, lakes, rivers, streams, ricefields, marshes, lagoons, mangroves and beaches. **Voice** Rapid, clear, very high-pitched, piercing *tsee-wee-wee* in flight. Alarm a drawn-out *sweeeee-eet*. [CD2:58]

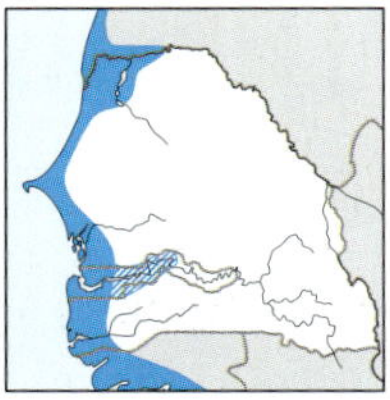

Ruddy Turnstone *Arenaria interpres*

21–25 cm; WS 43–49 cm

A. i. interpres. Stocky, coastal wader with stubby, wedge-shaped, black bill and ***short orange legs***. In flight, striking pied pattern with white wingbar, white stripe at base of wing and another on back, and white, black-banded tail. **Adult non-breeding** Mottled blackish-brown above and on breast, underparts white. **Adult breeding** Head and underparts white streaked black on crown with complex black pattern on face and breast. Upperparts a mix of bright rufous-chestnut and black. **Juvenile** Scaly upperparts. First-summer has variable mix of breeding and non-breeding plumages. **HH** Common Palearctic visitor (mainly Aug–Apr; records in all months). Frequents rocky coasts, shorelines, mangroves and muddy edges of lagoons; rarely inland waters. **Voice** Low, nasal *tuk* and rattling *tukatukatuk*. [CD2:59]

ad br
ad non-br
Green Sandpiper
ad non-br
ad br
ad non-br
Wood Sandpiper
ad br
ad non-br
juv
Common Sandpiper
ad non-br
Ruddy Turnstone
ad br

PLATE 51: PAINTED-SNIPE, SNIPES AND PHALAROPES

PAINTED-SNIPE Resembles true snipes, but brighter coloured, with shorter, slightly decurved bill and entirely different flight action. Sex roles reversed. Female is larger, polyandrous and leaves parental care to duller male.

Greater Painted-snipe *Rostratula benghalensis* 23–26 cm; WS 43 cm

R. b. benghalensis. Unobtrusive, rather plump, short-tailed, ***rail-like*** wader. In flight, ***rounded wings, slow wingbeats and dangling legs characteristic***. **Adult male** Golden-buff 'spectacles' and V on mantle, narrow median crown-stripe and spots on wing-coverts golden-buff. **Adult female** Head and neck chestnut, white 'spectacles', upperparts dark green. **Juvenile** As male but plumage less neatly patterned. **HH** Locally common. Singly or in pairs (occasionally small gatherings) in wetlands, incl. overgrown ricefields, edges of lakes, pools and streams, and swamps. Unobtrusive; mostly active at dawn and dusk. Seasonal, short-distance movements, responding to habitat requirements. **Voice** Usually silent. Breeding female utters low, sonorous *hooOOoo*. [CD6:19]

SNIPES Small to medium-sized with long, straight bill and cryptic plumage. Feed by probing rapidly into soft ground. ***PHALAROPES*** *are graceful waders that swim buoyantly and forage on water with spinning action. Sex roles reversed.*

Jack Snipe *Lymnocryptes minimus* 17–19 cm; WS 33–36 cm

Small and skulking with ***shortish bill*** and legs, ***buff, split supercilium***, dusky crescent around ear-coverts, ***no median crown-stripe*** and ***two yellowish stripes on upperparts***. Breast and flanks ***streaked*** (not barred) dark brown. In flight, note small size, shortish bill and dark, wedge-shaped tail. **HH** Rare to scarce Palearctic visitor (Nov–Mar). Singly in marshy areas and at muddy edges of pools, ricefields and small streams. Flushes silently from underfoot; does not usually fly far before dropping into cover. **Voice** Usually silent; quiet, hoarse *gatch* when flushed [CD2:41]

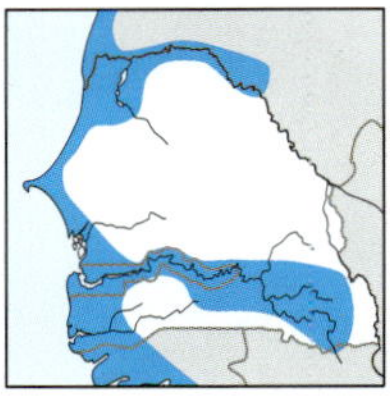

Common Snipe *Gallinago gallinago* 23–28 cm; WS 39–45 cm

G. g. gallinago. Cryptically patterned, with boldly striped head and very long, straight bill. Mainly brown, with buff median crown-stripe and buff stripes on mantle-sides and scapulars. Breast brown-buff densely marked with lines of dark brown spots, flanks barred dark brown, belly white. Flight fast and zigzagging, revealing ***narrow white trailing edge to inner wing*** and white edges to tail. **HH** Common to uncommon Palearctic visitor (mainly Oct–mid Apr). At muddy edges of pools, lakes and watercourses, and in marshes. Secretive and wary. Squats when threatened; flushes explosively at *c*.10 m distance, climbing rapidly in frantic, tilting flight and usually flying far before dropping again. **Voice** Hoarse *rretch* when flushed and in flight. [CD2:42]

Great Snipe *Gallinago media* 26–30 cm; WS 43–50 cm

Slightly bulkier than Common Snipe with slightly shorter bill, white wingbars, and entirely barred flanks, belly and underwings. Flight low and straight, distinctly slower and heavier than Common Snipe, with two narrow white lines delimiting dark midwing panel on greater coverts; broad white corners to tail conspicuous on take-off or landing. **HH** Rare Palearctic visitor (mainly Oct–Mar). Singly in marshy areas and at muddy edges of pools and small streams. Flushes silently but with audible sound of wings at *c*.5 m distance; usually dropping quickly into cover. Red List category: Near Threatened. **Voice** Hoarse, gruff *shrt* or *krrrt*, occasionally given in flight. [CD2:43]

Red-necked Phalarope *Phalaropus lobatus* 17–19 cm; WS 30–34 cm

Resembles Red Phalarope in shape and behaviour, but more dainty, with ***needle-thin, all-black bill***. In flight, more contrasting white wingbar. **Adult non-breeding** Slightly darker and less plain above than Red Phalarope. **Juvenile** Rusty-yellow V on dark brown upperparts. **HH** Pelagic; on land in various wetlands. Palearctic/N American vagrant. **Voice** Usually silent. [CD2:61]

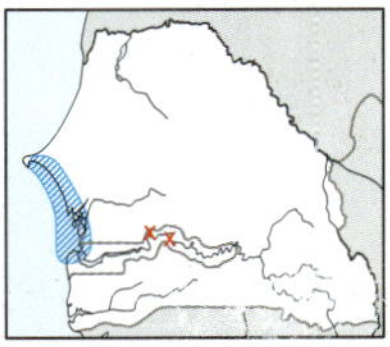

Red Phalarope *Phalaropus fulicarius* 20–22 cm; WS 36–41 cm

Small and active with shortish, blunt-tipped, straight bill. In flight, long white wingbar (broadest on inner wing). **Adult non-breeding** Plain, pale grey above with black eye patch, white below. **HH** Mainly pelagic. On land in various aquatic habitats. Palearctic/N American vagrant. **Voice** Sharp, high-pitched *kip*. [CD2:62]

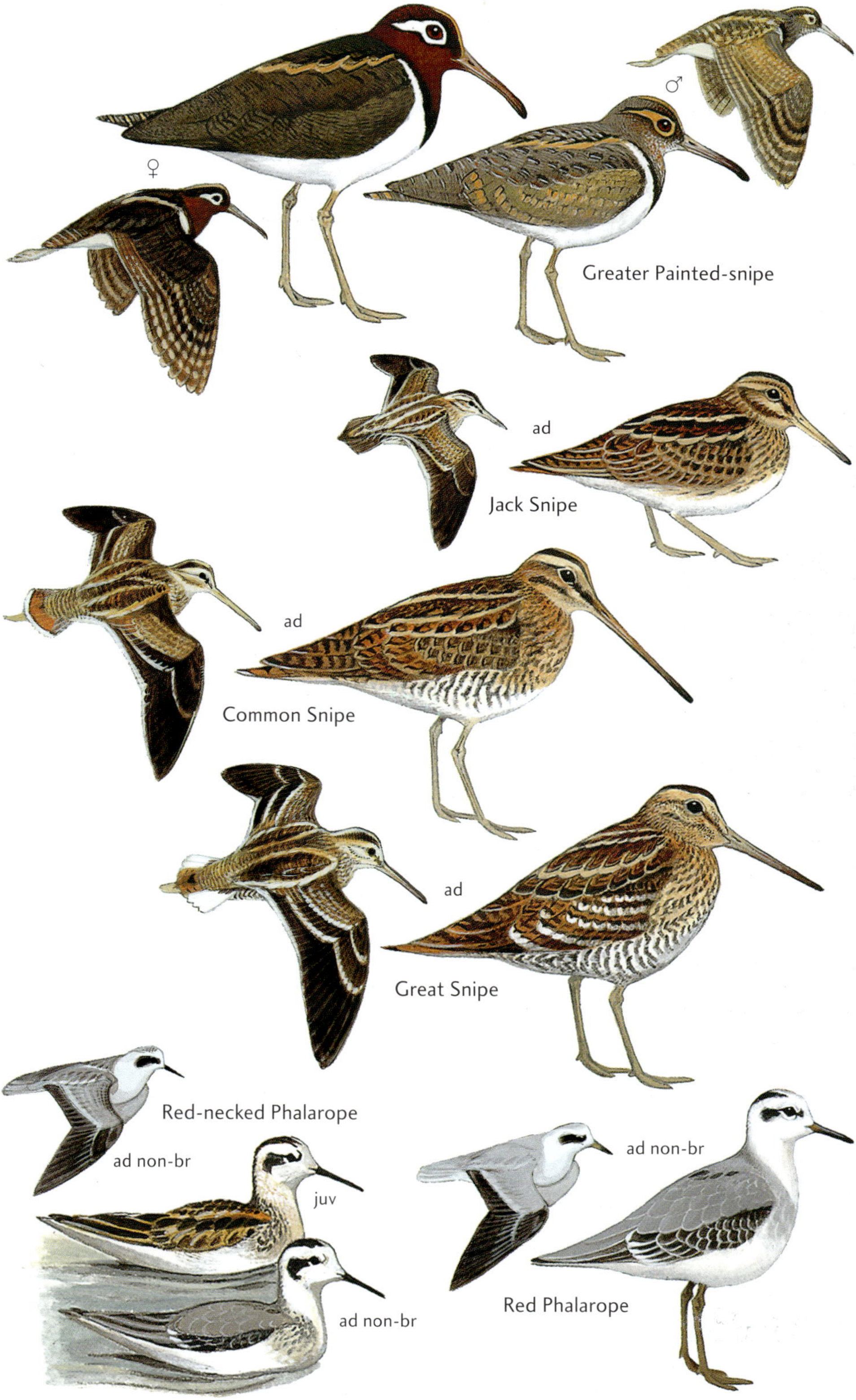
♀
♂
Greater Painted-snipe
ad
Jack Snipe
ad
Common Snipe
ad
Great Snipe
Red-necked Phalarope
ad non-br
juv
ad non-br
ad non-br
Red Phalarope

Greater Sand Plover *Charadrius leschenaultii* 20–23 cm; WS 44 cm

C. l. crassirostris/columbinus. Long-legged, medium-sized plover with relatively long and heavy bill. In flight, white wingbar and white sides to tail; feet project beyond tail. **Adult non-breeding** Long greenish-grey legs; no white nuchal collar. **HH** Palearctic vagrant, Senegal (Nov–Dec, Feb). Probably also Gambia (Jan 1991 & 1996, Dec 2005). On intertidal mudflats; also grassland. Associates with other waders. **Voice** Trilling *trrr* and *tirrirrilip*, often frequently repeated. [CD2:23]

Eurasian Dotterel *Charadrius morinellus* 20–23 cm; WS 46 cm

Adult non-breeding Rather plump and compact, recalling golden plovers, but has ***broad pale supercilia meeting in V on nape, narrow pale breast-band*** and ***yellowish legs***. In flight, plain upperwing; white shaft to outer primary forms narrow pale leading edge. **HH** Palearctic vagrant, Gambia (Cape St Mary, Mar 1985). Prefers open inland habitats. **Voice** Mostly silent. A rattling call in flight, rather reminiscent of Ruddy Turnstone. [CD2:24]

American Golden Plover *Pluvialis dominica* 24–28 cm; WS 54 cm

Noticeably smaller and slimmer than Grey Plover, with finer bill. In flight, indistinct wingbar and ***grey underwing***. At rest, 4–5 primary tips project beyond relatively short tertials; wings project clearly beyond tail. Has ***cold grey tones*** to non-breeding plumage. In breeding plumage has white 'shawl' from forehead to breast-sides. **HH** N American vagrant, Senegal (Oct 2005 & 2006, Dec 1983, Feb 2001, May 1979) and Gambia (Nov 1997, Dec 2005, Jan 1984). In various wetlands and farmland. **Voice** Rather mellow *tlu-ee*, with stress on first syllable, and *tlu-ee-uh*. [CD2:25]

Pacific Golden Plover *Pluvialis fulva* 22–26 cm; WS 49 cm (not illustrated)

Similar to American Golden Plover, with same wing pattern in flight, but has ***warm-buff tones*** to non-breeding plumage. Primaries shorter with, at rest, only 3 (rarely 4) tips projecting beyond tertials; tertials longer with tips approximately level with tail tip. Wings project slightly beyond tail. Legs slightly longer than American; toes project beyond tail in flight. In breeding plumage white 'shawl' extends to flanks, where barred with black. **HH** Vagrant from N Siberia, Senegal (Saloum delta, May 2012). **Voice** Clear, rapid *chu-it*, reminiscent of Spotted Redshank (and unlike American Golden Plover), and more drawn-out *chu-wee*; also *klu-wee-up*.

European Golden Plover *Pluvialis apricaria* 26–29 cm; WS 53–59 cm

Similar to American Golden Plover but ***white underwing*** visible in flight. Also slightly larger and more chunky, with shorter legs. At rest, wings project only just beyond tail; 3–4 primary tips project beyond tertials; tertial tips almost level with tail tip. In non-breeding plumage, has yellowish tones above, yellow-buff below. In breeding plumage white band extends to white undertail-coverts. **HH** Palearctic vagrant, Senegal and Gambia (Oct–May). On tidal mudflats. **Voice** Soft, melodious *tluuee*. [CD2:26]

Terek Sandpiper *Xenus cinereus* 22–25 cm; 42 cm

Rather stocky wader, readily identified by ***long, gently upturned bill*** and ***shortish, reddish-pink to orange-yellow legs***. In flight, distinctive wing pattern with dark leading edge and outer wing and whitish trailing edge; rump and tail grey. **Adult non-breeding** Rather plain pale grey above with dark carpal area, white below with pale grey breast-sides. Bill black with dull yellow or orange base. **HH** Palearctic vagrant, Gambia (Dec 1974). Frequents tidal mudflats, estuaries and muddy edges of lakes and rivers. Often forages fast and actively, running in low horizontal posture over mudflats, but also walks slowly. **Voice** Loud, melodious, ringing *eeb-eeb-eeb*. [CD2:57]

Buff-breasted Sandpiper *Tryngites subruficollis* 18–20 cm; WS 41 cm

Slightly smaller than female Ruff, with distinctly small, round head, short, fine black bill and yellowish legs. Sandy-buff below. In flight, no white above but white underwing has conspicuous ***dusky half crescent on greater primary-coverts***; feet do not project beyond tail. **HH** N American vagrant, Senegal (Lake Retba, Apr 1985) and Gambia (Kaur, Nov 1998; Pirang, Nov 2006). Frequents wetlands and grassy areas. Red List category: Near Threatened. **Voice** Usually silent.

ad non-br
Greater Sand Plover
ad non-br
Eurasian Dotterel
ad non-br
ad br
American
Golden Plover
ad non-br
ad non-br
ad br
European
Golden Plover
ad non-br
ad non-br
Terek Sandpiper
ad non-br
Buff-breasted
Sandpiper

Baird's Sandpiper *Calidris bairdii*

14–17 cm; ; WS 39 cm

Small, long bodied *Calidris* with long primary projection, ***wingtips extending clearly beyond tail, relatively short blackish legs*** and all-black, straight bill. In flight, faint white bar on long wings and dark-centred rump and uppertail-coverts. **Adult non-breeding** Grey-brown above with indistinct supercilium; breast washed buffish-brown and indistinctly streaked. **HH** Possible N American vagrant, claimed from N Senegal (Dec 1965) and Gambia (Nov 1976). **Voice** Soft, rolling *prrreet* or *kirrrp*, softer than Curlew Sandpiper.

Pectoral Sandpiper *Calidris melanotos*

19–23 cm; WS 38–44 cm

Slightly larger than Curlew Sandpiper and somewhat smaller than female Ruff, with relatively long neck and small head, ***densely streaked neck and upper breast sharply demarcated from white lower breast and belly***, bill slightly decurved and pale based, ***legs greenish-yellow to yellowish-brown***. In flight, indistinct narrow pale wingbar and white sides to dark-centred rump and uppertail-coverts. **HH** N American/E Palearctic vagrant, N Senegal (Apr 1993). Frequents muddy edges of pools, marshy places, wet grasslands, lagoons and mudflats. **Voice** Rolling, hoarse *krrrt*. [CD2:35]

Solitary Sandpiper *Tringa solitaria*

18–21 cm; WS 42 cm

Adult non-breeding As Green Sandpiper (Plate 50) but lacks white rump and is slightly smaller and slimmer. Best distinguished in flight, when ***dark band on centre of rump and tail*** is visible; tail-sides barred. **HH** N American vagrant, Gambia (Jan 1999, Jan 2005, Sep 2007). **Voice** High-pitched *tewit-weet* or *pleet-weet-weet*, resembling call of Green Sandpiper, but softer. [CD6:43]

Spotted Sandpiper *Actitis macularius*

18–20 cm; WS 33 cm

Adult non-breeding Very similar to Common Sandpiper (Plate 50) but with ***shorter tail***, which projects noticeably less beyond wingtips, and paler, yellowish legs. In flight, ***shorter wingbar***, not extending onto inner secondaries (in Common, wingbar extends across all secondaries), and usually less white on tail-sides. **Adult breeding** Underparts boldly spotted black, bill pinkish with black tip, legs pinkish. **HH** N American vagrant, Senegal (Gorée, Feb 2001). **Voice** A piping *peet!* or *tweet-weet-weet*, reminiscent of Green Sandpiper and lower pitched than Common Sandpiper.

Short-billed Dowitcher *Limnodromus griseus*

25–29 cm; WS 45 cm

Medium-sized with very long, straight, blunt-tipped bill, prominent supercilium and moderately long, greenish legs. In flight, whitish trailing edge to wing, white oval on back and barred tail; feet barely project beyond tail. **Adult non-breeding** Grey above with whitish supercilium and dusky eye-stripe, white below with grey upper breast and grey barring on flanks. Black bars in tail narrower than white bars. **Juvenile** Breast and upperparts tinged rusty-buff; upperpart feathers neatly fringed; centres of tertials, greater coverts and scapulars irregularly striped and barred buff. **Long-billed Dowitcher *L. scolopaceus*** very similar and best distinguished by call: a short, sharp, rather high-pitched *keek* or *kyip*, often quickly doubled or trebled when alarmed. Other differences include: black bars in tail broader than white bars. Juvenile/first-winter has plain (not 'marbled') dark centres to tertials, greater coverts and scapulars. **HH** N American species. A dowitcher was observed in Gambia in Dec 1978, but not specifically identified. **Voice** Soft, fast, rattling *tiu-tuk-tuk* or *chu-tu*, reminiscent of Ruddy Turnstone.

Lesser Yellowlegs *Tringa flavipes*

23–25 cm; WS 49 cm

Resembles Common Redshank (Plate 49) but distinctly slimmer, with finer, black bill and longer, ***yellow legs***. In flight, dark upperwing and square white rump; feet project well beyond tail. **HH** N American vagrant, Senegal (Mar 1997, Jan 1991) and Gambia (Jan 1976, Oct 2010). Frequents tidal mudflats, lagoons and various freshwater habitats. **Voice** Clear *tew* or *tew-tew*, reminiscent of Common Greenshank but more subdued.

ad non-br
Baird's Sandpiper
ad non-br
Pectoral Sandpiper
ad non-br
Solitary Sandpiper
ad br
ad non-br
ad non-br
ad non-br
Spotted Sandpiper
juv
ad non-br
Short-billed Dowitcher
tail
juv
tertial
ad non-br
tail
juv
tertial
Long-billed Dowitcher
for comparison
Lesser
Yellowlegs

SKUAS (Plates 54–55) Gull-like seabirds with powerful, steady and fast flight. Predatory and piratical. Feed mainly by harassing other seabirds, especially gulls and terns, to force them to disgorge or drop their catch. Plumages variable and identification often problematic; important features include structure, proportions and flight action; also note presence or absence of tail projection and of prominent pale areas in plumage.

Pomarine Skua *Stercorarius pomarinus* 42–50 cm; WS 115–125 cm

Largest and most powerful of the smaller skuas (slightly smaller than Lesser Black-backed Gull). In flight, ***noticeably heavier than Arctic Skua, with broader based wings and more rounded breast***. Bill slightly heavier and distinctly two-toned. Pale primary patch as Arctic but much smaller than in Great. **Adult breeding** Readily identified by ***elongated, spoon-shaped central tail feathers*** (projecting 5.5–11 cm). Pale morph has mottled breast-band. Rare dark morph entirely blackish-brown except pale primary patches. **Adult non-breeding** Variable. Most have head-sides darkish and upper- and undertail-coverts barred; tail streamers shorter and often lacking. **Juvenile** Blackish-brown to mid-brown with paler barring on underwings, uppertail-coverts and vent, ***which are typically paler than, and contrast with, rest of plumage***. Pale crescent at base of primary-coverts visible at fairly close range, creating ***distinctive double pale patch on underwing***. **HH** Common Palearctic migrant, Aug–Apr (records in all months). The commonest skua off Senegal (e.g. 119–6233 off Dakar in Oct–Nov 1995–2008). Occasionally well up the Gambia River estuary. Winters mainly at sea between 20° and 8°N. **Voice** [CD2:63]

Arctic Skua *Stercorarius parasiticus* 37–44 cm; WS 108–118 cm

Agile and rather elegant with ***fast, falcon-like flight***. Noticeably slimmer than Pomarine Skua, with ***more slender wings (narrower at base)***, smaller head and slimmer bill. Pale primary patch as Pomarine, but typically lacks pale crescent at base of primary-coverts. **Adult breeding** Distinguished by ***elongated, narrow, pointed central tail feathers*** (projecting up to 10 cm). Plumage variation as Pomarine. Pale morph has grey breast-band. Dark morph entirely blackish-brown to grey-brown except pale primary patches. **Adult non-breeding** Variable. Mostly with head-sides darkish and upper- and undertail-coverts barred; elongated tail feathers often absent. **Juvenile** Very variable. Blackish-brown to warm rufous-brown. Barring on underwings, uppertail-coverts and vent more irregular and wavy than Pomarine. Head narrowly streaked dark and ***often has warm rufous nuchal band*** (diagnostic); upperparts scaly, frequently tinged rusty; primary tips usually edged buffish (visible on closed wing). ***Central tail feathers pointed*** and only slightly projecting beyond rest of tail. **HH** Common Palearctic migrant, Aug–Apr (e.g. 476–2879 off Dakar in Oct–Nov 1995–2008); records in all months. Occasionally well up the Gambia River estuary. Winters mainly off Namibia and W South Africa. **Voice** [CD2:64]

Long-tailed Skua *Stercorarius longicaudus* 35–41 cm; WS 105–115 cm

Small and slim (about size of Black-headed Gull but often appearing larger) with buoyant, ***almost tern-like flight***. Head smaller than Arctic Skua, bill shorter, body and tail more elongated. Upperwing has two pale shafts on dark primaries, forming pale line on leading edge. **Adult breeding** ***Central tail feathers greatly elongated, very narrow and pointed*** (projecting up to 24 cm beyond rest of tail). Well-defined, neat black cap; head-sides and breast pale creamy becoming grey on belly; ***upperparts greyish with contrasting black flight feathers***. No dark morph. **Adult non-breeding** Variable. Most have darkish head-sides and dark barring or spotting on breast (creating band), flanks, and upper- and undertail-coverts; tail streamers much shorter and often absent. **Juvenile** Very variable. Palest have whitish head and belly, darkest mainly sooty-black. Typically cold brownish-grey, ***always lacking warm rufous-brown tones*** often shown by juvenile Arctic. Dark and pale barring on underwings, uppertail-coverts and vent. Upperparts scaly; fringes whitish (more contrasting and straighter than in Arctic). ***Central tail feathers narrow but rather blunt tipped and clearly projecting*** (projection up to 4 cm). **HH** Fairly common Palearctic migrant, Aug–Apr (e.g. 10–864 off Dakar in Oct–Nov 1995–2008). Winters mainly off Namibia and W South Africa. **Voice** [CD2:65]

ad br
pale morph
ad br
dark morph
ad br
pale morph
Pomarine Skua
juv
dark morph
juv
pale morph
ad non-br
juv
ad non-br
ad br
dark morph
ad br
pale morph
ad br
pale morph
Arctic Skua
juv
dark morph
juv
pale morph
juv
pale morph
ad br
Long-tailed Skua
ad br
juv
pale morph
juv
pale morph
juv
dark morph
ad non-br

PLATE 55: SKUAS AND LARGE GULLS

Great Skua *Stercorarius skua* 50–58 cm; WS 125–140 cm

Largest skua. ***Heavy, compact*** and entirely dark with relatively short tail and ***broad wings*** with ***conspicuous white patches at base of primaries above and below***. All plumages similar. **Adult** Coarsely streaked yellowish-buff. Sometimes has dark cap slightly contrasting with rest of head. Bill heavy, blackish. **Juvenile** Separable at close range. Darker and more uniform, less streaky than adult, generally more chestnut-brown below; white wing crescent often slightly narrower and less conspicuous above. **HH** Palearctic passage migrant and visitor. Main wintering area in Atlantic south to W Africa. Recorded off Senegal; status unclear due to confusion with South Polar Skua. **Voice** [CD2:66]

South Polar Skua *Stercorarius maccormicki* 50–54 cm; WS 125–135 cm

Very similar to Great Skua but slightly smaller and more slender, with relatively smaller head and slightly slimmer bill. Plumage variable but ***unstreaked***, and typically with paler nape (often hard to discern at sea) and lacking dark cap of Great. Occasionally has indistinct, small pale area at base of bill. **Adult dark morph** Wholly cold greyish-brown with paler nape. Darkest birds virtually impossible to separate from some juvenile Great Skuas. **Adult intermediate morph** Head and underparts pale brown; upperparts blackish-brown. **Adult pale morph** Distinctive. Head, mantle and underparts plain, pale buff-brown. **Juvenile** Similar to adult of respective morph. **HH** Breeds Antarctica. Immatures migrate to northern hemisphere in southern winter, performing clockwise migration. Southwestward migrants (presumably this species) observed in numbers off Senegal. Presumably regular passage migrant; status unclear due to confusion with Great Skua.

GULLS (Plates 55–58) *Robust, medium-sized to large birds of coast, oceans and inland waters. Distinct breeding, non-breeding and immature plumages. Adult plumage reached in 2nd–4th year; period of immaturity generally related to size. Tail white in adults; dark tail band indicative of immature. Sexes similar, male slightly larger. Flight strong and buoyant with slow wingbeats and much gliding and soaring.*

To identify gulls, note wing and head patterns, and bare-part coloration. Understanding moult greatly facilitates identification. First moult is post-juvenile: a partial moult, replacing head and body feathers and resulting in first-winter plumage. Subsequently, there are two moults per year: a partial one from winter to summer plumage, mainly replacing head and body feathers (flight feathers and all or most tail feathers thus still juvenile and much abraded in first-summer plumage), and a complete one after the breeding season, resulting in adult or immature winter plumages. The only exceptions to this rule are Sabine's Gull (Plate 57) and Franklin's Gull (Plate 58).

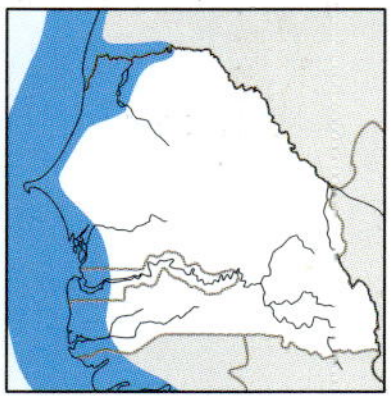

Lesser Black-backed Gull *Larus fuscus* 49–57 cm; WS 118–135 cm

Large but rather elegant with slightly narrower, longer looking wings than other large gulls. Adult plumage acquired in 4th year. **Adult breeding** Diagnostic combination of slate-grey (***graellsii***), blackish-grey (***intermedius***) or blackish (***fuscus***) upperparts and yellow legs. Black wingtips with 1–2 small white mirrors. In flight from below, dusky-grey band across flight feathers (darker than Yellow-legged Gull). **Adult non-breeding** Head and breast-sides with extensive grey-brown streaking (***graellsii***) or little or no streaking (***intermedius & fuscus***). **First-winter** Dark grey-brown, bill black, legs pink. In flight, outer wing and secondaries blackish, greater coverts dark. First-year Yellow-legged Gull has more contrasting upperwing pattern and less clear-cut tail band. **Second-winter** Acquires adult mantle colour and white head and body. **Third-year** As adult, but retaining some traces of immaturity on wings and tail; bare-part colours paler, bill with darkish spot on gonys. **HH** Common Palearctic migrant to entire coast and inland. Mainly Oct–Mar; some oversummer. **Voice** [CD2:74]

Kelp Gull *Larus dominicanus* 55–65 cm; WS 128–142 cm

L. d. vetula (Cape Gull). Large, heavy looking gull, with massive head, stout bill and dark eye. Adult plumage acquired in 4th year. **Adult** More robust than Lesser Black-backed Gull, bill heavier, eye dark, legs olive. **First-year** Mainly dark brown, bill black, legs pinkish-brown. In flight, very dark looking, with blackish outer wing and secondaries. **Second-year** Head largely white, dark brown 'saddle', white tail base. **Third-year** As adult, but retaining some traces of immaturity, esp. on wings. Immature Lesser Black-backed Gull very similar and best separated by structure. **HH** Uncommon migrant from southern hemisphere (possibly now resident), coastal Senegambia. Bred Senegal (first record Saloum, 1983). **Voice** [CD6:46]

Great Skua
ad
ad
dark morph
ad
pale morph
South Polar Skua
ad br
ad non-br
fuscus
1st-winter
1st-winter
ad non-br
graellsi
Lesser Black-backed Gull
2nd-winter
ad br
1st-winter
2nd-winter
ad
Kelp Gull
1st-winter
ad

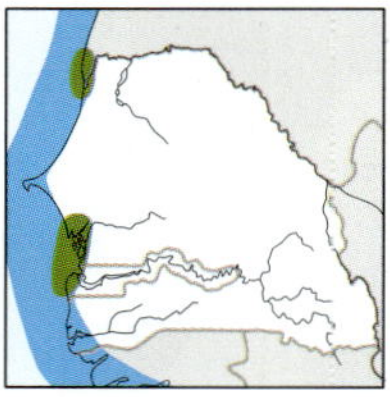

Slender-billed Gull *Larus genei* 40–44 cm; WS 100–110 cm

Medium-sized, white-headed and long-necked gull with distinctive ***elongated head***, produced by gently sloping forehead and longish bill. In flight, white leading edge to outer wing. Adult plumage acquired in 2nd winter. Compare Black-headed Gull. **Adult breeding** White head; pink flush to underparts. Eye pale (variable; sometimes looking dark depending on light). Dark reddish bill (looking black at distance) and legs. **Adult non-breeding** Faint dusky ear-spot. **First-winter** Brownish wing-coverts; pale bill and legs. In flight, pale brown carpal bar; blackish secondary bar and tail band. **HH** Locally not uncommon resident and intra-African migrant, breeding in Senegal (colonies Saloum delta, Langue de Barbarie). Also Palearctic winter visitor to Senegambian coasts. **Voice** [CD2:71]

Common Gull *Larus canus* 40–45 cm; WS 100–130 cm

L. c. canus. Medium-sized, fairly elegant gull with rounded head, slim bill and dark eye. Adult plumage acquired in 3rd winter. **Adult breeding** Dark blue-grey upperparts; yellow-green bill and legs. **Adult non-breeding** Head has dusky markings, bill pale with dark band. In flight, black wingtips with relatively large white 'mirrors'. **First-winter** Dusky head and breast; dusky-tipped pinkish bill; greyish-pink legs. In flight, blackish leading edge to outer wing, secondary bar and tail band. **Second-winter** Uniformly grey wings, black leading edge to outer wing, small white 'mirrors'. **HH** Palearctic vagrant, coastal Senegambia (Sep, Dec–Jan, Jul). Possibly under-recorded. **Voice** [CD2:73]

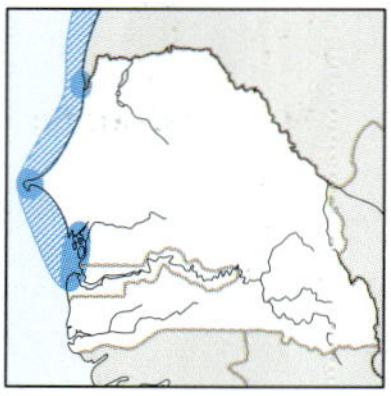

Audouin's Gull *Larus audouinii* 44–52 cm; WS 115–140 cm

Smallest, ***palest and most elegant of the large gulls*** in W Africa. Adult plumage acquired in 4th year. **Adult** Very pale with sharply contrasting black wingtips. In flight, ***obviously paler than Yellow-legged Gull*** (lacking prominent white leading and trailing edges to wing and white mirrors on wingtips), with ***dark eye*** and ***dark-looking bill*** (red, with black subterminal band and yellow tip) imparting less fierce expression. Legs olive-grey. **First-year** Whitish head contrasts with grey-brown upperparts; dark grey legs (diagnostic). In flight, blackish primaries and secondary bar; dark greater coverts; black tail; U-shaped band across rump. **Second-year** Upperparts obviously paler than corresponding Yellow-legged, with narrower tail band, dark legs and usually reddish-based bill. **Third-year** Much like adult, but black of wingtips extends onto greater primary-coverts and mirror on outer primary smaller or lacking. **HH** Rare to not uncommon migrant from Mediterranean to coastal Senegal and Gambia (Sep–May). Red List category: Near Threatened. **Voice** [CD2:72]

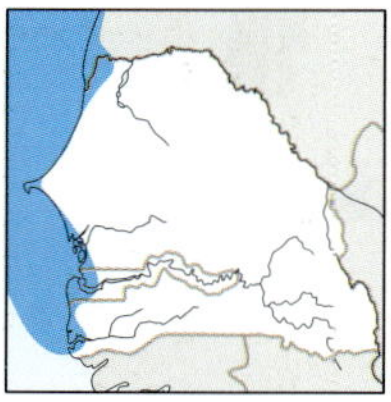

Yellow-legged Gull *Larus (cachinnans) michahellis* 52–58 cm; WS 120–140 cm

Large, powerful gull. Adult plumage acquired in 4th year. **Adult *atlantis*** Fierce looking, with ashy-grey upperparts, yellow bill with orange spot, yellow legs. In flight, black wingtips with small white 'mirrors'. In non-breeding plumage, head streaked, creating hooded appearance (more extensive than illustrated and similar in extent to first-winter; streaking lost from Dec/Jan onwards); legs sometimes yellowish-pink. ***L. m. michahellis*** is slightly paler grey above; black on wingtips usually less extensive; white 'mirrors' slightly larger. Non-breeding only very faintly streaked on head and neck. **First-winter *atlantis*** Mainly brownish, with head and underparts densely streaked or spotted, often rather uniform brownish; bill black; legs brownish-pink (often with darker spotting on upper tarsus and feet, a feature unique to this taxon and Kelp Gull). Many retain juvenile plumage through first winter moulting directly into second-winter plumage in late summer. Wings heavily worn from Feb/Mar. In flight, blackish-brown flight feathers contrast with rest of upperparts; broad dark tail band. ***L. m. michahellis*** has indistinct paler area on inner primaries, but in ***atlantis*** all primaries dark (thus resembling Lesser Black-backed Gull). **Second-winter** Acquires ashy-grey mantle and white head and body. **Third-winter** As adult, but traces of immaturity on wings and tail. **HH** Uncommon Palearctic visitor (mainly Oct–May; records in all months) to coastal Senegambia. Birds in Senegambia presumed to be *L. m. atlantis* (breeding on Atlantic islands) but *L. m. michahellis* (mainly Mediterranean) also possible; more research needed on distribution outside breeding season of respective races. **Voice** [CD2:76,77]

1st-winter
ad non-br
ad br
Slender-billed Gull
1st-winter
ad non-br
ad br
Common Gull
ad non-br
1st-winter
1st-win
2nd-winter
ad non-br
1st-winter
ad
ad
Audouin's Gull
1st-winter
1st-winter
michahellis
ad non-br
michahellis
1st-winter
atlantis
ad non-br
atlantis
Yellow-Legged Gull
1st-win
2nd-winter
3rd-winter
ad non-br

Little Gull *Larus minutus* 24–28 cm; WS 75–80 cm

Tiny gull with rounded-looking wings and buoyant, surface-picking feeding flight, recalling marsh tern. Adult plumage acquired in 3rd year. **Adult breeding** The only black-hooded gull lacking white eye-crescents; upperparts pale grey. **Adult non-breeding** Dusky cap and ear-spot; pale grey upperparts. In flight, no black on upperwing; ***underwings blackish with white trailing edge***. **First-winter** Dusky cap and ear-spot; black primaries and wingbar. In flight, blackish W on upperwings, underwings white, black tail band. **First-summer** Variably developed hood. **Second-winter/-summer** As adult breeding/non-breeding, but variable blackish markings on tips of outer primaries; summer plumage with hood often flecked white. **HH** Rare Palearctic visitor to Senegambian coast. **Voice** [CD2:68]

Sabine's Gull *Xema sabini* 27–35 cm; WS 90–100 cm

Rather small with ***tricoloured, triangular-patterned upperwing***, slightly forked tail (often hard to see), and elegant, tern-like flight. Unique among gulls in having a ***partial*** moult in Nov–Dec (in wintering areas) and a ***complete*** one in Feb–Apr (before northward migration). Adult plumage acquired in 2nd winter. **Adult breeding** Dark grey hood; yellow-tipped black bill. **Adult non-breeding** Head largely white with blackish nape patch of variable extent. **First-winter** Brownish wing-coverts; all-black bill. **Juvenile** Same pattern in flight as adult, but grey areas browner; black tail band. **HH** Highly pelagic Arctic migrant. Regularly recorded off Dakar in autumn, from Aug onwards (e.g. 6,724 in Oct 2007).

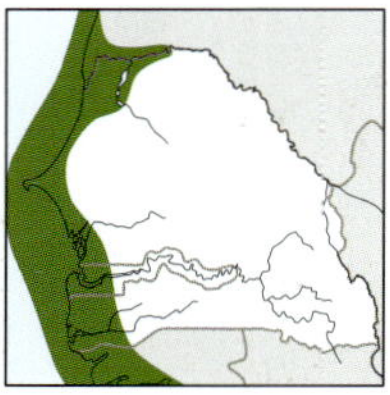

Grey-headed Gull *Larus cirrocephalus* 39–43 cm; WS 100–115 cm

L. c. poiiocephalus. Medium-sized gull with diagnostic upperwing pattern in adult and ***all-dusky underwing*** at all ages. Adult plumage usually acquired in 3rd winter. **Adult breeding** Pale grey hood; red bill and legs. In flight, extensive black wingtips with white mirrors, bordered by broad white diagonal bar. **Adult non-breeding** Head white, often with some greyish; faint dusky ear-spot. **First-winter** Head white with dusky markings; brownish wing-coverts. In flight, black outer primaries, secondary bar and tail band, brown carpal bar. **Second-winter** As adult non-breeding but black on wingtip more extensive, white mirrors small or lacking; dusky trailing edge. **HH** Common along whole coast in all months. Colonies in Senegal (Langue de Barbarie, Saloum delta, Basse Casamance); no recent breeding records in Gambia. **Voice** Harsh *kaarr* and short *kok-kok*. [CD2:69]

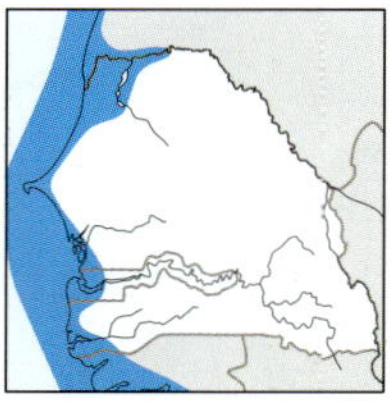

Black-headed Gull *Larus ridibundus* 34–37 cm; WS 90–100 cm

Rather small and slender. In flight slim, pointed wings with ***white leading edge to outer wing*** (most extensive on adults). Adult plumage acquired in 2nd winter. **Adult breeding** Dark-brown hood, dark red bill, legs and feet. **Adult non-breeding** Head white with dusky ear-spot and two faint dusky bands over crown ('headphones'). **First-winter** Brownish wing-coverts, dusky head markings. In flight, brown carpal bar, blackish secondary bar and tail band. **HH** Fairly common to common Palearctic visitor to coast and some inland aquatic habitats (mainly Oct–Apr; records in all months). **Voice** Harsh *krreeah* and *kwarr*, and short *kak, kok-kok-kok*. [CD2:70]

Mediterranean Gull *Larus melanocephalus* 36–40 cm; WS 92–102 cm

Slightly larger and heavier than Black-headed Gull, with stouter bill and longer legs. Adult plumage acquired in 3rd winter. **Adult breeding** ***Very white-looking*** with black hood extending on nape, white crescents above and below eye, blood-red bill (with black band near tip), and red legs. In flight striking: the only gull with ***all-white primaries and underwing***. **Adult non-breeding** White wingtips; dark wedge behind eye, extending diffusely over crown. **First-winter** Dusky patch on ear-coverts; blunt, blackish bill. In flight, pale saddle, contrasting upperwing pattern. **Second-winter** As adult non-breeding, but variable amount of black in tips of outer primaries. Bill colour variable. **HH** Scarce to rare Palearctic winter visitor, coastal Senegambia (Nov–Feb; also May); mainly immatures. **Voice** [CD2:67]

1st-winter
ad br
ad non-br
1st-winter
ad non-br
Little Gull
1st-winter
juv
ad non-br
ad non-br
ad br
Sabine's Gull
1st-winter
ad non-br
ad br
Grey-headed Gull
ad br
2nd-winter
1st-winter
1st-winter
ad br
ad non-br
Black-headed Gull
ad non-br
1st-winter
1st-winter
ad non-br
ad non-br
ad br
Mediterranean Gull
1st-winter

PLATE 58: VAGRANT GULLS

Black-legged Kittiwake *Rissa tridactyla* — 38–42 cm; WS 95–120 cm

R. t. tridactyla. Medium-sized, slender gull with rather pointed wings and a square or slightly notched tail. Adult plumage acquired in 2nd winter. **Adult breeding** Darkish grey and white, with diagnostic clear-cut black wingtips. Bill greenish-yellow. Legs black and rather short. **Adult non-breeding** Back of head and neck smudged grey; blackish ear-spot. **First-winter** Neatly defined, broad black W on wings (formed by black outer primaries and carpal bar); clear-cut black terminal tail band; blackish ear-spot; blackish collar (or traces of it) on hindneck. Bill black. Compare first-winter Little and Sabine's Gulls (Plate 57). **HH** Palearctic vagrant, Senegambia (Nov–Apr). Pelagic; probably regular offshore. **Voice** [CD2:81]

Bonaparte's Gull *Larus philadelphia* — 28–31 cm; WS 80–90 cm

Small, elegant version of Black-headed Gull (Plate 57), with white underwing and mainly black bill. Plumage sequence as Black-headed; adult plumage acquired in 2nd winter. **Adult non-breeding** As Black-headed Gull but smaller; ear-spot more distinct. In flight, neat black edge to primaries. **First-winter** Bill all dark; legs pinkish. In flight, pattern of primary coverts different from Black-headed Gull, black trailing edge more distinct. **HH** N American vagrant, N Senegal (Jan 1986).

Franklin's Gull *Larus pipixcan* — 32–36 cm; WS 85–92 cm

Small and rather dumpy, with shortish wings, tail, bill and legs, and conspicuous white eye-crescents in all plumages. In winter, darker headed than any other similar-sized gull (but beware transitional stages of moult). The only gull to have two complete moults per year (only post-juvenile moult is partial); adult plumage normally acquired in 2nd summer. **Adult breeding** Differs from Laughing Gull in more rounded head, shorter bill, broader eye-crescents. In flight, black on wingtips surrounded by white. **Adult non-breeding** ***Blackish 'half-hood'*** and ***broad white eye-crescents*** diagnostic. **First-winter** Differs from first-winter Laughing Gull in dark half-hood, white neck and underparts, paler inner primaries, and narrower tail band, which does not extend to outermost tail feathers. **HH** N American vagrant, coastal Senegambia (Oct–Apr). One paired with Grey-headed Gull (Saloum delta, Senegal, May 1983). **Voice** [CD6:45]

Laughing Gull *Larus atricilla* — 38–41 cm; WS 100–120 cm

Rather small and slim (only slightly larger than Black-headed Gull, Plate 57) with relatively long wings, legs and bill, the latter often appearing drooping. Adult plumage acquired in 3rd winter. **Adult breeding** Black hood; white eye-crescents; dull red bill and legs. In flight, ashy-grey upperparts (darker than Common Gull); black wingtips (***lacking mirrors***). **Adult non-breeding** Dusky-marked head; longish, blackish bill. **First-winter** Rear of head to breast and flanks dusky brownish-grey; blackish primaries, secondary bar and tail band. **First-summer** Variably developed hood; wing markings faded. **Second-winter/-summer** As adult breeding/non-breeding, but black on wing more extensive, and often have traces of secondary bar and/or tail band. **HH** American vagrant, coastal Senegambia (Oct–Apr). One may have paired with Grey-headed Gull (Saloum delta, Senegal, Mar 1985). **Voice** [CD6:44]

Ring-billed Gull *Larus delawarensis* — 41–49 cm; WS 115–135 cm

Intermediate in size and structure between Common and Yellow-legged Gulls (Plate 56), with rather flat head and stout, 'parallel', blunt-looking bill. Adult plumage acquired in 3rd winter. Plumage sequence and patterns similar to Common. **Adult non-breeding** Yellow bill with black subterminal band. In flight, white mirrors smaller than in Common Gull. **First-winter** Typically heavily spotted on head, neck and breast-sides; pinkish, dark-tipped bill. In flight, grey on upperparts paler than in Common Gull; tail band not clear-cut. **Second-winter** Single small white mirror (or none); tail often with traces of subterminal band. Eye usually pale. **HH** N American vagrant, Senegal (Saloum delta, Oct 1985) and Gambia (Tanji, Dec 2008).

1st-winter
1st-winter
ad non-br
ad br
Black-legged Kittiwake
1st-winter
1st-winter
ad non-br
ad non-br
Bonaparte's Gull
1st-winter
ad non-br
1st-winter
Franklin's Gull
ad br
ad non-br
1st-winter
Laughing Gull
ad non-br
1st-winter
ad br
1st-winter
1st-winter
ad non-br
Ring-billed Gull
ad non-br

TERNS (Plates 59–62) *Slender, small to medium-sized marine and freshwater birds with long, pointed wings, usually long, forked tail and slender, pointed bill. Distinct breeding and non-breeding plumages. Sexes similar. Flight graceful and buoyant. Gregarious. Most immatures of Palearctic species spend first, sometimes also second, summer in winter quarters. Adult plumage usually acquired durirg 2nd or 3rd winter.*

Main identification features include size, wing pattern, bill shape and colour, and call.

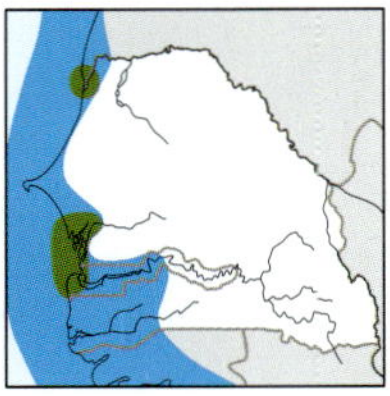

Caspian Tern *Sterna caspia* 47–54 cm; WS 130–145 cm

Largest tern (about size of large gull) with ***massive red bill***, thick neck, conspicuous ***dark wedge on tip of underwing*** (formed by blackish outer primaries) and short, shallowly forked tail. Flight gull-like with slow, heavy wingbeats. **Adult breeding** Large black cap ending in short rough crest, bill blood-red with yellow tip and black subterminal band, legs black. **Adult non-breeding** Dark cap densely streaked white (appearing greyish). **Juvenile** Scaly pattern on upperparts; bill dull reddish tipped dusky. **Immature** in flight has dusky leading edge and secondary bar on upper wing. **HH** Common resident and (probably) Palearctic migrant. Colonies, Senegal (Langue de Barbarie, Saloum delta, Kalissaye). Mainly on coasts, lagoons, large rivers and lakes. Often with other terns. Flies high, occasionally soars. Catches prey by plunge-diving, submerging completely. **Voice** Deep, hoarse *kraah-ap* or *rrha-ak*. [CD2:83]

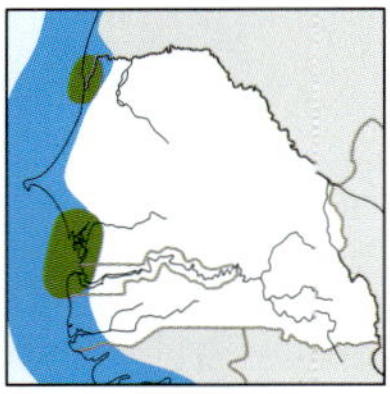

Royal Tern *Sterna maxima* 45–50 cm; WS 125–135 cm

S. m. albididorsalis. Second largest tern. **Adult breeding** Black cap ending in short, shaggy crest, ***deep orange bill***, black legs. In flight from below, ***dark trailing edge to outer primaries***. Black cap retained for only short period, in early breeding season. **Adult non-breeding** Black on head reduced to 'shawl' from eyes across hindcrown (sometimes abrading to leave head almost white); bill pale orange. **Juvenile** Head pattern similar to adult non-breeding, upperparts indistinctly scaled, prominent carpal and secondary bars, dull yellowish legs. In flight, boldly patterned upperwing with dark outer wing and pale grey inner wing marked with three contrasting dark bars (on leading edge, greater coverts and secondaries); tail tipped dark. **Immature** Initially as juvenile but with uniformly grey upperparts; gradually more like adult non-breeding. **HH** Common resident and intra-African migrant. Colonies, Senegal (Langue de Barbarie, Saloum delta, Kalissaye). Has attempted to breed in Gambia. Marine; mainly on shores, lagoons and in harbours. **Voice** High-pitched, shrill *kee-err* and a loud, sharp *krryuk*, resembling Sandwich Tern but deeper. [CD2:84]

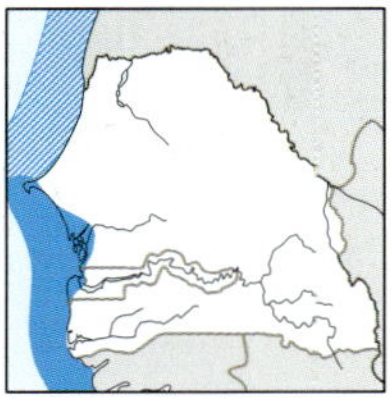

Lesser Crested Tern *Sterna bengalensis* 35–37 cm; WS 92–105 cm

S. b. emigrata. Fairly large, elegant crested tern with slender ***orange bill***. Legs black. Royal Tern is larger, with stouter bill. Similar-sized Sandwich Tern (Plate 61) has yellow-tipped black bill. Both have slightly paler upperparts and white rump and tail. **Adult breeding** Black cap ending in shaggy crest, silvery-white primaries contrasting with rest of grey upperparts, underwing white with narrow dark trailing edge to outer primaries, rump and tail grey. Bill orange-red. **Adult non-breeding** Forehead and most of crown white; bill paler orange. **Immature** Similar to adult non-breeding but with dark outer primaries and dusky secondary bar. Bill dusky-orange. **HH** Uncommon Mediterranean migrant. Most records from Gambia, but probably under-recorded in Senegal. Marine; mainly occurring inshore. Gregarious, often foraging with other terns. Catches prey by plunge-diving, usually submerging completely. **Voice** Rasping *errik* and *krrr-eep*, resembling Sandwich Tern but higher pitched. [CD2:85]

juv
ad br
Caspian Tern
ad non-br
1st-winter
ad non-br
1st-winter
Royal Tern
juv
ad non-br
ad br
ad non-br
ad br
1st-winter
ad non-br
Lesser Crested Tern
ad non-br

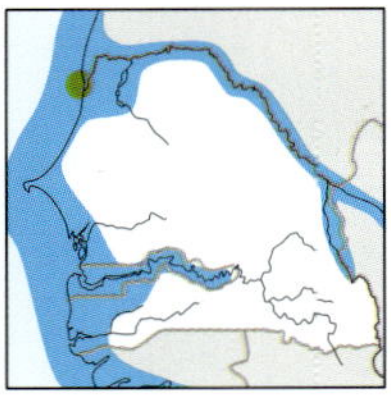

Gull-billed Tern *Gelochelidon nilotica* 35–43 cm; WS 86–103 cm

G. n. nilotica. Fairly large, comparatively thickset and white-looking tern with ***stout, all-black bill***, no crest, short shallowly forked tail and longish black legs. **Adult breeding** Black cap (from Mar); stout black bill. Upperparts and tail very pale grey; upper- and underwing with dark trailing edge to primaries. **Adult non-breeding** Head white with black streak behind eye. Outer primaries gradually darken through wear. **Juvenile** Faint brownish tinge to crown and upperparts; indistinct V-shaped markings above. **Immature** In flight, primaries worn and darker than in adult. **HH** Fairly common Palearctic visitor (mainly Sep–Mar); also resident, Senegal (colony Langue de Barbarie). Frequents larger lagoons and saltpans; also lakes and rivers. Not on open sea. Attracted to bush fires. Does not habitually plunge-dive (unlike Sandwich Tern). **Voice** Harsh, raucous *ger-vik* and *khaak*. Juvenile utters high-pitched *pe-eep* or fast *pe-pe-eep*. [CD2:82]

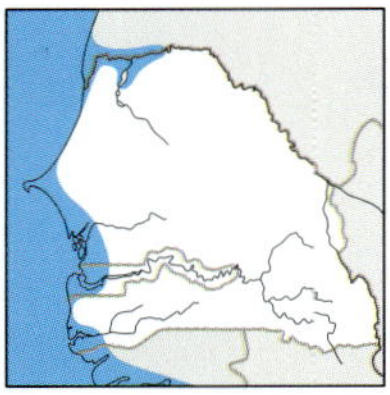

Sandwich Tern *Sterna sandvicensis* 36–41 cm; WS 95–105 cm

S. s. sandvicensis. Fairly large, white-looking tern with long, slender black bill (yellow-tipped in adult), long wings and short, forked tail. **Adult breeding** Black cap with shaggy crest; yellow-tipped black bill. In flight, very white looking, with outer primaries forming darkish wedge. **Adult non-breeding** Head white with black 'shawl'. **Immature** Initially differs from adult non-breeding in darkish carpal bar, secondaries and outer tail feathers, but these lost by late winter. **HH** Common Palearctic migrant. Marine; mainly occurring inshore. Gregarious, often with other terns. Typically plunge-dives from great heights (5–10 m), but also from lower. **Voice** Loud, rasping *kerrik* or *keerr-wit*. [CD2:86]

Roseate Tern *Sterna dougallii* 33–38 cm; WS 72–80 cm

S. d. dougallii. Medium-sized, slender sea tern. Underwing lacks dark trailing edge to primaries in all plumages. **Adult breeding** Very pale grey, ***appearing almost white at distance*** (recalling Sandwich Tern), with black cap, extremely long tail streamers (moulted Aug–Sep) and black bill with red base. Underparts white tinged pink. On upperwing, 2–4 outer primaries gradually darken to form narrow dark wedge; ***underwing all white***. Legs bright red. **Adult non-breeding** Much paler overall than Common and Arctic Terns; no tail streamers; legs dusky-red. **First-winter** In flight, dusky carpal bar, whitish secondaries. **HH** Fairly common Palearctic passage migrant, Senegal; rare (probably under-recorded), Gambia. Bred Senegal (one pair, 1980). Marine; more pelagic than Common Tern. Gregarious, often with other terns. Plunge-dives; also dips to surface. **Voice** A shrill *cher-vrik* (reminiscent of Sandwich Tern). Also a guttural, rasping *rrraakh*. [CD2:87]

Common Tern *Sterna hirundo* 31–35 cm; WS 77–98 cm

S. h. hirundo. Medium-sized, elegant sea tern with, in all plumages, diagnostic underwing pattern of ***dusky-tipped outer primaries forming diffuse trailing edge***. **Adult breeding** Bill red ***with black tip***; tail not projecting beyond wingtips. White line between cap and gape broader than in Arctic Tern. In flight, upperwing never uniformly grey, but with dusky wedge on outer wing. **Adult non-breeding** Bill dark with reddish base, forehead white, no tail streamers. Inner wing with ***dusky carpal and fainter secondary bar***, outer wing more uniformly pale once worn outer primaries moulted early winter. **Juvenile** Gingery tinge to forehead and scaly upperparts; blackish carpal bar. **Immature** As adult non-breeding. **HH** Common Palearctic migrant. Breeding recorded, Senegal (Saloum delta). Mainly coastal. Gregarious. Forages by hovering and plunge-diving from low height. **Voice** Shrill, drawn-out *KEEEarrh* and a sharp *kip*. [CD2:88]

Arctic Tern *Sterna paradisaea* 33–35 cm; WS 75–85 cm

Medium-sized, very elegant sea tern with, at all ages, diagnostic ***well-defined narrow black trailing edge to primaries*** on underwing. **Adult breeding** Bill all red, underparts pale grey, tail projecting beyond wingtips. In flight, uniform upperwings, ***underwings pure white (looking wholly translucent against light)*** with tip bordered by neat black line. **Adult non-breeding** Bill blackish (shorter than Common Tern), forehead white, no tail streamers. **First-winter** In flight, dusky carpal bar (less obvious than in first-winter Common Tern); very pale hindwing with broad white trailing edge. **HH** Palearctic passage migrant. Regularly recorded off Dakar in autumn, from Aug onwards (e.g. 1446 in Oct 1996). Rare, Gambia. Both inshore and offshore; more pelagic than Common Tern. Hovers and plunge-dives from low height. **Voice** Strident *keeAARRGH* resembling call of Common Tern, but harsher and with stress on second syllable; rapid *ki-ki-ki-ki*. [CD2:89]

juv
ad non-br
1st-winter
Gull-billed Tern
ad br
ad br
ad non-br
1st-winter
ad br
Sandwich Tern
ad non-br
ad br
Roseate Tern
ad non-br
1st-winter
ad br
juv
ad br
Common Tern
ad non-br
ad br
1st-winter
ad br
Arctic Tern
ad br
1st-winter
ad non-br

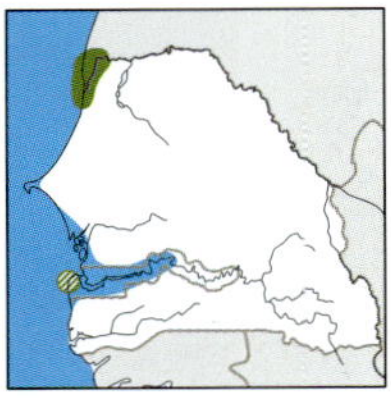

Little Tern *Sterna albifrons* 22–24 cm; WS 48–55 cm

Small with narrow pointed wings, short tail and ***white forehead in all plumages***. Flight fluttering. **Adult breeding** African race ***guineae*** has ***yellow bill*** with little or no black, yellow legs and white rump. In flight, dark outer primaries form narrow wedge on leading edge of wing. Palearctic ***S. a. albifrons*** has ***yellow bill with clear black tip*** and grey rump. **Adult non-breeding** Bill black; black on head recedes to form 'shawl' from eyes across nape, crown streaked grey, legs brownish. In flight, dark carpal bar and leading edge to outer wing. **Juvenile** Upperparts with buffish fringes. In flight, upperwing has dusky leading edge to entire wing contrasting with pale grey mid-wing panel and almost white secondaries and inner primaries. **Immature** Much as adult non-breeding. **HH** Fairly common to locally common resident and Palearctic migrant (mainly Nov–Apr; records in all months). Breeds N Senegal; possibly breeds, Gambia. Coastal; inland along major rivers. Forages by hovering and plunge-diving from moderate height, also by surface-picking. **Voice** Short, shrill *kitik* and *kree-ik*. [CD2:92]

MARSH TERNS Small and compact, with shallowly forked tail. Non-breeding and immature plumages may present identification problems. Adult plumage usually acquired during 2nd summer. Do not plunge for food but forage low over water or aquatic vegetation, hovering and surface-picking; also catch insects in flight.

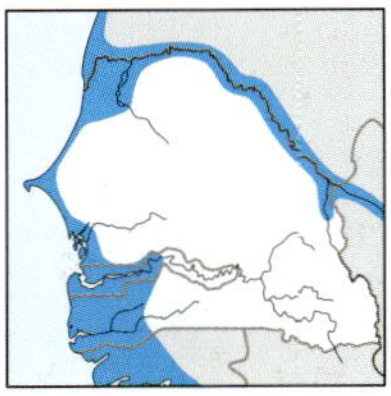

Whiskered Tern *Chlidonias hybrida* 23–25 cm; WS 74–78 cm

C. h. hybrida. Largest marsh tern, with distinctive *Sterna* jizz. **Adult breeding** (mainly early Apr–Jun/Jul) Black cap; contrasting white streak on head-sides. Underparts dark grey becoming blackish-grey on belly. Bill blood-red. **Adult non-breeding** Very pale. Black on head reduced to U-shaped band from eyes across nape and some streaking on rear of crown. Upperparts and tail uniformly pale grey, underparts white. Bill black. **Juvenile** Dark brown 'saddle' (soon replaced), dark bill and legs. **Immature** Gradually more like adult non-breeding (much individual variation; post-juvenile moult usually completed Oct–Jan). **HH** Uncommon to fairly common Palearctic migrant (Aug–May; records in all months). Frequents coastal areas and inland reservoirs. **Voice** Low grating and rattling sounds, incl. harsh *kirrk* and nasal *airkh*. [CD2:93]

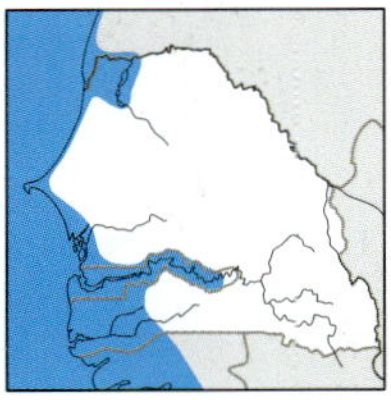

Black Tern *Chlidonias niger* 22–24 cm; WS 64–68 cm

C. n. niger. Small. Unmistakable in breeding plumage; in other plumages ***small dark breast patches*** diagnostic. **Adult breeding** Black head and underparts, slate-grey upperparts, black bill, darkish red legs. In flight, contrasting pale grey wings and white undertail-coverts. **Adult non-breeding** Black cap with large 'headphones'; rest of head and underparts white. In flight, dark patch on breast-sides; grey rump and tail. Moulting birds blotched below. **Juvenile** As adult non-breeding but with faintly contrasting, dark grey-brown 'saddle' and broad, dark carpal bar. **Immature** Post-juvenile moult generally started in winter quarters. Gradually more like adult non-breeding (much individual variation). **HH** Common Palearctic migrant (mainly Jun–Feb; records in all months). Coastal. Some inland records. **Voice** Nasal *kyay* or *kleeah* and *kyek*, grating *kerr* and short *kik*. [CD2:94]

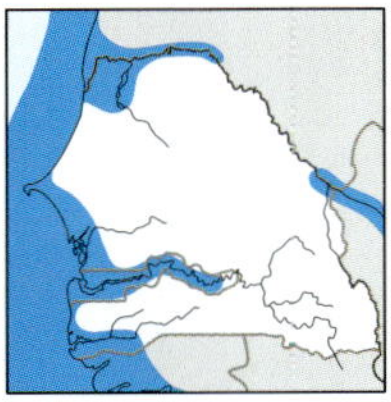

White-winged Tern *Chlidonias leucopterus* 20–23 cm; WS 63–67 cm

Small. Unmistakable in breeding plumage; ***no breast patches*** in other plumages. **Adult breeding** Black with contrasting pale grey wings and white rump, vent and tail. Black underwing-coverts contrast with pale grey remiges. **Adult non-breeding** Black confined to head, where less extensive than on Black Tern and reduced to small patch behind eye and streaking over rear crown (forming 'headphones'). Upperparts pale grey (paler than Black) with contrasting white rump. Often retains traces of breeding plumage late into moult (esp. on wing-coverts). **Juvenile** Dark brown 'saddle' contrasts strongly with pale grey wings (paler than juvenile Black) and white rump. Carpal bar narrower and less conspicuous than Black. Head as adult non-breeding. **Immature** Post-juvenile moult generally started in winter quarters. Gradually more like adult non-breeding (much individual variation). **HH** Fairly common to common Palearctic migrant (mainly late Oct–early May). Frequents coastal areas and inland reservoirs. Occasionally also far from water. **Voice** Various squealing and grating sounds, incl. *khee-ep*, a harsh *krrek* and a softer *kverr-kek*; also a fast *kikikik*. [CD2:95]

ad br
albifrons
juv
1st-winter
ad br
guineae
Little Tern
ad non-br
ad br
guineae
ad non-br
ad br
1st-winter
Whiskered Tern
ad non-br
ad non-br
juv
ad non-br
Black Tern
ad br
ad moulting
juv
ad non-br
ad br
ad non-br
White-winged Tern
ad br
ad non-br
ad moulting
juv

Bridled Tern *Sterna anaethetus*

30–32 cm; WS 77–81 cm

S. a. melanoptera. Medium-sized sea tern with ***dark upperparts*** and very buoyant, graceful flight. Long, slender black bill. Legs black. **Adult** Black cap and ***white forehead extending beyond eye as narrow supercilium***. Upperparts and tail dark grey-brown, tail long and deeply forked with white outer tail feathers, underparts white washed greyish on breast and belly. White underwing-coverts contrast with dark flight feathers; grey collar usually separates black crown from dark upperparts. **Juvenile** Paler and browner above, upperparts scalloped, underparts white with grey wash on breast-sides and flanks, forehead dusky. Immature mostly as adult. **HH** Breeds in Senegal (a few pairs, Îles de la Madeleine and Langue de Barbarie). Intra-african migrant/vagrant, Senegal (few records) and Gambia (Mar–Aug 2005, July 2007). Breeding colonies Mauritania, São Tomé and Annobón. Pelagic. Flies low over sea, hovering frequently and surface-picking; also plunge-dives. **Voice** Various harsh, grating calls. [CD2:90]

Sooty Tern *Sterna fuscata*

33–36 cm; WS 82–94 cm

S. f. fuscata. Medium-sized slender sea tern with ***blackish upperparts*** and long forked tail. Legs and feet black. **Adult** Highly contrasting pattern of blackish upperparts and tail, and pure white underparts. ***Forehead white, not extending as supercilium***. Underwing white with contrasting blackish flight feathers, upperparts all black without neck collar. **Juvenile** Wholly blackish-brown with whitish lower belly and undertail-coverts; upperparts have white fringes. **Immature** Gradually becomes paler on head and underparts. **HH** Small numbers breed erratically, Senegal. Afrotropical and Nearctic vagrant, Gambia (Aug & Dec 2006, Mar 2007). Pelagic. Does not plunge-dive but flies low over water and snatches prey from or just below surface. **Voice** Nasal *wekawek* and *kwèè-eh*; also a low *krrrr*. [CD2:91]

NODDIES *Distinctive pelagic terns with wholly dark brown plumage, contrasting pale cap, and spatulate, slightly forked tail. Bill and legs black. Rarely soar high above ocean or plunge-dive; normally fly c.3 m above sea.*

Black Noddy *Anous minutus*

35–39 cm; WS 66–72 cm

A. m. atlanticus. **Adult** Blacker than Brown Noddy, bill longer and more slender, cap whiter, more extensive and more sharply defined. In flight, uniformly blackish-brown. More graceful, with faster wingbeats than Brown Noddy. **Juvenile** White restricted to forehead and sharply demarcated; upperparts and coverts fringed buffish. **HH** Afrotropical vagrant, Gambia (Oct–Nov 1977, Nov 1984). Pelagic. **Voice** Mainly silent. [CD6:48]

Brown Noddy *Anous stolidus*

38–40 cm; WS 77–85 cm

A. s. stolidus. **Adult** Contrasting greyish-white forehead and forecrown becoming ash-grey on rear crown and nape. White crescent below eye. ***Underwing greyish-brown bordered by contrasting dark brown band***, narrow on leading edge, broad on trailing edge. Upperwing-coverts contrast slightly with darker flight feathers (visible only in good light). **Juvenile** Almost entirely dark; upperparts and coverts with pale fringes. **Immature** Grey restricted to forehead. **HH** Afrotropical vagrant, Gambia (Aug 1974). Highly pelagic, occurring mainly offshore. **Voice** Low grating *krrrah*. Mainly silent. [CD6:49]

SKIMMER *Black-and-white tern-like bird with unique bill shape and feeding technique. Fishes by flying close to surface with tip of much longer lower mandible slicing through water. Wings long and narrow, tail short and forked. Sexes similar, male slightly larger. Flight fast and agile. Breeds colonially.*

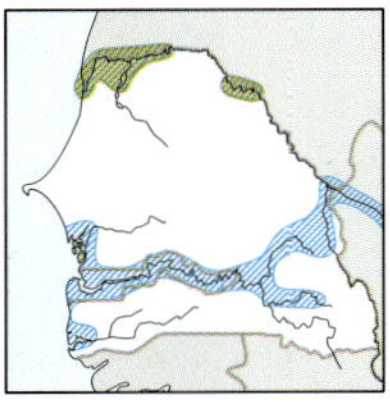

African Skimmer *Rynchops flavirostris*

36–42 cm; WS 125–135 cm

Adult breeding Black above, white below; ***scissor-like bill and legs orange-red***. In flight, has white trailing edge to inner primaries and secondaries on both surfaces; underwing greyish-white with blackish primaries. **Adult non-breeding** Browner above; whitish collar on lower hindneck. **Juvenile** Upperparts browner, fringed buffish; bill has dusky tip. **HH** Uncommon to rare. Breeding recorded, N Senegal. Curious feeding action requires large expanses of quiet water; breeding dependent on large sandbars in broad rivers and lakes. Disperses when not breeding, then also on estuaries, coastal lagoons, dams and saltpans. Spends much of day resting on sandbars. Red List category: Near Threatened. **Voice** Loud, shrill *kik* and *kree*, often repeated. [CD6:50]

juv
imm
ad
Bridled Tern
ad
juv
imm
ad
Sooty Tern
ad
juv
ad
Black Noddy
ad
ad
imm
Brown Noddy
juv
ad br
ad non-br
juv
African Skimmer
ad br

AUKS *Small to medium-sized seabirds with elongated bodies and strong bills. Plumages typically dark above and pale below. Wings short and narrow, used for both flight and swimming underwater; tail short. Feed by diving. Flight direct and low with rapid, whirring wingbeats.*

Black Guillemot *Cepphus grylle* — 32–38 cm; WS 49–58 cm

Adult non-breeding Variably barred above, rump and underparts pure white. Upperwing black with ***large, oval white panel***. Bill pointed, black. **Adult breeding** (unlikely in Senegambia) All black with large, oval white upperwing panel and bright red feet. **HH** Palearctic vagrant, Senegal (one off Dakar, Oct 2008). **Voice** Silent when not breeding.

SANDGROUSE *Cryptically plumaged, terrestrial birds of arid habitats. Wings and tail pointed, bill and legs short. Sexually dimorphic, females less distinctly marked. Flight swift and direct, reminiscent of pigeons. May gather in flocks at favourite water holes, usually at dawn or dusk. Calls are a good identification clue.*

Chestnut-bellied Sandgrouse *Pterocles exustus* — 29–33 cm

P. e. exustus. ***Long-tailed***, pale yellow-buff with ***dark underwing*** and ***dark belly***. Above, blackish remiges contrast with pale upperwing-coverts. **Adult male** Head and throat plain; ***belly dark chestnut***. **Adult female** Densely barred dark brown above. **HH** Common to not uncommon resident, Senegal. Southward movements in dry season, with records from Gambia. **Voice** In flight, a guttural *kwit-gurut, kwit kwit-gurut*… [CD6:51]

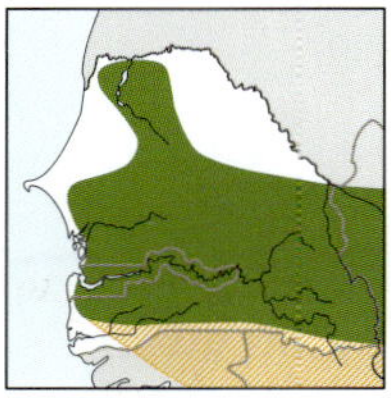

Four-banded Sandgrouse *Pterocles quadricinctus* — 25–28 cm

Short-tailed. In flight, blackish flight feathers contrast with yellowish-buff upperwing-coverts and grey underwing-coverts. **Adult male** Black-and-white forehead; chestnut, white and black breast-bands. **Adult female** Plain face, throat and upper breast. **HH** Locally common to not uncommon in dry wooded grassland, cultivation, open sandy scrub. Resident and dry-season breeding visitor in Gambia and S Senegal; wet-season non-breeding migrant in north. Flocks gather at dusk and fly to waterholes. Apparently largely nocturnal. **Voice** In flight, a far-carrying whistled twittering *kik-krrr-reee*. Also an aerial song *whup whip twhu turuptu-wrree*, at night. [CD6:52] **Note** Name 'Four-banded' is a misnomer, as male has only three breast-bands.

Lichtenstein's Sandgrouse *Pterocles lichtensteinii* — 22–26 cm

P. l. targius. Stocky, smallish, ***finely and densely barred*** sandgrouse. In flight, blackish flight feathers contrast with paler upperwing-coverts; underwing dark grey-brown, with little or no contrast; tail short. **Adult male** Forehead white traversed by black band; white supercilium with black spot above eye. Breast unbarred traversed by two narrow black bands; bill orange. **Adult female** Finely and densely barred. **HH** Vagrant, N Senegal (Jul and Sep). Frequents rocky and scrubby desert with scattered bushes. Gathers at freshwater source at dusk. **Voice** Clear, liquid, sharp *k-kwio k-kwio*…. or *kliuw kliuw*… (also transcribed as *KWEtal* or *kuitl*). Alarm a dry croaking *krre-krre-krre-krre-*… [CD3:3a]

Black Guillemot

Chestnut-bellied Sandgrouse

Four-banded Sandgrouse

Lichtenstein's Sandgrouse

PLATE 64: WOOD DOVES AND NAMAQUA DOVE

PIGEONS AND DOVES (Plates 64–66) *Arboreal and terrestrial species with compact body, small head and strong, fast flight. Feed on fruit or seeds. Calls distinctive and an excellent identification clue.*

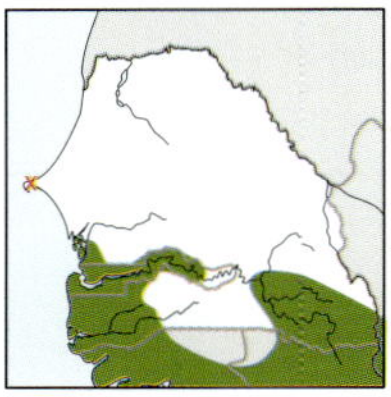

Blue-spotted Wood Dove *Turtur afer* — 20 cm

Small with ***dusky red, yellow-tipped bill***. In flight, ***conspicuous rufous primaries***, two blackish bands on lower back and broad dark subterminal tail band. **Adult** Top of head blue-grey; 5–6 dark metallic blue wing spots. **Juvenile** Duller and browner, with scaly upperparts and dusky bill. **HH** Common resident. Singly or in pairs at forest edges and in clearings, gardens, cultivation, gallery forest and dense woodland. Often foraging on ground in open situations and near habitation. **Voice** Two muffled, plaintive syllables followed by a series of *hoo*s, initially rather hesitant and irregular, then accelerating, finally fading *oh-wuh oh-wuh oh-ho-ho-ho-hohohohoho*. [CD6:58]

Black-billed Wood Dove *Turtur abyssinicus* — 20 cm

Adult Very similar to Blue-spotted Wood Dove but slightly paler and greyer with ***black bill***. Top of head blue-grey becoming whitish on forehead. **Juvenile** Very similar to juvenile Blue-spotted, but overall paler, with greyer head. **HH** Common resident. Singly or in pairs in dry savanna woodland, open scrub, cultivation, thickets and second growth. **Voice** A series of soft, plaintive *hoo*s, similar to Blue-spotted Wood Dove, but initial notes more hesitant and drawn out. [CD6:60]

Namaqua Dove *Oena capensis* — 22–26 cm

O. c. capensis. Small and slim with ***long, steeply graduated, pointed tail***. In flight, ***conspicuous rufous primaries*** and two black bands on lower back. **Adult male** Black mask; bill dark red broadly tipped yellow. **Adult female** Face greyish; bill dusky. **Juvenile** Upperparts scalloped. **HH** Common resident/Afrotropical migrant throughout. In thornbush, dry grassland and cultivation. Forages on bare ground; often on roads and tracks. **Voice** Short, soft note, followed by plaintive, drawn-out, emphatic and slightly rising syllable: *oh-whooooah...* [CD3:3b]

Tambourine Dove *Turtur tympanistria* — 20–22 cm

Small with ***contrastingly white underparts***. In flight, rufous flight feathers and two blackish bands on lower back. **Adult male** ***White face***; dark brown upperparts. **Adult female** Face and breast greyish. **Juvenile** Duller and browner, with scaly upperparts. **HH** Rare, SW Senegal (Casamance). In various types of forest, dense woodland and plantations. Forages on ground; occasionally venturing onto tracks. **Voice** A series of *hoo*s, starting rather slowly, then accelerating *ho, ho, ho ho hohoho hohohohohohohohoho*; similar to Blue-headed Wood Dove, but 'flatter' (not modulated), less 'bouncing'. [CD6:57]

Laughing Dove *Streptopelia senegalensis* — 23–25 cm

S. s. senegalensis. Slender and small with mainly rufous-brown upperparts, ***blue-grey outer wing-coverts***, and ***black-speckled rufous necklace***. In flight, blackish flight feathers separated from rufous shoulders by blue-grey panel; tail has white corners; undertail white. **Adult male** Head and breast vinous-pink. **Adult female** Paler head and breast, and less prominent necklace. **Juvenile** Duller and browner, without necklace. **HH** Common resident. Singly or in pairs, in villages and towns, farmland, bush and open woodland. Feeds on ground. Not shy. **Voice** A single, emphatic note followed by a series of 5–7 hurried notes, first rising, then falling *hoo koHUHUhu-hoo*, likened to gentle laugh. [CD3:9]

ad
juv
ad
Blue-spotted
Wood Dove
ad
juv
Black-billed
Wood Dove
ad
juv
♀
Namaqua Dove
♂
♂
Tambourine Dove
♀
juv
ad
Laughing Dove
♂

African Green Pigeon *Treron calvus* 25–28 cm

T. c. sharpei. **Adult** Plump, heavily built green pigeon with ***green head and underparts***, purplish patch on shoulder and pale bill with ***large red base***. Greater coverts and secondaries tipped and edged yellow; undertail-coverts chestnut edged buff. **Juvenile** Duller and lacking purplish shoulder patch. **HH** Fairly common to locally common resident. Usually in flocks, in forest and wooded savanna. Feeds on figs and other fruit, clambering among branches like a parrot, occasionally hanging upside-down. Rarely on ground, but may locally visit saltpans. Easily overlooked when perching quietly in thick foliage. **Voice** Curious and not pigeon-like, starting with a rather soft, fluty trill, suddenly rising, then descending, *hru-hru-wrih-hu-hrruu hrruuu* and ending in throaty chuckling, creaking, barking and growling notes. [CD6:53]

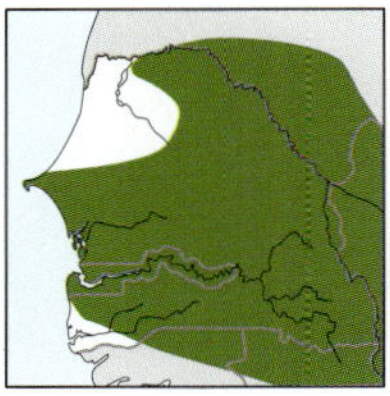

Bruce's Green Pigeon *Treron waalia* 28–30 cm

Adult Plump green pigeon with ***pale grey head and breast, bright yellow belly***, purplish patch on shoulder, and whitish bill with ***bluish cere***. Vent mainly whitish streaked olive-grey; undertail-coverts chestnut edged buff. **Juvenile** Duller with no or smaller purplish shoulder patch. **HH** Common in small flocks in dry wooded savanna and thorn scrub. Partial to figs. Easily overlooked when perched in leafy canopy. Local movements, related to rains or ripening of figs. **Voice** Curious and not pigeon-like, starting with a rapid series of hard, cracking notes (resembling slow opening of a creaking door), followed by fluty whistles, suddenly rising, then descending and ending in some abrupt yapping grunts *k k k-k-k-kkkkkkrrrrrrr whuuuuu-wiwhuwheeew errrr whrreh errrr whrreh whrreh whrreh*. [CD6:55]

Speckled Pigeon *Columba guinea* 35–40 cm

C. g. guinea. Large grey-and-chestnut pigeon with ***triangular white spots on wing-coverts***. In flight, pale grey rump and uppertail-coverts contrast with dark upperparts; tail grey with broad black terminal band. **Adult** Head and neck grey with conspicuous bare red patch around yellow eye, and vinous-chestnut streaking on neck. Upperparts vinous-chestnut; underparts grey. **Juvenile** Duller and browner; bare eye patch brownish. **HH** Common to frequent in open country, incl. savanna, cultivation, gardens; also villages and towns, where often perches on buildings. Feeds on ground. Local seasonal movements probable, as numbers increase in dry season in some areas. **Voice** A fast, rising series of *oo* notes, increasing in pace and volume *oo-oo-oo-oo-oo-oo-oo-OOH-OOH-OOH-*…. Also *rroh rroh rroh*. [CD6:68]

Rock Dove *Columba livia* 31–34 cm

C. l. gymnocycla. **Adult** ***Dark slate-grey with pure white rump***, two broad black wingbars and white underwings. Neck and upper mantle glossed metallic green and purple. Eye yellowish with narrow red orbital ring. Bill dark grey with white cere. Feet dark red. **Juvenile** Duller. Ancestor of domestic pigeon; feral population of latter highly variable, some very similar to genuine Rock Dove. **HH** Small populations occurred on coastal cliffs and rock ledges, N Senegal; also observed SE Senegal (occurring on nearby rocky hills?). Situation at present unclear due to interbreeding with feral birds. Feeds on ground. **Voice** Low, drawn-out, emphatic *rrooh* repeated several times. [CD3:6]

ad
African Green Pigeon
ad
Bruce's Green Pigeon
ad
juv
Speckled Pigeon
Rock Dove
ad

***'COLLARED' DOVES** Medium-sized grey-brown and pinkish-grey doves with black half-collar on hindneck. Juveniles as adults but duller. Feed mainly on the ground. Note in particular eye-colour, tail pattern and voice.*

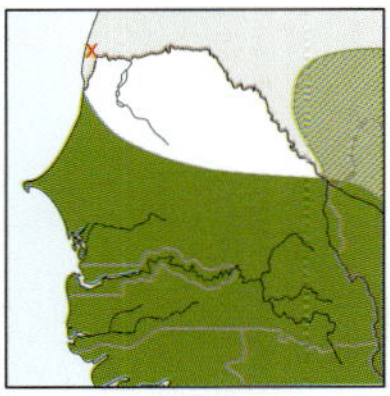

Red-eyed Dove *Streptopelia semitorquata* — 30–34 cm

Adult Large and dark with grey crown, dark grey-brown upperparts and vinous-pink underparts. In flight, dark band across central tail. Eye dark red with narrow dark red orbital ring (visible at close range). **Juvenile** Upperparts with rufous-buff fringes; half-collar indistinct. **HH** Common throughout, except N Senegal, where rare or absent. In various habitats, incl. woodland, gardens, clearings, forest edge and mangrove. Not in forest. Feeds mainly on ground, sometimes in trees. **Voice** One of the most familiar sounds of the Senegambian bush. Rapid series of six short, rather nasal coos, with stress on 5th, *ho-hu ho-hu HOO-ho*. Also *rroorr-huhu rroorr-huhuhu…* and *rroorr-rrhuh rroorr-rrhuh…* Occasionally a nasal *ehèèh* on landing. [CD6:69]

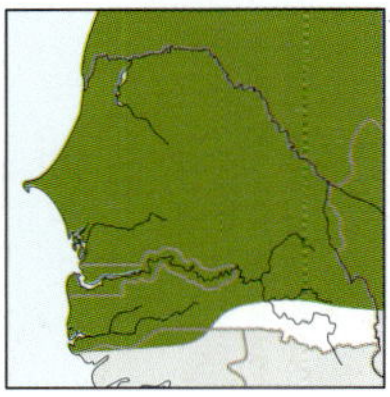

African Mourning Dove *Streptopelia decipiens* — 28–30 cm

S. d. shelleyi*. Adult** Dry-country 'collared' dove with grey face and crown, ***yellow to pale orange eyes and ***narrow but conspicuous red orbital ring***. In flight, tail has white corners. **Juvenile** Browner, esp. on head; eye pale brownish. **HH** Common in arid wooded grasslands near water. **Voice** Emphatic note followed by quavering coo, *whoh! kho-o-o-o-o-o* and rapidly repeated *whoh whoh-hoo*. Quavering gargle *arh-r-r-r-r-rw* on landing. [CD6:70]

Vinaceous Dove *Streptopelia vinacea* — 25 cm

Smallest 'collared' dove. **Adult** Head and underparts pale vinous-pink; eye dark. In flight, white tail corners; pale grey underwing. **Juvenile** Similar but with pale fringes to wing-coverts. **HH** Common. Singly, in pairs or flocks in various savanna habitats, incl. cultivation. Seasonal movements recorded, north with rains, south in dry season. **Voice** One of the most characteristic sounds of the Senegambian bush. Rather high-pitched, far-carrying, fast, 3-syllable *wheh heh-ho wheh heh-ho…*, throughout the day and even at night. Also fast series of rolling *horrrh-horrrh-horrrh-horrrh-…* Nasal call on landing. [CD6:71]

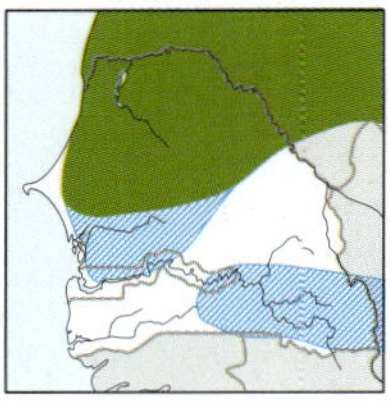

African Collared Dove *Streptopelia roseogrisea* — 27–29 cm

S. r. roseogrisea*. Adult** Pale, fairly uniform, medium-sized 'collared' dove with grey-pink head and underparts, and dark wine-red eye. In flight, pale wing-coverts contrast with dark primaries; tail has white corners; ***underwings white. **Juvenile** Duller, with indistinct collar and pale fringes to feathers of upperparts. **HH** Common, N Senegal; uncommon to scarce further south. In pairs or flocks in thornbush and arid farmland. Feeds on ground, often with other doves. Congregates in large numbers at water and roosts. **Voice** Single, emphatic note followed, after a brief pause, by rolling purr *whooh rrrwhrrrooh* or *ooh krrruuuuu…* [CD6:73]

***TURTLE DOVES** Medium-sized doves with dark neck patch and chequered upperparts.*

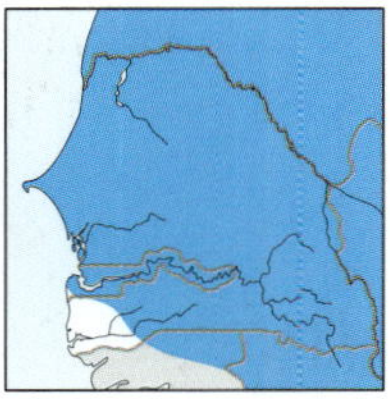

European Turtle Dove *Streptopelia turtur* — 26–28 cm

Slender, medium-sized dove with blue-grey top of head and hindneck, ***black-and-white streaked patch on neck-sides*** and ***chequered upperparts***. In flight, blackish flight feathers separated from rufous shoulders by blue-grey band; tail has contrasting white terminal band. N African ***S. t. arenicola*** paler than European ***S. t. turtur***. **HH** Common to fairly common Palearctic migrant (Aug–May) in savanna woodland, ricefields and farmland. *S. t. arenicola* recorded N Senegal. Has seriously decreased since 1970s (when probably several millions in Senegal delta). **Voice** Deep purring *rrrurrr rrrurrr…* Not recorded on migration. [CD3:8]

Adamawa Turtle Dove *Streptopelia hypopyrrha* — 30–31 cm

Adult Fairly large and ***dark*** with blue-grey head to upper breast, ***broad black patch on neck-sides*** and ***chequered upperparts***. Lower underparts dark cinnamon-pink becoming grey on vent. In flight, dark overall; pale grey terminal tail band. **Juvenile** Paler and duller. **HH** Uncommon (probably overlooked), E Gambia to SE Senegal; first recorded in 1990. In various wooded habitats, incl. open woodland on rocky, hilly terrain, riparian woodland, cultivation and gardens. **Voice** Low, hard purring *rrrurrr rr-rrurr*, resembling European Turtle Dove, but sharper and deeper. [CD6:74]

ad
ad
Red-eyed Dove
juv
ad
African Mourning Dove
ad
Vinaceous Dove
ad
ad
African Collared Dove
ad
ad
ad
European Turtle Dove
Adamawa Turtle Dove

PARROTS *Mostly arboreal species with compact body and stout, strongly hooked and powerful bills, also used in climbing. Legs short, feet strong and zygodactyl, used like hands in feeding. Flight strong, fast and direct. Noisy, especially in flight. Calls include high-pitched screeches. Rather gregarious. Feed mostly on fruits and seeds. Nest in tree holes.*

Rose-ringed Parakeet *Psittacula krameri* 38–42 cm

P. k. krameri. Slender, all-green parrot with ***long pointed tail*** and red bill. **Adult male** Head encircled by narrow ring. **Adult female** Head all-green; tail shorter. **Juvenile** As adult female, with even shorter tail and pinkish bill. **HH** Common. In small flocks in savanna-like habitats, incl. wooded grassland, riparian forest, farmland, gardens and mangroves. **Voice** Vocal. Loud, screeching calls, incl. shrill *kee-ak*. [CD6:84]

Senegal Parrot *Poicephalus senegalus* 23 cm

P. s. senegalus. **Adult** Rather small, noisy, green parrot with grey head and conspicuous yellow belly darkening to orange on centre. Breast and thighs green; underwing-coverts yellow. Eye yellow. **Juvenile** Mostly greenish below; eyes dark brown. **HH** Common. In pairs or small groups in savanna woodland and farmland with scattered trees. **Voice** Various harsh shrieks. [CD6:80]

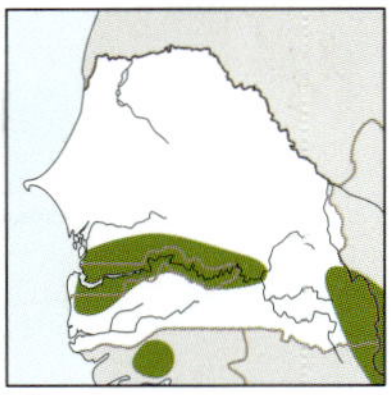

Brown-necked Parrot *Poicephalus robustus* 28–33 cm

P. r. fuscicollis. **Adult male** Large, green parrot with ***brownish-grey head*** and conspicuously ***large, horn-coloured bill***. Leading edge of wing and thighs red. Flight and tail feathers blackish. **Adult female** Forehead and forecrown red. **Juvenile** Duller; head brownish-olive; no red on wings and thighs. **HH** Very local, uncommon to frequent, and declining. Singly, in pairs or small groups in savanna woodland. Undertakes regular daily flights from favoured roosts to feeding areas. **Voice** Strident, harsh *zzkeek*. [CD6:77] **Note** Sometimes treated as a separate species, *P. fuscicollis* (with *P. robustus* split off as Cape Parrot).

Rose-ringed Parakeet
♀
♂
♂
Senegal Parrot
ad
juv
ad
Brown-necked Parrot
♀
♀
♂

TURACOS *Arboreal birds with crested heads and long tails. Flight weak, over short distances only and consisting of several flaps followed by a glide. Agile in trees, characteristically running and bounding along branches (except* Crinifer*). Feed principally on fruit. Endemic to Africa. Vocal.*

Green Turaco *Tauraco persa* 40–43 cm

T. p. buffoni. **Adult** Head, mantle and breast bright green; back, wing-coverts and tail glossy violet-black. Crest entirely green. Small white patch in front of eye. Bill dusky orange-red; red orbital ring. Flashing ***crimson flight feathers striking in flight***. **Juvenile** As adult but crimson on wing more limited. **HH** Locally common to fairly common resident in lowland and gallery forest, S Senegal; very local, Gambia. In pairs or family parties. **Voice** Two or three rising notes *woop-woop* followed by a regular series of loud, raucous *khaw* notes; emitted throughout the day by both sexes, often in duet. Also a repeated *rroooh* followed by *keh* in duet: *rroooh-keh rroooh-keh…* A calling bird often incites all others within earshot to join a rapidly swelling chorus. [CD6:86]

Violet Turaco *Musophaga violacea* 45–50 cm

Adult All-purple and blue-black turaco with ***heavy, deep yellow and reddish-tipped bill***. Crown crimson; bare orbital skin scarlet. White line over ear-coverts, bordering orbital skin and crown. Bright crimson in wings conspicuous in flight. **Juvenile** Much duller, with all-dark head and blackish bill. **HH** Fairly common to locally common resident in gallery forest and forest edge. In pairs or small parties. **Voice** A melodious rolling series of far-carrying *koorroo* notes; often joined by second bird in asynchronous duet and eliciting response from others. [CD6:94]

Western Grey Plantain-eater *Crinifer piscator* *c.*50 cm

Adult Mainly ***grey***, long-tailed turaco with ***lemon-yellow bill*** and ***spiky nuchal crest***. Lower breast to undertail-coverts white, streaked blackish. In flight has striking ***white wing patch***. **Juvenile** As adult but crest shorter. **HH** Common resident in gallery forest along watercourses, savanna woodland and cultivated areas. In pairs or small parties, frequenting treetops. Feeds almost exclusively on wild and cultivated fruits, but not known to eat plantains or cultivated bananas in the wild. **Voice** A series of loud, high-pitched and rolling *kow-kow-kow-…* ending in chatter; cackling *kak-kak-kak-kalak-kalak…* [CD6:98]

ad
Green Turaco
Violet Turaco
ad
Western Grey
Plantain-eater
ad

TYPICAL CUCKOOS ***(Plates 69–71)*** *Arboreal with long, pointed wings and long tails, reminiscent of a small raptor in flight. Feed on insects and hairy caterpillars. Brood parasites. Soon after hatching, young cuckoos usually throw host's eggs or young out of nest. Many rather inconspicuous and some quite difficult to observe, but all have distinctive calls.*

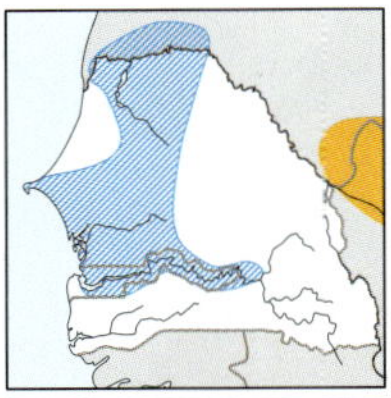

Jacobin Cuckoo *Oxylophus jacobinus* — 33 cm

O. j. pica. **Adult** Pied, crested cuckoo, black above and white below. Long graduated tail black, broadly tipped white. ***White wing patch conspicuous in flight.*** **Juvenile** Duller, with shorter crest and buffish underparts. **HH** Scarce intra-African migrant. No confirmed breeding in Senegambia. In dry savanna woodland; not in forest. Moves south in dry season, crossing equator. Parasitises mainly Common Bulbul, also babblers. **Voice** A series of clear *keew* notes (as Levaillant's Cuckoo but slightly higher pitched) ending in staccato chatter. [CD7:1]

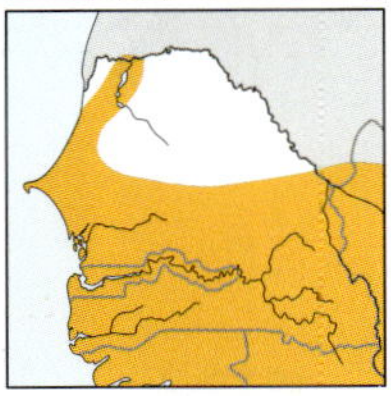

Levaillant's Cuckoo *Oxylophus levaillantii* — 38–40 cm

Adult As Jacobin Cuckoo, but larger and with ***heavily streaked throat and breast. White wing patch conspicuous in flight.*** **Juvenile** Duskier, with shorter crest, finer streaks and buffish underparts. **HH** Fairly common intra-African migrant. Frequents dense savanna woodland, forest edge, second growth and adjacent cultivated areas in forest zone. Wet-season breeder, moving south in dry season. Solitary and unobtrusive, foraging within cover. Parasitises babblers; also Chestnut-bellied Starling. **Voice** A series of clear, loud *KEEow* notes, ending in staccato chatter (like machine gun burst). [CD7:2]

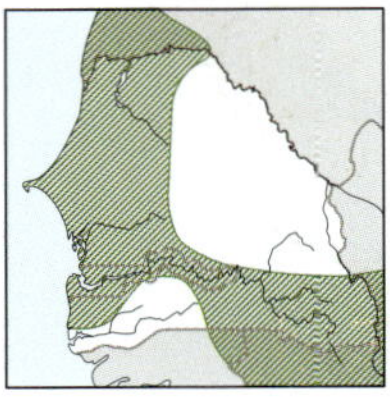

Great Spotted Cuckoo *Clamator glandarius* — 35–40 cm

Large, elongated cuckoo, boldly spotted white above. **Adult** ***Pale grey crest***; upperparts grey-brown boldly spotted white. Long graduated tail blackish, all feathers except central pair broadly tipped white. Throat and breast creamy-buff; rest of underparts white. **Juvenile** ***Black cap*** with small crest; upperparts darker brown; ***primaries rufous*** (conspicuous in flight); underparts richer coloured. **HH** Scarce to fairly common resident and Palearctic winter visitor. Mainly in semi-arid open woodland. Easily overlooked when not calling, despite size and open savanna-type habitat. Breeds during rains, moving south in dry season, when Palearctic migrants winter south of Sahara (south to 10°N). Parasitises mostly Pied Crow, also starlings. **Voice** Vocal in breeding season. Loud *kweeow kweeow kweeow*, excited *kiu-ku-ku-ker* and variety of harsh and chattering calls. [CD3:10]

ad
juv
ad
Jacobin Cuckoo
juv
ad
ad
Levaillant's Cuckoo
juv
Great Spotted Cuckoo
ad

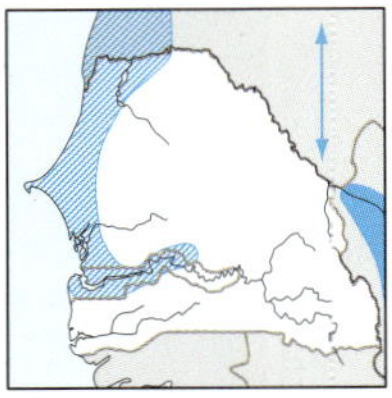

Common Cuckoo *Cuculus canorus* — *32–36 cm*

C. c. canorus/bangsi. Very similar to African Cuckoo, but ***bill mainly dusky with yellowish base***; outer tail feathers with incomplete white bars (visible from below). **Female** similar but with rusty-brown tinge. Rare female hepatic form is entirely rufous barred black above. **Immature** as young African Cuckoo but has crown partially (not entirely) barred; rump and uppertail-coverts plain (not barred). **HH** Scarce to uncommon Palearctic passage migrant in wooded and derived savanna. **Voice** Silent in Africa. [CD3:11]

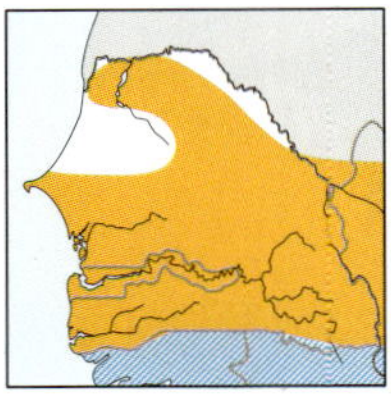

African Cuckoo *Cuculus gularis* — 31–33 cm

Slender, grey-backed cuckoo with long pointed wings and long, graduated tail. **Adult *Bill mainly orange-yellow with black tip*,** upperparts and breast grey; rest underparts white barred grey; outer tail feathers with complete white bars (visible from below). Female may have more buffish underparts, esp. throat and breast. **Juvenile/immature** Grey upperparts finely barred white; throat and breast dirty buff. **HH** Fairly common intra-African migrant in wooded savanna. Unobtrusive, like congeners, except when breeding. Often calls from bare branch, perching upright; occasionally at night. Flies with rapid shallow wingbeats, wings mostly held below horizontal (unlike small raptors) while head, body and tail are very straight, with bill pointing forward. May recall small hawk. Parasitises mainly Fork-tailed Drongo; also Yellow-billed Shrike. Wet-season breeder, moving south in dry season. **Voice** Frequently repeated, far-carrying *oo-OO*, with stress on second, slightly higher pitched syllable (as flat, inverted song of Common Cuckoo). Female utters rapid bubbling *bibibibibibi*. [CD7:6]

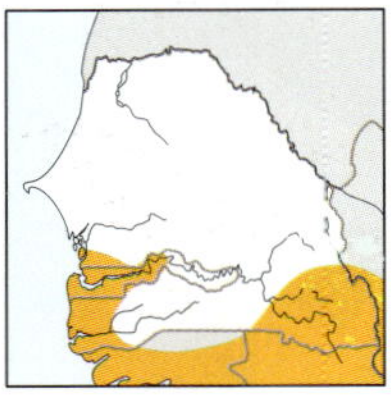

Red-chested Cuckoo *Cuculus solitarius* — 28–31 cm

C. s. solitarius. **Adult** Mainly slate-grey with chestnut breast (occasionally barred), and buffish-white belly and undertail-coverts boldly barred blackish. Tail tipped white. Female often has chestnut of breast paler, less extensive and barred. **Juvenile** Very dark. Upperparts blackish-brown, feathers narrowly fringed white; white spot on nape; belly and undertail-coverts boldly barred on buffish-white. **HH** Uncommon Afrotropical wet-season visitor. In forest; also wooded savanna. Highly vocal when breeding, usually calling from hidden perch. Parasitises mainly Turdidae. **Voice** Loud and descending three-note *WHIT-whit-teew* (*'LET-it-rain'*), frequently uttered. Also calls at night. [CD7:4]

Black Cuckoo *Cuculus clamosus* — 28–31 cm

C. c. clamosus. **Adult** Typically all black with graduated, narrowly white-tipped tail. Underparts may be partially or wholly barred with white to pale rufous and have some chestnut on throat and breast. **Juvenile** Entirely dull sooty-black. **HH** Afrotropical wet-season vagrant, S Senegal; no confirmed recent records, Gambia. Usually solitary and unobtrusive in forest patches in savanna and woodland, remaining in dense cover, often high. In S Africa recorded to parasitise mainly shrikes. **Voice** A series of three hesitant notes, each with more emphasis and higher pitched than the preceding, final note sometimes repeated after short pause: *who whuu whee, wheee*. Also a wild, bubbling trill, rising and falling and gradually dying away *lulululululuWHIRlulululuWHIRluluWHIRlu...* Calls also at night. [CD7:5]

♀
hepatic form
Common Cuckoo
ad
♂
♀
African Cuckoo
juv
♂
juv
ad
Red-chested Cuckoo
juv
ad
Black Cuckoo

CHRYSOCOCCYX *CUCKOOS* *Small, slender cuckoos with, in adult males, glossy upperparts. Males readily identified, but females and juveniles may be confusing. Calls distinctive.*

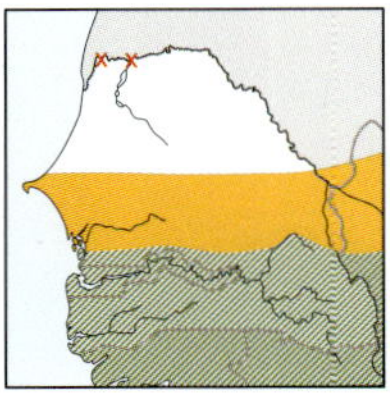

Klaas's Cuckoo *Chrysococcyx klaas* 18 cm

In all plumages, ***white outer tail conspicuous in flight*** (poorly seen Klaas's may thus recall small honeyguide). **Adult male** Head and upperparts bright metallic green with ***small white streak behind eye***. Underparts clean white with ***green patch on breast-sides*** forming incomplete collar. **Adult female** Variable, but typically has upperparts more bronzy and flanks finely barred olive-brown. Post-ocular patch buffy, less conspicuous than in male and sometimes indistinct. Some females similar to males. **Juvenile** Barred green and russet above; finely barred olive-brown below; slight post-ocular patch (female/juvenile Emerald Cuckoo very similar but always lack post-ocular patch and have bolder barring on underparts). **HH** Fairly common resident in south, moving north with rains. In various wooded habitats, incl. savanna woodland, forest (mainly edges and clearings) and cultivated areas. Solitary and unobtrusive; female rarely seen. Parasitises warblers, sunbirds, flycatchers and white-eyes. **Voice** Plaintive *huee-ti huee-ti* or *whee-ee chew whee-ee chew*. [CD7:15]

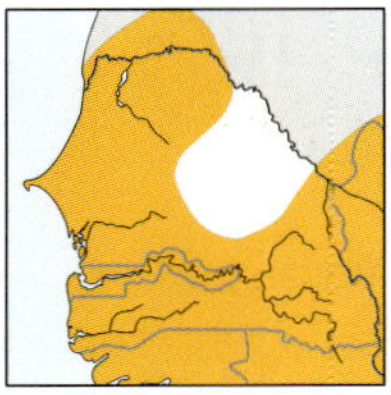

Didric Cuckoo *Chrysococcyx caprius* 19 cm

Adult male Head and upperparts metallic green with bronze reflections. ***White streak in front and behind eye***; short green malar stripe; white spots on wings and outer tail feathers. Underparts white barred bronzy-green on flanks and thighs. ***Orbital ring and eye red***. **Adult female** Variable, with upperparts more bronzy than male and white areas sometimes tinged buff. **Juvenile** Highly variable, with ***coral-red bill***. Upperparts dull green (some with rusty barring) or russet. Underparts white with streaked throat and breast and lightly barred flanks, or spotted from breast to belly. **HH** Common intra-African migrant in various open and wooded habitats (not closed forest). Conspicuous (unusual for a cuckoo). Migrates north from forest zone in rainy season to breed. Parasitises mainly weavers. **Voice** Plaintive *deea deea deedrik* (name is onomatopoeic). Also rapid *di-di-di-di-di*. Female a *deea-deea-deea*. [CD7:16]

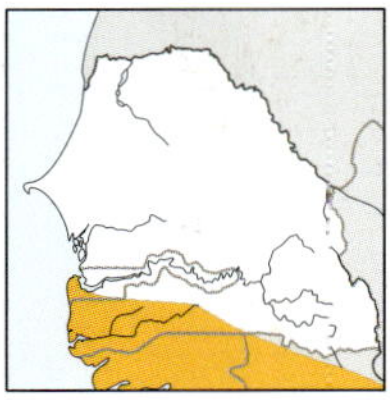

African Emerald Cuckoo *Chrysococcyx cupreus* 23 cm

Adult male Unmistakable. Strikingly beautiful, with bright glossy green upperparts, throat and breast and ***deep golden-yellow belly***. Bill bluish; blue-green orbital ring. **Adult female** Barred russet and green above; boldly barred bronzy-green below. **Juvenile** As female, but crown and nape feathers green (not bronzy-brown) fringed white. **HH** Scarce intra-African migrant in lowland and gallery forest, and dense woodland. Often heard, but much less seen, as usually well hidden in canopy. Occasionally forages lower. Known hosts include bulbuls, flycatchers, sunbirds and weavers. **Voice** Far-carrying, clear and melodious *ptiu, tiu-ut*; a characteristic sound of African rainforest. Also an explosive, stuttering *tiuw tu-tu-tu...* Female utters clear *tiuw* and *huu tu-tu*. [CD7:13]

♂
♀
juv
Klaas's Cuckoo
Didric Cuckoo
♂
♀
juv
juv
♂
African Emerald Cuckoo
♀

YELLOWBILL *Arboreal and mainly insectivorous, non-parasitic cuckoo related to Asian malkohas. Sexes similar. Endemic to Africa.*

Yellowbill *Ceuthmochares aereus* 33 cm

C. a. flavirostris*.** **Adult** Wholly slate-grey with ***long, graduated tail and ***conspicuous yellow bill***. Upperparts have purplish-blue gloss. Eye reddish; orbital ring yellowish. **Juvenile** Duller; upper mandible brownish. **HH** Uncommon resident, S Senegal; rare and local, Gambia. In forest (even where heavily degraded). At all levels, mostly mid-stratum and lower canopy, unobtrusively creeping through dense leafy vegetation with creepers. Frequently joins mixed-species flocks. **Voice** A series of loud, explosive *kuk* notes, first uttered slowly, but rapidly gathering speed *kuk, kuk, kuk kuk kukkukkukkukukukkkkkkrrrrrr*. Single *kuk* notes resemble those of Western Nicator. Also soft, plaintive *mweeeew*. [CD7:17]

COUCALS *Semi-terrestrial, stoutly built non-parasitic cuckoos with rounded wings and long, broad tails. Flight clumsy and not sustained. Sexes similar; females larger than males. Mainly insectivorous, but also take small vertebrates. Utter rapid series of deep, hollow hoots and liquid bubbling notes (like water poured from a bottle).*

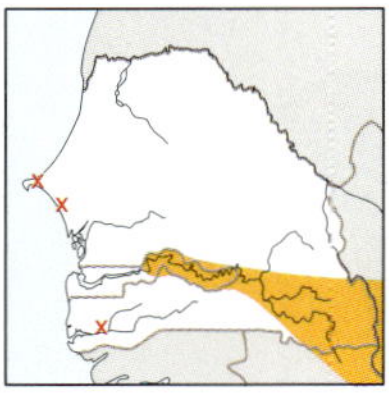

Black Coucal *Centropus grillii* 30–35 cm

Adult breeding Rather small, ***all-black coucal with contrasting chestnut wings***. **Adult non-breeding** Pale, with top of head to mantle dusky streaked buffish; rest of upperparts and tail tawny-chestnut barred blackish; rump and uppertail-coverts buff narrowly barred black. Underparts buffish; flanks and undertail-coverts with some dusky barring. **Juvenile** Similar to adult non-breeding but more heavily barred. Full adult plumage attained in *c*.2 years; young birds with fully or partly barred wings and tail may breed. **HH** Local and uncommon wet-season breeding visitor. Frequents moist, rank grasslands and marshy areas; occasionally shrub and cultivation. **Voice** In breeding season, a long series of variable speed *wok-wok, wok-wok, wok-wok, …* or *po-op po-op po-op…* and faster *popopopopopopop…*, from exposed perch. Also 'water-bottle' song, similar to Senegal Coucal but higher pitched, faster and not rising at end. [CD7:21]

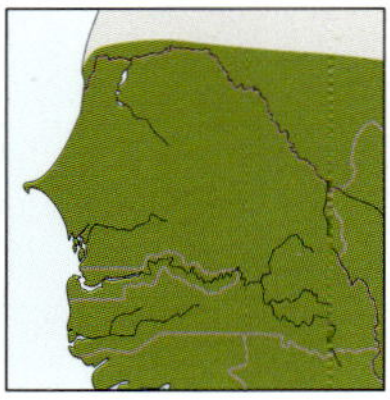

Senegal Coucal *Centropus senegalensis* 36–40 cm; WS *c*.50 cm

***C. s. senegalensis*.** Commonest coucal. **Adult** Head and tail glossy black; upperparts rufous-brown; underparts whitish washed buff. **Juvenile** Strongly barred upperparts; head dark and streaky; underparts tawny. **HH** Common resident. Singly or in pairs in farmland, roadsides, gardens and variety of open habitats with tall grass and bushes. Clambers in low vegetation, often in thick cover, runs and hops on ground, or flies clumsily for short distance, crash-landing into bush. **Voice** A characteristic, accelerating series of hollow *hoots*, first rising in pitch, then dying away. Often in duet, both songs delivered in similar rhythm but different pitch. Speed and pitch variable, slowest and deepest resembling song of Black-throated Coucal. Calls include nasal *gook* (alarm) and fast, dry *k-t-k-t-k-t-k-t-k-t-k-…* [CD7:22–23]

Black-throated Coucal *Centropus leucogaster* 48–58 cm

C. l. leucogaster*.** **Adult** Large forest coucal with ***blue-black head, mantle, throat and upper breast; rest of underparts white. Wings deep chestnut; back to uppertail-coverts black barred buff. **Juvenile** Duskier, with barred upperparts and tail. **HH** Scarce, SW Senegal (Casamance), where presumed resident. Singly or in pairs in lower strata of lowland forest and at forest edges. Rather shy and retiring; difficult to see. **Voice** A series of 10–20 deep, resonant *hoots*; similar to Senegal Coucal's in structure, but lower and slower, never accelerating at end. Songs of duetting pair have similar pitch but different rhythm. Call a fast *dookdookdookdook…* [CD7:18]

Yellowbill

Black Coucal

Senegal Coucal

Black-throated Coucal

BARN OWLS *Medium-sized nocturnal birds of prey with a distinctive heart-shaped, pale facial disc. Flight buoyant and silent. Perch upright.*

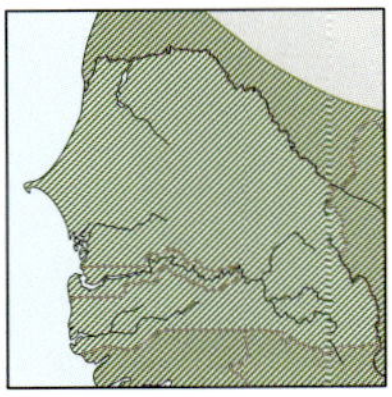

Barn Owl *Tyto alba*

33–36 cm; WS 80–95 cm

T. a. poensis. **Adult** ***Pale owl*** with conspicuously ***whitish, heart-shaped face***, black beady eye and no ear-tufts. Upperparts pale grey mixed with golden-buff and speckled white and black. Underparts white sparsely speckled black and variably washed pale golden-buff. Flight feathers and tail barred dark grey. **Juvenile** Upperparts darker grey, underparts washed golden-buff. **HH** Fairly common to locally common in various habitats (not in closed forest); often associated with man. Particularly numerous, N Senegal (where breeding almost year-round); dry-season breeder, Gambia. Nocturnal and crepuscular. Takes mainly rodents and shrews; locally also birds, bats, reptiles and insects. **Voice** Hoarse, drawn-out screech *shreeeeeeeee*, in flight. [CD3:12]

TYPICAL OWLS (Plates 73–75) *Small to large, mainly nocturnal birds of prey with characteristic large, rounded heads and large, forward-facing eyes within flat facial disc. Plumage mostly cryptic, many species with ear-tufts. Calls distinctive; usually most vocal just after dark and before dawn.*

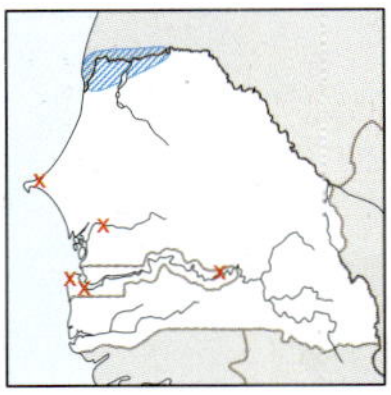

Short-eared Owl *Asio flammeus*

35–40 cm; WS 95–110 cm

A. f. flammeus. Sandy-buff marked dark brown and tawny above; whitish face narrowly edged dark; yellow eyes. Small, short ear-tufts in centre of forehead usually hard to see. In flight, long wings with dark carpal patch and contrasting sandy-buff patch at base of primaries; underwing pale with black tips and dark carpal patch. Marsh Owl is darker and plainer above and has dark brown eyes. **HH** Rare to uncommon Palearctic migrant (Oct–Apr) in various open habitats. Crepuscular, nocturnal and often diurnal. Hunts low; buoyant flight with rigid wingbeats alternating with gliding and banking, and wings held in shallow V, recalling harrier. Roosts on ground. Takes mainly small mammals; also birds and insects. **Voice** Usually silent in Africa. [CD3:18]

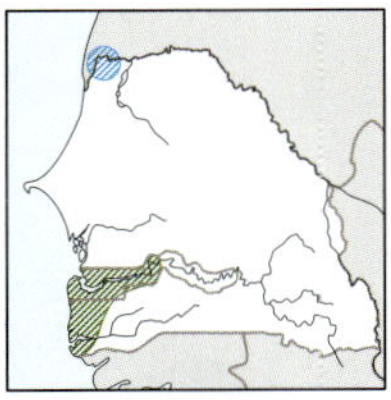

Marsh Owl *Asio capensis*

30–35 cm; WS 80–95 cm

A. c. capensis. **Adult** Rather plain, dark brown owl with round head and large dark eyes emphasised by blackish surround. Small, short ear-tufts in centre of forehead rarely visible. In flight, contrasting large pale patch at base of primaries; underwing pale with dark carpal patch and black tips to primaries. **Juvenile** Face darker and bordered black. **HH** Local resident and partially intra-African migrant. Uncommon to rare, coastal Gambia and Senegal delta. Also claimed from coastal S Senegal. Singly, in pairs or small groups in open grassland. Mainly crepuscular and nocturnal, occasionally diurnal. Hunts from low perch or in low, buoyant flight with wings held in shallow V and sudden banking, recalling harrier. Roosts on ground. Inquisitive, often circling around intruder when flushed. Takes mainly rodents; also other small mammals, birds, lizards, frogs, scorpions and insects. **Voice** Various harsh croaks and rasping calls, uttered in flight and on ground. [CD3:19]

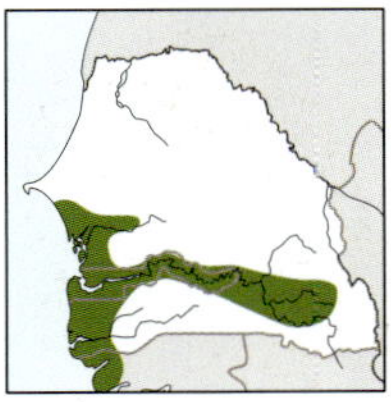

African Wood Owl *Strix woodfordii*

33–35 cm

S. w. nuchalis. The commonest forest owl. **Adult** Medium-sized, with ***round head***, barred underparts and large dark eyes. Dark brown above, speckled white; scapulars and wing-coverts spotted white. Face pale brownish, with dark eyes emphasised by dark brown surround and broad whitish eyebrows. Underparts barred rufous-brown and white. Bill yellowish. **Juvenile** White facial markings less distinct. **HH** Scarce (Senegal) to rare (Gambia) resident in evergreen and gallery forests, dense woodland and thickets. Often in pairs. Nocturnal. Takes mainly insects; also small mammals, birds, reptiles and frogs. **Voice** Characteristic, single, drawn-out *whoOOow* or *oowhEEoo* and rhythmic *hu-hoo, hu-hoo hoo-hu-hoo*. Duets; female higher pitched. Male may answer female *whoOOow* by low *hooo*. [CD7:44]

Barn Owl

Short-eared Owl

Marsh Owl

African Wood Owl

Greyish Eagle Owl *Bubo (africanus) cinerascens* 43–48 cm

Adult Large, greyish owl with prominent, two-toned ear-tufts, finely barred underparts and dark brown eyes. Face darkish bordered black; breast with dark blotches. **Juvenile** Browner, with shorter ear-tufts. **HH** Uncommon resident in various habitats, from wooded grassland to forest edge; often near villages and towns. Often on dirt roads at night. Calls mainly at dawn and dusk. Takes wide variety of prey, but mainly insects, spiders, scorpions, small mammals, birds and reptiles. **Voice** Male utters short, abrupt hoot, followed after short pause by second, lower and drawn-out hoot *HO!, hoooo*. Female has trisyllabic *ho, hohooo*, middle note higher. Often duets. [CD7:31]

Verreaux's Eagle Owl *Bubo lacteus* 58–65 cm

Adult ***Very large***, greyish owl with pink eyelids. Face whitish bordered black at sides; broad, bushy ear-tufts seldom raised; eyes dark. **Juvenile** Browner, with ear-tufts limited to small bumps. **HH** Locally fairly common resident in wooded savanna, often in riverine woodland with large trees. Calls mainly at dusk; also not infrequently during day, particularly in early morning. Takes wide range of prey, but mainly mammals and birds. **Voice** A far-reaching, short, irregular series of low, discontented grunts. [CD7:35]

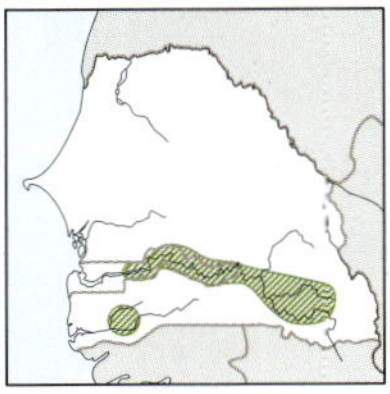

Pel's Fishing Owl *Scotopelia peli* 55–63 cm

Adult ***Very large, strikingly orange-rufous owl*** with large round head, plain face and no ear-tufts. Upperparts and tail ***narrowly barred blackish***; underparts paler, variably marked with dark brown drop-shaped spots, streaks and bars. Eyes large, dark brown. **Juvenile** Paler. Initially covered in whitish to pale buffish down, with pale rufous wing and tail feathers narrowly barred blackish. Retains pale yellowish head until attaining adult plumage at *c.*15 months. **HH** Scarce and local resident, frequenting large forested rivers. Largely nocturnal. Roosts by day in shady spot in large tree near water. Flushes with noisy wingbeats. Catches fish from low perch. **Voice** Far-carrying, deep sonorous hoot, sometimes preceded and often followed by low grunt *hooommm-hut*. Pair may duet, male starting with low grunting *uh-uh-uhu...* reaching higher *hoommm*, answered by deeper hoot of female. Young utter single, initially rising, then falling scream *weeeeaaow*; when hungry, may continue to call throughout the day. [CD7:36]

Desert Eagle Owl *Bubo ascalaphus* 46–50 cm

B. a. desertorum. Large owl with distinct ear-tufts and orange or deep yellow eyes. **Adult** Mainly tawny-rufous to sandy-ochre above with dark and pale blotches and streaks. **HH** Vagrant, Senegal (one found dead on Serpent islet, near Dakar, Mar 1974; also an unlikely claim of a pair in Basse Casamance, SW Senegal, Mar 1983). Occurs in desert and sub-desert, esp. with rocky outcrops and in oases. Seasonal movements recorded. Mainly nocturnal. Calls at dusk. Takes mainly smaller mammals; also birds, reptiles, larger insects and scorpions. **Voice** Far-carrying, deep, sonorous hooting *WHOO-oo*, second syllable lower pitched; repeated at intervals of 10–15 seconds. During courtship, trisyllabic *WHOO-hoo-hoo*. Duets; female higher pitched. Calls at dusk. [CD3:14]

Greyish Eagle Owl

Verreaux's Eagle Owl

Pel's Fishing Owl

Desert Eagle Owl

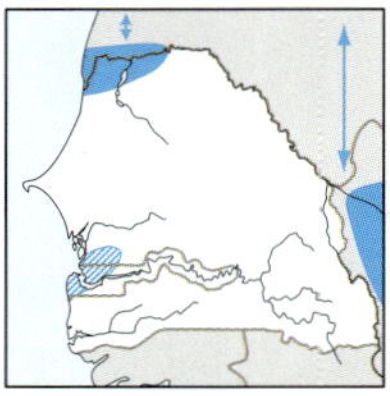

Eurasian Scops Owl *Otus scops* 18–20 cm; WS 47–54 cm

O. s. scops. As African Scops Owl; tends to be browner. In hand: outermost (10th) primary longer than 5th. **HH** Regular Palearctic visitor in small numbers, N Senegal (Sep–Apr); rare, Gambia. In various bushy and wooded habitats. Takes mainly insects; occasionally small mammals, small birds, lizards and frogs. **Voice** Pure, whistled *pyuu*, rarely uttered in Africa. [CD3:13]

African Scops Owl *Otus senegalensis* 16–18 cm

O. s. senegalensis. ***Cryptically coloured small owl with prominent ear-tufts*** (occasionally flattened). Short tail does not project beyond wingtips at rest. Mainly grey, with fine blackish streaks and vermiculations; row of bold white spots on shoulder. Face outlined with black; eyes pale yellow. **Juvenile** as adult. **HH** Fairly common to uncommon resident, Senegal; locally fairly common, Gambia. In wooded grassland. Generally starts calling at dusk, but may do so up to one hour before. Takes mainly insects; occasionally rodents, small birds, lizards, frogs, scorpions and spiders. **Voice** Single, short, vibrant *prr-u-u-p*, often repeated, with pauses of 5–10 seconds. [CD7:28]

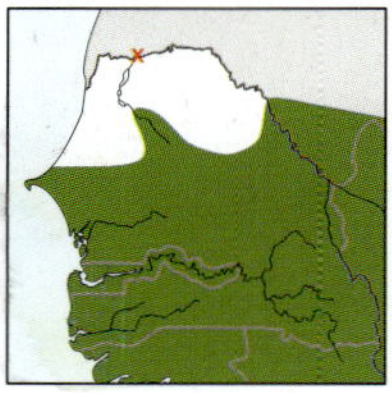

Pearl-spotted Owlet *Glaucidium perlatum* 19–21 cm

G. p. perlatum. **Adult** Small owl with rounded head, no ear-tufts and rather long tail. Two dark 'eye-spots' on nape. Brown spotted white above; row of bold white spots on shoulder; underparts white heavily streaked rufous-brown. Eyes yellow. **Juvenile** Lacks white pearly spots on crown and mantle. **HH** Locally common to uncommon resident in wooded savanna. Partially diurnal; regularly calls by day. Takes wide range of prey, incl. insects, reptiles, snails, small birds and small mammals. **Voice** A distinctive, rhythmically rising series of short notes followed by loud whistling climax *hu-hu-hu-hu-HU-HU-HU TEEEUW TEEEUW TEEEUW…* Also a high-pitched *siiiu*. [CD7:39]

Northern White-faced Owl *Ptilopsis leucotis* 23–28 cm

Adult Smallish owl with ***contrasting white face boldly outlined with black, large orange eyes***, and ***distinct ear-tufts***. Mainly soft grey, with long, narrow blackish streaks and fine vermiculations. **Juvenile** Paler and browner; eyes yellow. **HH** Locally fairly common to uncommon resident in various woodlands, mainly wooded grassland. Nocturnal and crepuscular. Roosts by day in trees and bushes, often in rather open situations. Takes small rodents, shrews, birds, insects, scorpions and spiders. **Voice** Two rather low, melodious, fluting syllables *whoh whoow*, repeated at intervals of 4–8 seconds. Also a lower and slightly faster *whoh thohoow*. [CD7:30]

NIGHTJARS (Plates 75–77) Crepuscular and nocturnal with very cryptic plumage, long wings, moderately long tail, tiny bill and huge gape for catching insects on the wing. Normally not seen during the day, except when flushed from underfoot. Identification of many species problematic; important features include overall coloration, markings on scapulars and wing-coverts, presence of nuchal collar, and amount of white (male) or buff (female) in wings and tail, but best identified by voice. Typically sing at or just after dusk and just before dawn; often also throughout moonlit nights. Often on dirt roads at night; orange reflection of eyes in headlights of car visible at considerable distance.

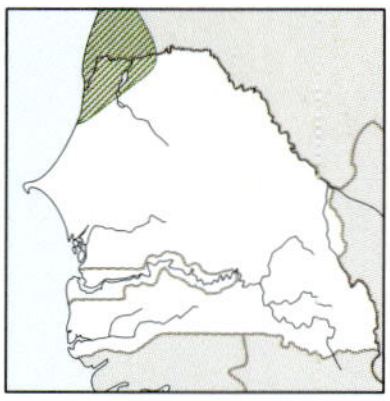

Golden Nightjar *Caprimulgus eximius* 23–25 cm

C. e. simplicior. Unique, golden-buff nightjar of arid country. **Adult male** Entirely cinnamon-buff covered with irregular white spots, each bordered and speckled black. In flight, conspicuous large white patch towards wingtip and white corners to tail. **Adult female** Wing patch and tail corners smaller and buffish. **HH** Rare to uncommon resident in Sahel zone, N Senegal. Favours stony ground and rocks. **Voice** Song a monotonous, sustained, hard reeling *rrorrrrrrrrrrrrrrrr… rrerrrrrrrrrrrrrrrrrrrrrr…* identical to that of Plain Nightjar. Also a rhythmic *kodu-kodu-kodu-…*

Egyptian Nightjar *Caprimulgus aegyptius* 24–27 cm; WS 53–58 cm

C. a saharae. Rather uniform, pale greyish desert nightjar. May resemble greyish forms of Plain Nightjar, but has white on throat and, in flight, ***no white in wings and tail***; dark brownish outer wing. **HH** Uncommon to fairly common Palearctic migrant, N Senegal (mainly Dec–Feb). Favours dry, open sandy habitats. **Voice** Mostly silent in Senegal. Song a long, very rapid, regular series of *krow* notes, *krowkrowkrowkrowkrowkrowkrowkrow…*, resembling sound of running engine. [CD3:20]

Eurasian Scops Owl
ad
African Scops Owl
ad
ad
head facing away
Pearl-spotted Owlet
ad
Northern White-faced Owl

♂
♀
Golden Nightjar
ad
ad
Egyptian Nightjar

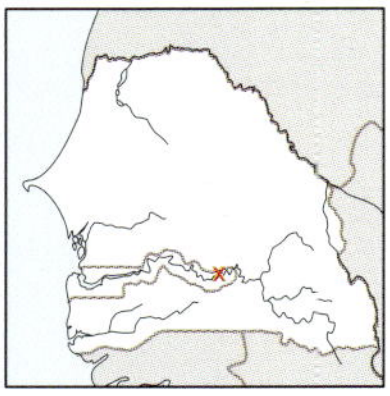

Swamp Nightjar *Caprimulgus natalensis* 20–24 cm

C. n. natalensis. Medium-sized, mainly grey-brown, heavily marked and spotted nightjar of wet grasslands. Tail rather short. ***Head-sides dark brown bordered by long, pale supercilium and moustachial stripe***. Large blackish markings on scapulars and wing-coverts; white, triangular throat-patch; ***round buff spots*** on breast. In flight, **male** has ***white patch towards wingtip*** and ***white-edged tail***; in **female**, white replaced by buff. **HH** Status unknown; no sight records. Only tape-recorded near Basse, Gambia, Mar 1975. Favours damp or locally dry grasslands and edges of swamps. **Voice** Song a long, regular series of rapid notes *chuk-chuk-chuk-chuk-chuk-chuk-…* or *tuk-tuk-tuk-tuk-tuk-…*, resembling knocking on dry wood, and faster *chukukukukukukukuk*. Also a sharp note, often repeated a few times, *kip kip-kip-kip*; a highly distinctive, variable series of melodious, tremulous notes *whoa-whulululu whoa-whuwhu whoa-whulululu whoa!* in flight, and fast *rrrrukukukukukukuk-whoalulululu*. [CD7:47]

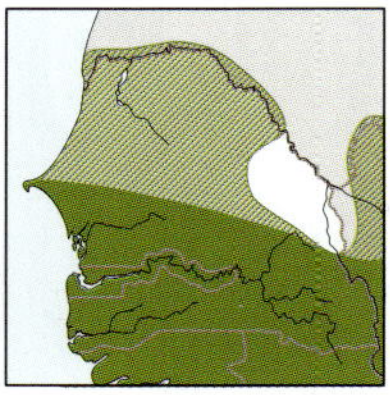

Long-tailed Nightjar *Caprimulgus climacurus* 28–43 cm

C. c. climacurus. Slim, small nightjar with ***long, graduated tail*** and variable, relatively contrasting plumage. Above, mainly grey-brown to pale buff-brown with ***clear white bar on wing-coverts***. Head-sides dark rufous bordered by narrow white moustachial stripe and, usually, pale supercilium; white triangular throat patch. Broad tawny or buffish collar on hindneck. In flight, **male** has white patch towards wingtip and ***white trailing edge to wing***. **Female** has shorter tail and buffish wing patch. **HH** Locally common in various open and lightly wooded habitats. Entirely or partially migratory; some probably also sedentary. Movements poorly understood. Faithful to (often communal) roost sites. **Voice** Song a monotonously sustained, hard reeling with single, fast rhythm *rrerrrrrrrrrrrrrrrrrrrrrr…* Female has similar, slightly lower pitched *rrorrrrrrr…* In flight, a repeated *chiong-chiong…* [CD7:48]

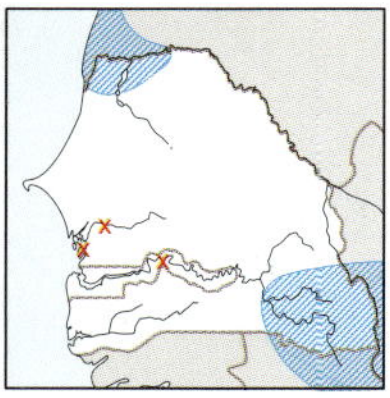

Plain Nightjar *Caprimulgus inornatus* 22–23 cm

Rather slender nightjar with fairly ***uniform plumage*** lacking distinctive markings. Plumage coloration varies from cinnamon-rufous or sandy-buff to grey-brown, grey or dark brown (matching soil colour). ***Row of small black spots on scapulars***. In flight, **male** has small white patch towards wingtip and white corners to tail; **female** has buff wing patch and tail corners. **HH** Status inadequately known. Reportedly fairly common, SE Senegal, but possibly irruptive; few records, Gambia. In various open and wooded habitats. **Voice** Song a monotonous, sustained, hard reeling *rrorrrrrrrrrrrrrrr… rrerrrrrrrrrrrrrrrrrrrr…* similar to but noticeably lower pitched and slower than that of Long-tailed Nightjar. In flight, chuckling *kwakow*. On ground, low *chuk*. [CD7:55]

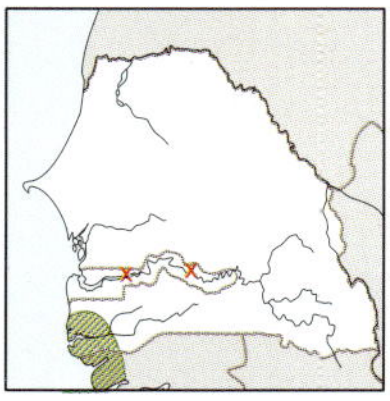

Black-shouldered Nightjar *Caprimulgus nigriscapularis* 23–25 cm

Rufous-brown with ***contrasting blackish shoulder***; rusty-buff line on scapulars. Head-sides and hindneck collar dark rufous; throat with white patch on either side or single triangular patch, bordered below by black; breast dark rufous-brown mixed with grey and blackish. In flight, obvious small white patch towards wingtip and white corners to tail. **Female** has slightly smaller white wing patch and tail corners. **HH** Status uncertain. A few identified by voice in Basse Casamance, SW Senegal (Jan, Feb and Jun). One in mangroves at Tendaba, Gambia (Nov–Dec 1998). Frequents woodland, gallery forest and forest edge. **Voice** Song a clear, melodious drawn-out note followed by tremulous whistling *kiiiuu-iup kiiurrrr*, sometimes preceded by fast *whoap-whoap-whoap-whoap-…* Flight call a low *chuk* often repeated rapidly 2–3 times. [CD7:52]

ad
Swamp Nightjar
♀
♂
♂
♀
Long-tailed Nightjar
♂
♀ greyish individual
Plain Nightjar
greyish individual
♂ rufous individual
rufous individual
♂
Black-shouldered Nightjar
ad
♀

Red-necked Nightjar *Caprimulgus ruficollis* 30–34 cm; WS 60–65 cm

Large, grey-brown nightjar with ***broad rufous collar*** on hindneck. Head-sides dusky-rufous, bordered below by white moustachial stripe; throat rufous with white patch on either side or single triangular patch. In flight, **male** has white patch towards wingtip and white tail corners; **female** has buffish wing patch and tail corners. **HH** Scarce Palearctic migrant in various open and wooded habitats. Most records known from road kills. **Voice** Song a series of rapid, mechanical-sounding *kotok-kotok-kotok-kotok-kotok-kotok-…* accelerating in long *kotokotokotokotokotokoto…* then slowing down again. [CD3:21]

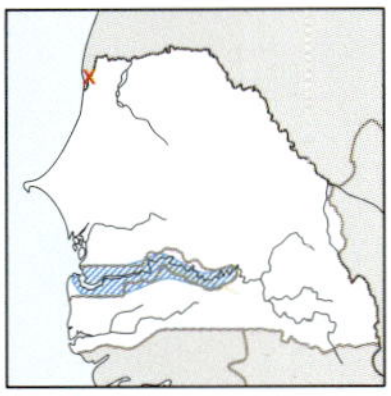

European Nightjar *Caprimulgus europaeus* 24–28 cm; WS 52–60 cm

Rather large and dark greyish nightjar. Broad pale upperwing-covert bar, ***pale line on scapulars***; whitish moustachial stripe. In flight, **male** has white patch towards wingtip and white corners to tail; **female** has no white in wings and tail. N African ***C. e. meridionalis*** overall smaller and paler than ***C. e. europaeus***. **HH** Rare Palearctic migrant, NW Senegal (one record of *C. e. meridionalis*, Nov); scarce/uncommon, Gambia (late Oct–Feb). In various open and lightly wooded habitats. **Voice** Normally silent in W Africa. Flight call a short *kweek*. [CD3:22]

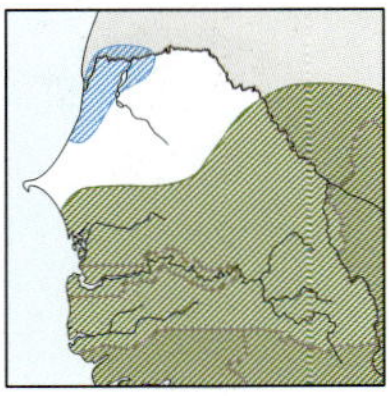

Standard-winged Nightjar *Macrodipteryx longipennis* 21–22 cm

Smallish, richly coloured nightjar with ***broad rufous nuchal collar*** and ***no white in wings and tail***. Buffish line on scapulars; buffish throat. **Adult male breeding** Unmistakable. In flight, ***extremely elongated 2nd innermost primaries with bare shaft and blackish vane at tip*** ('standards'), gives impression of bird being closely pursued by two small fluttering bats. **Adult male non-breeding/ adult female** Similar, but without 'standards'. **HH** Fairly common to locally common in various open, wooded habitats, incl. wooded grassland, coastal plains, farmland, dry scrub, and open bush in and around villages and towns. Moves south in dry season. **Voice** Song a fast, high-pitched, shrill, insect-like *tsikitsikitsikitsikitsiki…* and slower *tseepetseepetseep etseepetseepe…* Flight call a low *kuk*. [CD7:58]

Pennant-winged Nightjar *Macrodipteryx vexillarius* 24–28 cm

Medium-large, brownish nightjar with broad rufous nuchal collar. **Adult male breeding** Unmistakable. In flight, ***uniquely shaped wings*** with black remiges, ***broad white band on primaries and extremely elongated whitish 2nd innermost primary*** ('pennant'). Inner primaries and secondaries tipped whitish forming contrasting trailing edge. **Adult male non-breeding** Similar, but pennants broken or short. **Adult female** Very similar to adult female/male non-breeding Standard-winged Nightjar but upperparts slightly darker, nuchal collar broader and darker rufous, throat patch white, lower underparts paler buff and more heavily barred, primaries with larger rufous bars; also larger. In flight, no white in wings and tail. **HH** No certain records; claimed from Gambia (one examined in hand near Soma, Nov 1977). Intra-African migrant occurring as non-breeding visitor from E Nigeria east, with vagrants west to SE Burkina Faso and Togo. In various types of wooded grasslands and bush. May start foraging well before dusk. **Voice** Probably silent in W Africa. Song a fast, high-pitched, shrill, insect-like *tsitsitsitsitsitsitsitsitsitsi…* recalling Standard-winged Nightjar. Flight calls *wheeeo* and *chup*. [CD7:59]

♂
♀
Red-necked Nightjar
ad
♂
♀
European Nightjar
ad
Standard-winged Nightjar
♂
♀/ ♂ non-br
Pennant-winged Nightjar
♂
♀
♀

SWIFTS (Plates 78–79) *Highly aerial, anchor-shaped birds with mostly dark plumage. Flight swift and effortless on long, stiff wings. Sexes similar; juveniles similar to adults or duller. Catch insects on the wing; never perch. Often gregarious and vocal. Some difficult to identify; useful features include size, silhouette, rump pattern, tail shape, habitat and locality.*

SPINETAILS *have distinctively shaped, notched wings, with curved wingtips, bulging midwings and shorter secondaries pinched-in at body, and square tails (with bare shafts extending beyond feather tips, hence name, but invisible in the field).*

Mottled Spinetail *Telacanthura ussheri* 13–14 cm

T. u. ussheri. Blackish with ***broad white rump*** patch and narrow white band of variable extent on lower belly (usually difficult to see in field). Throat pale mottled brownish. Tail square and relatively long (imparting *Apus*-like appearance). Superficially similar to Little Swift; see below for differences. **HH** Fairly common to uncommon resident in wooded savanna (esp. with *Borassus* palms and baobabs). Usually in pairs or small groups. Associates with other swifts. **Voice** Rasping twittering, *chrwit chrwitt-itit chrrr-r-r-r-itititttt* and variations. [CD7:62]

TYPICAL SWIFTS *(all other species on this and the following plate) have sickle-shaped wings and generally forked tails.*

Little Swift *Apus affinis* 12–13.5 cm; WS 32–34 cm

A. a. aerobates. Small, rather stocky, blackish swift with ***square tail*** (sometimes appearing slightly notched) and ***broad white rump patch*** extending onto rear flanks; throat white. Superficially similar to Mottled Spinetail, but has different jizz, with less protruding head, shorter wings and more fluttery flight; also well-defined white throat, no white on lower belly and different call. **HH** Common resident. Highly gregarious; mainly in towns, but also forages over adjacent open areas. Very partial to bridges and other man-made structures for nesting. Associates with other swifts. **Voice** Rapid, high-pitched twittering, frequently uttered. [CD3:26]

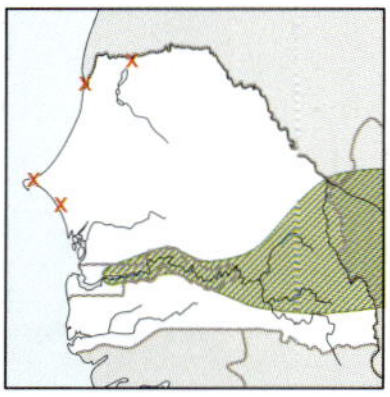

White-rumped Swift *Apus caffer* 14–15.5 cm; WS 33–37 cm

Rather small, slender black swift with long outer tail feathers forming ***deeply forked tail*** (usually held closed, tapering to thin point), ***relatively narrow U-shaped white rump patch*** extending to rear flanks, and white throat. **HH** Locally and seasonally scarce to fairly common intra-African migrant, mainly in wet season. Singly, in pairs or small flocks in savanna and farmland. Flight fast and graceful; also fluttering. Associates with other swifts and swallows. **Voice** Rather silent. Low, harsh, twittering notes. [CD3:25]

African Palm Swift *Cypsiurus parvus* 16 cm, incl. tail of up to 9 cm

C. p. parvus. Distinctive, very slender, ***all-greyish*** swift (sometimes appearing brownish) with ***very long and narrow wings and tail***. Deeply forked tail often held closed, appearing thin and pointed. Juvenile has shorter tail. **HH** Common resident. In pairs or small groups, typically foraging low around palm trees. Frequently in towns. Flight fast and graceful, with quick wingbeats and sudden twists and turns. Occasionally in mixed flocks with other swifts. **Voice** Vocal. Rapid, high-pitched, sibilant twittering *srrit-itititititit*, frequently uttered. [CD7:66]

Little Swift
ad
Mottled Spinetail
ad
ad
African Palm Swift
ad
White-rumped Swift

PLATE 79: SWIFTS II

Pallid Swift *Apus pallidus* — 16–18 cm; WS 39–44 cm

A. p. brehmorum. Medium-sized, strongly resembling Common Swift but ***paler***, more greyish-brown; ***paler inner wing contrasts with darker mantle, forewing and outer wing***; whitish throat patch larger and extending over lores to forehead, making dark eye patch stand out. Slightly broader and blunter wings, slightly shallower tail fork and less agile flight with slower wingbeats impart subtly different jizz. **HH** Seasonally common to uncommon Palearctic migrant (Jul–Feb). Over various habitats, often at great height. Associates with other swifts. **Voice** As Common Swift but often slightly deeper and disyllabic. [CD3:23]

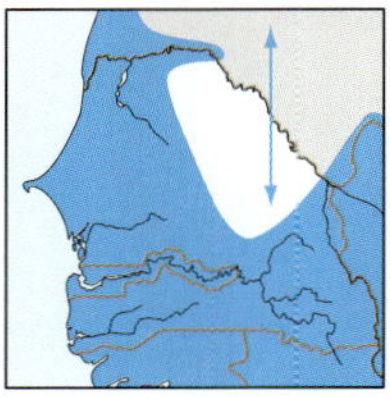

Common Swift *Apus apus* — 17–18.5 cm; WS 40–44 cm

A. a. apus. Medium-sized, mainly uniformly blackish-brown swift with narrow pointed wings, forked tail and small, ill-defined whitish throat patch. Compare Pallid Swift. **HH** Common Palearctic migrant (mainly on passage, but records in all months). Highly gregarious, occurring over all habitats, often at great height. Flight strong and fast. Very mobile, covering long distances in search of good feeding areas and to escape adverse weather. Attracted to centres of low atmospheric pressure. Presumed to remain on the wing, even for roosting, during entire sub-Saharan stay. Often in mixed groups with other swifts and hirundines. Some immatures assumed to remain in winter quarters during northern summer. **Voice** High-pitched, shrill, screaming *srreeee*. [CD3:24]

Mottled Swift *Tachymarptis aequatorialis* — 23 cm

T. a. lowei. Very large, robust, dark swift with forked tail. Upperparts dark brownish; throat greyish-white to white; rest of underparts scalloped whitish, appearing greyish in field. Single or distant Common Swifts separated by different jizz imparted by smaller size, less bulky body, and faster wingbeats; closer birds show uniform underparts. **HH** Afrotropical vagrant or rare visitor, SE Senegal (Niokolo Koba). Gregarious, but also occurring singly. Frequents crags, rocky outcrops and escarpments but may travel some distance to forage and disperses widely after breeding; then recorded over variety of habitats. Associates with other swifts, esp. Common Swift. **Voice** High-pitched *titititititititi*, rising and falling in patch; similar to Alpine Swift but delivered more slowly. Also a brief, dry call, infrequently uttered. [CD7:69]

Alpine Swift *Tachymarptis melba* — 20–23 cm; WS 51–58 cm

T. m. melba* (& *tuneti*?)**. ***Very large, mainly brown swift with ***white throat and belly separated by dark brown breast-band***; tail shallowly forked. Size and powerful flight may recall small falcon. **HH** Rare Palearctic (and Afrotropical?) migrant. Over a variety of habitats. Gregarious. Flight fast and more powerful than Common Swift, with slower wingbeats. Associates with other swifts. Usually at great height. **Voice** Rapid, rolling, twittering *chit rititititititit chetetet*… rising and falling in pitch. Reminiscent of Little Swift but slower, lower pitched and less harsh. [CD3:27]

ad

Pallid Swift

ad

Common Swift

ad

Mottled Swift

ad

Alpine Swift

KINGFISHERS (Plates 80–81) Characterized by compact silhouette with large head, long dagger-shaped bill, short body and very short legs. Most species brightly coloured. Occur in a variety of habitats. Aquatic species feed mainly on fish captured by plunge-diving; terrestrial species subsist on insects and reptiles on which they swoop from a perch. Flight fast and direct. Nest in tree or earth holes.

African Pygmy Kingfisher *Ceyx pictus* 12 cm

C. p. pictus**. Adult** ***Very small*** with brilliant violet-blue upperparts, orange-rufous underparts and coral-red bill and feet. Head mainly orange-rufous with ***blue-black crown***, broad orange-rufous supercilium, ***ear-coverts strongly washed violet*** and white patch on neck-sides. Throat white. **Juvenile** Duller, with black, white-tipped bill, blackish moustachial stripe and lores, and mottled upperparts. **HH** Fairly common resident and partial migrant; presumed breeding visitor, N Senegal. In various habitats, incl. woodland, cultivation, gardens, forest edge and clearings; not usually near water. Hunts for insects from low perch. **Voice** Single sharp, high-pitched *tseet*, in flight. [CD7:83]

Malachite Kingfisher *Alcedo cristata* 13 cm

A. c. galerita**. Adult** Small, ***aquatic*** kingfisher with glossy violet-blue upperparts, orange-rufous underparts and coral-red bill and feet. Forehead and crown feathers long and violet-blue barred black, ***occasionally raised and forming crest***; no supercilium. **Juvenile** Duller and duskier, esp. on ear-coverts and breast, with black bill (turns red from c.3 months). **HH** Common resident, but partially migratory in some areas. In aquatic habitats, incl. lakes, pools, marshes, lagoons and slow-running rivers. Dives for fish, shrimps and insects from low perch. Not shy. **Voice** Shrill *pseek*, in flight. Song in duet *tsrrr-tsrrrr tsip-tsip tsrr tsip t-t-t...* [CD7:84]

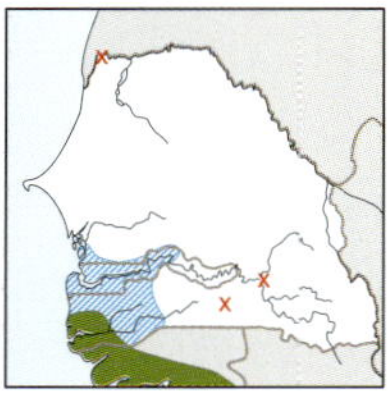

Shining-blue Kingfisher *Alcedo quadribrachys* 16 cm

A. q. quadribrachys**. Adult** Unobtrusive ***aquatic*** forest kingfisher with ***black bill***. Head, upperparts and patch at breast-sides brilliant, deep ultramarine-blue with white patch on neck-sides and rufous loral spot. Throat white; rest of underparts deep orange-rufous. Feet red. Female has some dark red on base of lower mandible. **Juvenile** Duller, with dusky mottling on breast and bill tipped whitish. **HH** Rare to seasonally uncommon along forested rivers, lakes, lagoons and mangroves. Apparently only occurs in wet season; no confirmed breeding records. Easily overlooked; mostly seen in low, fast flight over water. **Voice** Shrill, high-pitched *tseep*, in flight. [CD7:86]

Striped Kingfisher *Halcyon chelicuti* 17 cm

H. c. chelicuti**.** Small, ***rather drab savanna kingfisher*** with distinctive song. **Adult** Crown and underparts streaked; black streak through eye; head-sides and throat whitish. Bill dusky above, reddish below. Underwing white with dark carpal patch, broad grey trailing edge and broad black subterminal band. In female, underwing has smaller carpal patch and lacks subterminal black band on remiges. **Juvenile** Bill blackish tipped whitish, underparts more buffy, blue areas of wings and tail paler and more limited, or lacking. **HH** Fairly common to common in dry savanna woodland and bush. Perches unobtrusively on branch at mid-height. Feeds on insects (mainly grasshoppers) and, occasionally, small vertebrates, nearly all taken on ground. Returns to perch to consume prey. Displaying birds face each other perching upright, flicking wings open and shut, and calling excitedly. **Voice** Song a series of loud, far-carrying *KEE-RRRUU KEE-RRRUU KEE-RRRUU...* [CD7:81]

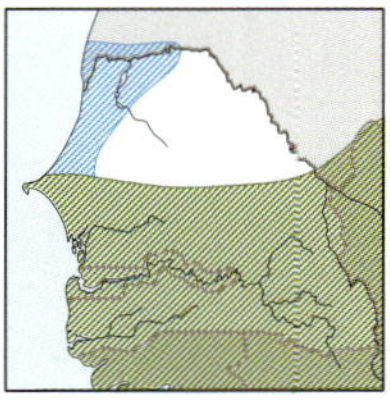

Grey-headed Kingfisher *Halcyon leucocephala* 22 cm

H. l. leucocephala**. Adult** Diagnostic combination of pale grey head and breast, ***bright chestnut belly and undertail-coverts*** and black mantle, scapulars and wing-coverts. Bill and feet bright red. **Juvenile** Duller, with dusky bill and buff or pale chestnut belly. **HH** Seasonally fairly common in woodland, bush and cultivated areas. On migration in a variety of habitats, except dense forest and very arid areas. Hunts from perch, sitting quietly for long periods, occasionally pouncing on insects on ground. Displaying pairs circle above trees, calling. Migrates north in rains, south to open areas in forest zone in dry season. No breeding records from far north. **Voice** Call a loud, scolding and explosive *CHeK!* or *KHE!*, frequently uttered in short series. Song a weak descending trill *chichichichi-chi-chiu*, usually given in flight. [CD7:78]

ad
juv
African Pygmy Kingfisher
juv
Malachite Kingfisher
ad
juv
♂
♀
Shining-blue Kingfisher
ad
Striped Kingfisher
juv
Grey-headed Kingfisher
ad

Blue-breasted Kingfisher *Halcyon malimbica* 25 cm

H. m. forbesi. **Adult** Large, mainly bright azure-blue kingfisher with ***black back, scapulars and upperwing-coverts*** and ***pale azure-blue breast*** becoming grey or whitish on rest of underparts. Crown grey. Black streak through eye produces stern appearance. Bill heavy, upper mandible red, lower mandible black. **Juvenile** Duller, with head and underparts tinged buff. **HH** Fairly common in mangroves, gallery forest, densely wooded savanna, and forest. Mainly keeps in deep shade below canopy. Feeds primarily on insects, also small vertebrates, hunted from perch. **Voice** Song starts with single, abrupt *chiup!*, followed after short pause by a series of piping *pu* notes, initially accelerating and ascending slightly, then descending and slowing down: *chiup! pu-pupupuPUPUUpuu-puu puu puu*; uttered throughout the day, mostly from perch, but also in display flight. Both sexes sing. Alarm a raucous *chup, chup-chup-chup* or *kiah, kiah-kiah*. [CD7:79]

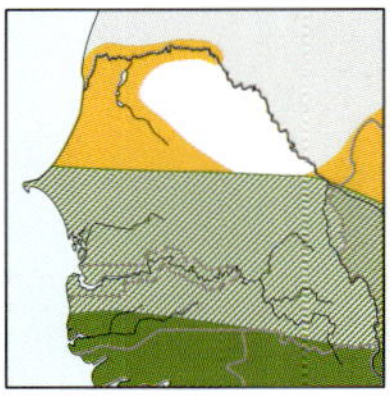

Woodland Kingfisher *Halcyon senegalensis* 22 cm

H. s. senegalensis. **Adult** Widespread, medium-sized kingfisher with grey head and mantle, and rest of upperparts bright azure-blue with contrasting ***black upperwing-coverts***, and greyish-white underparts. **Juvenile** Duller, with dusky bill. **HH** Common wet-season breeding visitor (mainly Jun–Nov; uncommon in other months); resident in S Senegal (Casamance). In woodlands, farmland, mangroves, gardens and forest clearings. Perches on high branch or other, often exposed, vantage point, hunting insects and small vertebrates. Conspicuous, attracting attention by loud song uttered throughout the day (mostly in mid-morning; occasionally after dark). In display flicks wings open while singing, revealing contrasting underwing pattern, and turns around repeatedly. Often interacts aggressively with other birds. **Voice** Distinctive and easily learnt song vigorous and explosive, consisting of single sharp initial note, followed after short pause by hard descending trill *PTIK TIRRRRrrrrrrr*; uttered throughout the day. Alarm a fast *kee-kee-kee-kee-*... [CD7:80]

Pied Kingfisher *Ceryle rudis* 25–27 cm

C. r. rudis. Conspicuous and unmistakable black-and-white aquatic kingfisher. **Adult male** has double breast-band; **adult female** a single incomplete breast-band (limited to patch on breast-sides). **Juvenile** as adult female but feathers of lores, throat and breast fringed brown; breast patches greyish-black; gape pink. **HH** Common resident. In pairs or small parties in open areas on rivers, lagoons, mangroves, lakes and creeks. Feeds almost exclusively on fish, caught by plunge-diving from low perch or hovering flight. **Voice** Vocal. Calls include high-pitched, chattering *kwik-kwik* uttered in flight or on perch, *chikrr-chekrr*... (threat call) and *kittle-te-ker* (when flying from perch). [CD7:88]

Giant Kingfisher *Megaceryle maxima* 42–46 cm

M. m. maxima. Largest kingfisher, with massive, black bill, bushy crest, and white-speckled, slate-grey upperparts. Unmistakable, but beware of confusing flying bird with Green-backed Heron (Plate 8), which has similar size, upperpart coloration and call. **Adult male** Breast chestnut, belly white barred black. **Adult female** Breast white very densely spotted black, belly chestnut. **Juvenile male** as adult male, but breast-sides speckled black; flanks with some chestnut feathers. **Juvenile female** as adult female, but breast less densely spotted; spots crescentic and mainly at sides. **HH** Generally uncommon but locally fairly common in wet season. Frequents wooded rivers, lagoons and lakes, mangroves; also rocky and sandy seashores. Perches quietly and often concealed over water, searching for fish and crabs. Rather shy. Flies low over water when disturbed. **Voice** Loud *KEK!* uttered singly or in series. [CD7:87]

juv
Blue-breasted Kingfisher
ad
juv
ad
hovering
Woodland Kingfisher
♀
♂
♀
Giant Kingfisher
Pied Kingfisher
♂

BEE-EATERS (Plates 82–83) *Slender, mostly brightly coloured species with long, pointed, slightly decurved bills and triangular wings. Feed mainly on wasps and bees, either caught in short sallies from a perch or in lengthy hawking flights. Some migratory and gregarious, others sedentary and occurring in pairs or family parties. Nest in burrows excavated in banks or level ground. Most species in this distinctive and beautiful family are easy to identify.*

Little Bee-eater *Merops pusillus* 15–17 cm

M. p. pusillus. Smallest bee-eater. **Adult** Top of head and upperparts grass-green. Throat rich yellow bordered by black mask and gorget; breast chestnut becoming cinnamon to buff on belly and vent. In flight, wings mainly cinnamon with broad black trailing edge; square or slightly notched tail cinnamon with black tips and green central feathers. **Juvenile** Paler, with pale greenish breast, no gorget. **HH** Common. In pairs or family parties in grassland, open savanna, plantations and various open-country habitats; fond of moist areas. Perches on long grass stem or low shrub, occasionally higher. **Voice** Quiet *slip* or *sip* and hard, clipped *tsip* or *klip*, uttered singly or in series. Also high-pitched, sibilant *siddle-iddle-ip, d'jee* in high-intensity greeting display. [CD7:92]

Swallow-tailed Bee-eater *Merops hirundineus* 20–22 cm

M. h. chrysolaemus. Medium-sized greenish bee-eater easily distinguished by ***strongly forked, pale blue tail***. In flight appears mainly green with forked, white-tipped tail; wings have some rufous and black trailing edge. **Adult** Throat yellow bordered by black mask and narrow blue gorget. **Juvenile** Throat greenish-white; no gorget. **HH** Locally fairly common. In pairs or small parties in open savanna woodland. Rather silent and unobtrusive. Feeds in short sallies from perch. Often moves from one spot to another when foraging. May roost in tightly packed ranks of 6–10 (rarely up to 30) when not breeding. At least partially migratory, with some moving south in dry season. **Voice** Subdued *weeerp-weeerp, tip-tip*, or *diddle-diddle-ip*, typically bee-eater-like in quality. [CD7:95]

Red-throated Bee-eater *Merops bulocki* 20–22 cm

M. b. bulocki. **Adult** The only bee-eater in its range with green upperparts, ***bright red throat*** (rarely yellow) and deep blue undertail-coverts. In flight, wings green above and buff below with black trailing edge; square tail green with some buff. **Juvenile** Duller. **HH** Common. In small to large groups in open savanna woodland and bushy pastures. Breeds colonially in stream banks, erosion gullies and quarries. **Voice** Vocal. Calls include *wip*, querulous *kirrup, kwirrup* or *krrip*, delivered in trilling series in greeting, and nasal *kweep*, uttered in flight and from perch. [CD7:96]

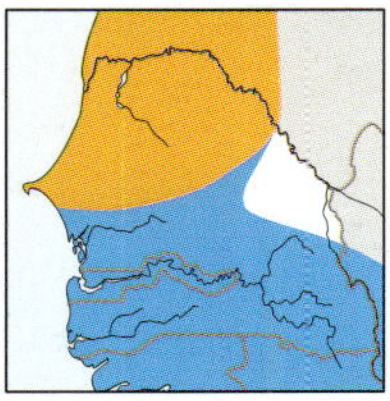

White-throated Bee-eater *Merops albicollis* 19–21 cm

Adult Graceful, conspicuous bee-eater with largely pale green upperparts, ***long tail streamers*** (projecting up to 12 cm) and striking ***black-and-white head pattern***. Black crown (unique among bee-eaters), mask and gorget contrast with white forehead, supercilium and throat. In flight, green and pale rufous wings (buffish below) have black trailing edge. **Juvenile** Duller, with yellowish-tinged throat and no streamers. **HH** Seasonally common intra-African migrant. Breeds during rains in narrow band in Sahel, moving south in dry season. Migration and non-breeding range include moist savanna and rainforest zones. Highly gregarious. Usually hunts from medium-high to high perch in non-breeding zone. **Voice** Vocal. Pleasing, melodious and far-carrying *pruuee, trooee* or *pruik*, reminiscent of European Bee-eater but distinctly higher pitched and delivered in longer series. Sometimes interspersed with shorter notes (e.g. *kwik kwik kwik kearrlup kearrlup…*). Frequently uttered in flight and on perch. [CD7:98]

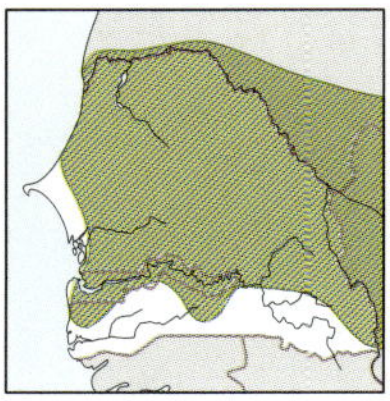

Little Green Bee-eater *Merops orientalis* 16–18 cm

M. o. viridissimus. **Adult male** Small, ***all-green*** bee-eater with very long tail streamers (projecting up to 9.5 cm), black mask and thin black gorget. In flight, shows rufous flight feathers above and below, and black trailing edge (like many bee-eaters). **Adult female** Gorget usually even narrower and tail streamers shorter. **Juvenile** Duller, lacking gorget and streamers. **HH** Fairly common to uncommon resident and partial migrant, dispersing mostly through south of range in dry season, after breeding. Singly, in pairs or small parties in thornbush and woodland. Makes short flycatching sallies from low perch. **Voice** Hard, rapid and slightly buzzing trill *trrri-trrri-trrri* or *kree-kree-kree-…* Alarm a staccato *ti-ti-ti* or *ti-ik*. [CD8:1a]

juv
ad
Little Bee-eater
juv
ad
ad
Swallow-tailed Bee-eater
juv
ad
yellow-throated
variant
ad
ad
Red-throated Bee-eater
juv
ad
White-throated Bee-eater
ad
juv
ad
ad
Little Green Bee-eater

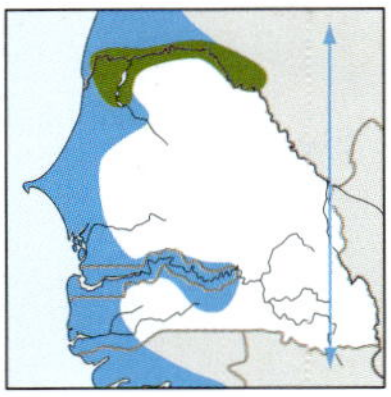

Blue-cheeked Bee-eater *Merops persicus* 24–28 cm; WS 35–39 cm

M. p. chrysocercus**. Adult** Large, slender, bright green bee-eater with very long tail streamers (projecting up to 11 cm). Black mask bordered with blue, and yellow chin merging into chestnut of throat and upper breast. In flight, golden-green above, underwing rufous with ***narrow*** dusky trailing edge. In worn plumage appears paler, more bluish. **Juvenile** Duller, lacking streamers. **HH** Seasonally common Palearctic visitor (Oct–May; records in all months). Breeds locally, N Senegal (Jun–Sep). Winters in bushy grassland and open farmland. Gregarious. Perches on power lines and low bushes. **Voice** Far-carrying *prri-ip*, *dirrip* or *pririk*, very similar to European Bee-eater but harder, less liquid and shorter. Alarm *kuk*. [CD3:29]

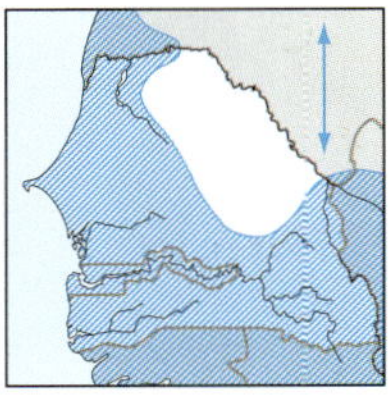

European Bee-eater *Merops apiaster* 23–25 cm; WS 36–40 cm

Large, vocal, uniquely coloured bee-eater with chestnut crown and mantle, golden scapulars, yellow throat, bluish underparts and short tail streamers. Female has less chestnut on wing and scapulars suffused with green. In flight, scapulars and rump form golden V; wings chestnut and green with ***broad*** black trailing edge. **HH** Locally fairly common Palearctic migrant (Sep–May) in various open habitats. Gregarious. Flight graceful, wheeling; rapid wingbeats alternating with gliding. **Voice** Melodious, liquid, remarkably far-carrying *prruuip*, *pru-ik* or *kwirip* (calls of Blue-cheeked Bee-eater very similar and only separable with practice). Also *pik-pik*. [CD3:30]

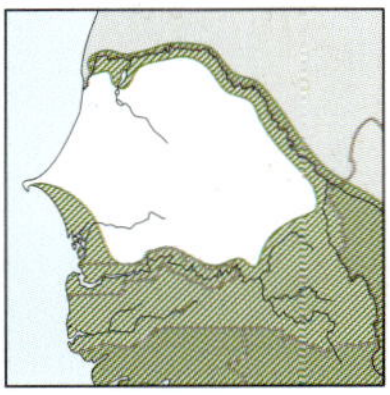

Northern Carmine Bee-eater *Merops nubicus* 24–27 cm

Adult Large, conspicuous, ***carmine-and-pink*** bee-eater with ***blue head*** and ***long tail streamers***. Rump, upper- and undertail-coverts pale blue, strongly contrasting with rest of plumage. Underwing buffy. Tail grey below. **Juvenile** Duskier, with mainly brownish upperparts and short streamers. **HH** Seasonally fairly common to locally common, in dry and open savanna (but records in all months). Often near water. Gregarious. Forages in 'sailing', twisting flight. Also feeds from perch. Partial to grasshoppers and locusts and therefore attracted to bush fires. Disperses south in dry season, after breeding. **Voice** Rather unmusical *klienk*, *klunk* or *terk*, often repeated and sometimes followed by *ki-ki-ki-ki-*… [CD8:3a]

MOUSEBIRDS Buffish-brown or greyish with long, stiff, graduated tails and short, erectile crests. Always in small parties. Flight on short, whirring wings, fast and direct with long glides. Climb well. Feed mainly on fruit and leaves. Endemic to Africa.

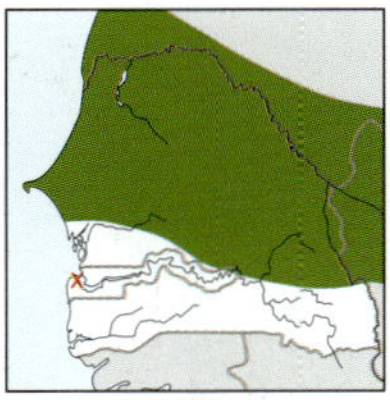

Blue-naped Mousebird *Urocolius macrourus* 33–38 cm

U. m. macrourus**. Adult** Greyish with conspicuous ***azure-blue nape***, very long, slender tail (of up to 28 cm), red on bill and bare red patch around eye. **Juvenile** Paler, without blue on nape, shorter crest; dull-coloured bare parts. **HH** Fairly common to *c.*15°N, more local and uncommon further south, Senegal; vagrant, Gambia. In small, vocal parties in arid open country with scrub, thorn trees and cultivated areas. Sedentary or locally nomadic. **Voice** Far-carrying, long, clear whistle *pwheeeeeee*; also shorter notes including *kwee*, *pwee-u* and *kruw*. [CD7:71]

TROGONS Arboreal forest birds with brightly coloured plumage and long, broad tails. Feed on insects, which they catch on vegetation during swift, swooping flight, reminiscent of drongos. Nest in tree-holes. Easily overlooked despite their brilliant coloration as they perch motionlessly for long periods at mid-level in forest shade.

Narina's Trogon *Apaloderma narina* 29–34 cm

A. n. constantia**. Adult male** Upperparts, throat and breast bright green; rest of underparts crimson. Tail white below, mainly blackish-blue above. Wing-coverts finely barred grey and white. Bill yellow. Head-sides with vivid yellow bare skin at gape, below eye and above eye. **Adult female** Duller, with forehead, throat and breast grey washed green; belly to undertail-coverts pink; bare skin bluer. **Juvenile** Resembles adult female, but secondaries and wing-coverts tipped with large white and buff spots. **HH** Rare and very local resident, SE Senegal (where discovered in 2010) in gallery forest. When singing, throat swells, revealing blue skin, while tail pumps rhythmically down. **Voice** A series of paired, moaning, soft and dove-like hoots, with stress on second syllable and repeated up to 14 times, starting hesitantly, then slightly rising in volume and giving impression of being forced out *who-ot-WHO poe-WHO poe-WHO poe-WHO*…; easily passing unnoticed. [CD7:74]

juv
ad
Blue-cheeked Bee-eater
juv
ad
European Bee-eater
ad
juv
ad
Northern Carmine Bee-eater
♂
♀
Blue-naped Mousebird
♀
♂
Narina's Trogon

ROLLERS *Robust, conspicuous birds with colourful plumage, large heads and stout bills, slightly hooked at tip. Vocal and pugnacious. Voice harsh. Have spectacular display flights. Breed in tree holes.* Coracias *species hunt from perch, swooping to ground to catch large insects or small vertebrates.*

Rufous-crowned Roller *Coracias naevius* — 35–40 cm

C. n. naevius. **Adult** Large, stocky and comparatively dull roller with ***conspicuous white forehead and supercilium, dull brownish-pink underparts heavily streaked white*** and square tail. Upperparts mainly olive-brown; rump pale purple; upperwing-coverts pinkish-brown; flight feathers purplish. Tail dark purple-blue with brownish-green central feathers. **Juvenile** Duller, more olive. **HH** Fairly common, in open woodland. Usually singly, perching on high vantage point. Less vocal than other rollers. Seasonal movements recorded (north with rains). **Voice** Various cackling, muffled and nasal notes, uttered singly or in rapid series, less harsh than other rollers and sometimes reminiscent of Green Wood-hoopoe. [CD8:4a]

Blue-bellied Roller *Coracias cyanogaster* — 28–30 cm

Adult Strikingly coloured roller, with pale buffish head and breast, black back, purplish-blue wings, rump and belly, and azure-blue, forked tail with streamers. In flight, shows broad azure-blue wing bar. **Juvenile** Duller, lacking streamers. **HH** Common in wooded savanna, edges of gallery forest, derived savanna and cultivation. Usually in pairs. Perches conspicuously atop trees, electricity poles, etc. **Voice** Vocal. Fast, sharp *keh-keh-keh-keh-k-r-r-r-r-r* and an accelerating, descending, scolding 'laugh' *HEhehehehheh…*, in flight and from perch. [CD8:4b]

Abyssinian Roller *Coracias abyssinicus* — 28–30 cm

Adult Slender and brightly coloured roller with ***very long outer tail feathers*** (projecting up to 12 cm). Mainly bright turquoise-blue with chestnut mantle, scapulars and tertials, deep violet-blue rump and uppertail-coverts. Forewing violet-blue, forming shoulder patch (sometimes concealed) when perched; flight feathers deep violet-blue. **Juvenile** Duller, without tail streamers. **HH** Common in various dry wooded habitats, incl. *Acacia* savanna, wooded grassland, farmland and large gardens. Noisy and conspicuous, perching on exposed vantage point. Singly or in pairs, gathering in small groups at bush fires. Moves south during dry season and north at beginning of wet season. **Voice** Harsh, explosive, scolding screech *kwrèèèèh* or *kèèèèèhh* uttered from perch; sharp *kek*, mainly in flight. In display flight a rapid series of similar harsh notes. [CD8:5a]

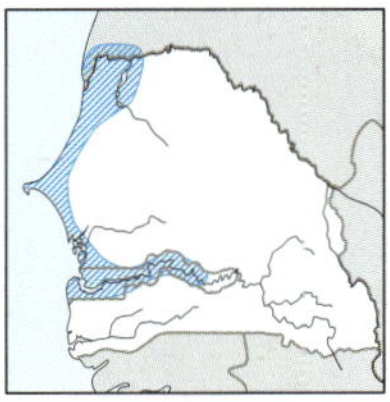

European Roller *Coracias garrulus* — 29–32 cm; WS 52–57 cm

C. g. garrulus. **Adult** Mainly turquoise-blue with chestnut mantle, scapulars and tertials, ***blackish flight feathers*** (deep violet below) and ***square tail***. Back to uppertail-coverts and forewing purplish-blue. Adults usually appear very washed out and worn on arrival in Africa. **Juvenile** Paler and duller until moult (Nov–Dec). Juvenile and moulting adult Abyssinian Roller differ in having more extensive white or pale buff on forehead and face, uniformly pale blue wing-coverts and upperside of flight feathers dark blue (not blackish); they appear bluer and more slender, are noisier and have a more agile flight. **HH** Uncommon to rare Palearctic migrant (late Aug–Jun), in open wooded habitats. Singly or in groups. Perches on bare branches, power lines and poles. Flight strong and direct. Red List category: Near Threatened. **Voice** Mostly silent in W Africa. A harsh, short *wrek*, often developing into *wrek-kekekek-erek…* Also a drawn-out *wrèèèèèh*. [CD3:31]

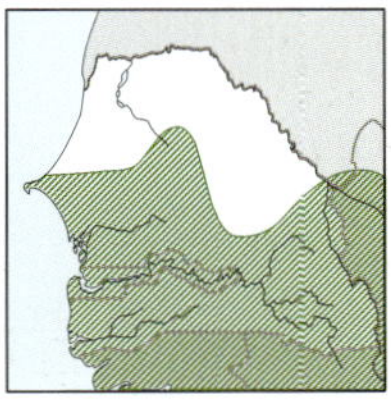

Broad-billed Roller *Eurystomus glaucurus* — 29–30 cm

E. g. afer. **Adult** Relatively small, thickset and dark roller with ***bright yellow, triangular bill***. Chestnut above, deep lilac below, tail and undertail mainly azure-blue. Flight feathers and greater coverts dark blue. **Juvenile** Duskier; underparts mottled brown and bluish; bill partly dusky. **HH** Seasonally common, in various wooded habitats, from forest to dry savanna. Moves north during rains. Perches on tall treetops, occasionally hawking insects high in the air. Singly or in pairs, but assembles in wheeling feeding parties in late afternoon and at dusk. Pointed wings combined with powerful and aerobatic flight may recall falcon. **Voice** Vocal. Sharp, nasal *kek*, sometimes developing into harsh 'laughing' *kekekekekek-k-k-k-k-r-r-r-r*; also variety of other guttural, grating and growling notes, uttered in series. [CD8:7a]

juv
ad
Rufous-crowned Roller
Blue-bellied Roller
ad
ad
Abyssinian Roller
juv
ad
European Roller
juv
ad
Broad-billed Roller

WOOD-HOOPOES Slender birds with mainly dark glossy plumage, slender, decurved bills and long graduated tails. Arboreal and agile, often hanging upside-down while probing bark for insects. Nest in tree holes. Endemic to Africa.

Green Wood-hoopoe *Phoeniculus purpureus* 35–40 cm

P. p. guineensis* & *senegalensis*. Adult** Large savanna wood-hoopoe with all-black plumage glossed violet-green and long, graduated tail ***with white subterminal spots on outer 2–3 rectrices. Double white wingbar conspicuous in flight. Bill bright red or black with red tip. Feet red. **Juvenile** Smaller and duller, with black bill and feet. **HH** Common resident in open wooded habitats. Noisy and gregarious, typically in groups of 3–12 (or more), flying in single file between trees. Striking group display, with flock members perching close together, rocking back and forth, wings partly extended, raising and depressing tail, calling loudly. Breeds cooperatively. **Voice** Single, loud, cracked *whak* or *kuk* (male) and higher-pitched *keek* (female), usually in alarm, and loud, high-pitched cackling, usually started by one and soon accelerating and developing into resonant 'laughter' when other group members join in. [CD8:8b]

Black Scimitarbill *Rhinopomastus aterrimus* 23 cm

R. a. aterrimus*. Adult *Small*,** violet-black savanna wood-hoopoe with ***black bill and feet. Double white wingbar obvious in flight. **Juvenile** Duller, with shorter bill and tail. **HH** Locally fairly common to scarce resident in open wooded habitats. Singly or in pairs in trees and bushes, occasionally low, even on ground. Joins mixed-species flocks. **Voice** A series of plaintive, fairly loud or subdued notes *kwheep-kwheep-kwheep-...* recalling raptor. [CD8:9b]

HOOPOES Very distinctive, cinnamon-buff and black-and-white birds with slender decurved bills, broad rounded wings and erectile, fan-shaped crests. Feed on ground. Highly conspicuous in flight when pattern and irregular flapping of broad wings suggest huge butterfly. Surprisingly easily overlooked when on the ground. Crest not often raised, except on landing and when alarmed. Nest in holes.

Hoopoe *Upupa epops* 25–29 cm; WS 44–48 cm

Two races occur. Palearctic **nominate** is palest (cinnamon-buff or sandy) with broad white band across primaries and four relatively neat white bands across coverts and secondaries. Afrotropical ***senegalensis*** is slightly darker, more cinnamon-rufous, and has more white on secondaries, forming large white patches. **Females** slightly duller and smaller. **Juveniles** duller still, with shorter crest and bill. **HH** Fairly common, in wooded savanna and dry scrub. Forages by walking briskly about, vigorously probing ground with long bill. Palearctic winter visitor *epops* (late Aug–early Apr) overlaps with Afrotropical *senegalensis*; local status and distribution of subspecies inadequately known. **Voice** Low and soft, but far-carrying *hoop-oop-oop* (sometimes 2 or 4 notes). [CD3:32] **Note** African breeding races occasionally treated as separate species (African Hoopoe), but morphological, behavioural and vocal similarities do not suggest specific distinctness.

ad
Green Wood-hoopoe
juv
ad
Black Scimitarbill
ad
ad ♂
senegalensis
ad
epops
Hoopoe

HORNBILLS (Plates 86–87) *Medium-sized to very large, with large decurved bills surmounted by horny casque of variable size. Tail long. Sexes separable by shape and pattern of casque, coloration of bare parts and/or plumage. Juveniles as adults but with smaller bills and no casque. Most species arboreal, frugivorous and insectivorous. Many may make long daily and/or seasonal movements in response to fruit availability. Fly with neck outstretched. Nest in tree holes; entrance sealed from within by female (except in ground hornbills), leaving a narrow slit through which she is fed by her mate throughout incubation and until the young are partially grown. Most species vocal, with distinctive calls.*

(Western) Red-billed Hornbill *Tockus (erythrorhynchus) kempi* 40–48 cm

Adult male Medium-small, black-and-white savanna hornbill with ***slender red bill***. Upperparts blackish-brown with white streak on central back and ***white spots on wing-coverts***. Orbital ring and bare facial skin black. **Adult female** Bill smaller. **Juvenile** Bill duller and smaller still. **HH** Common resident in open savanna woodland and thorn scrub. In pairs or small parties. May congregate in large feeding flocks outside breeding season. Forages mostly on ground. **Voice** A very common sound of the Senegambian bush: a long series of clucking notes, increasing in tempo and volume (as if getting more excited) *uk uk uk uk-uk-uk-UK-UK-UK-UK-UK-uhWUK uhWUK UK-UK-uhWUK...* Several birds may join in excited clucking chorus. [CD8:12b]

African Grey Hornbill *Tockus nasutus* 45–51 cm

T. n. nasutus. Medium-sized, dull greyish savanna hornbill with prominent white supercilium. In flight, dull brownish-grey, with outer tail feathers broadly tipped white. **Adult male** Bill black with triangular creamy patch at base of upper mandible. **Adult female** Bill smaller and tipped dark red; rest of upper mandible pale yellow. **Juvenile** Bill small, entirely dusky-grey. **HH** Common. In pairs or small family parties in savanna woodland. Arboreal. Flight undulating and flapping with glides on broad, rounded wings. Moves north with rains, after breeding. **Voice** Rather melancholy *pee-o*, often uttered in series, and a series of far-carrying, rhythmic, piping notes *pee-pee-pee-pee-PEE-PEE PEE-pyew PEE-pyew PEE-PEE-pyew-pee...* [CD8:15]

GROUND HORNBILLS *Unmistakable, huge, ground-dwelling hornbills, sometimes placed in a separate familiy (Bucorvidae).*

Abyssinian Ground Hornbill *Bucorvus abyssinicus* 90–110 cm

Huge, black, ***terrestrial hornbill***, with long, black bill and stout black legs. Reveals ***all-white primaries in flight***. **Adult male** Bill large with short, high, open-ended casque and reddish patch at base of upper mandible; eye surrounded by blue skin; inflatable red and blue skin on throat. **Adult female** Smaller, with bare skin entirely dark blue. **Juvenile** Dark sooty-brown with smaller bill, undeveloped casque and grey-blue bare skin. **Immature** Gradually as adult. In male, bare skin, incl. around eye, becomes dull pinkish. Adult coloration acquired in c.3 years. **HH** Uncommon and probably declining resident in open grassland. Usually in pairs or small family groups. Walks in search of prey; occasionally flies and perches in trees. **Voice** Deep, far-carrying, booming *uu-uh, uh-uh-uh*, from ground or perch, esp. at dawn. [CD8:10a]

♀
♂
juv
(Western) Red-billed Hornbill
♂
♀
♂
juv
African Grey Hornbill
♂
♂
Abyssinian Ground Hornbill
♀
juv

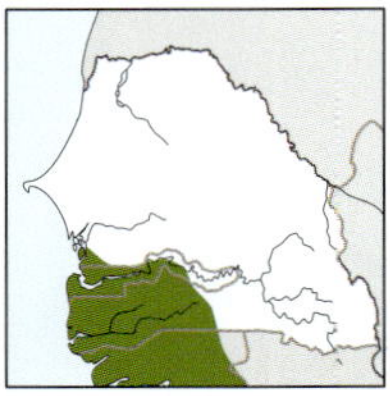

African Pied Hornbill *Tockus fasciatus* 48–55 cm

T. f. semifasciatus*.** **Adult male** Medium-sized, slender hornbill with ***cream-yellow bill tipped black. Black, with lower breast to undertail-coverts white. Outer tail feathers 3 and 4 broadly tipped white. ***Flight flapping and buoyant on broad, rounded wings*.** **Adult female** Smaller, with slightly smaller bill. **Juvenile** Bill smaller and entirely creamy. Similar-sized Piping Hornbill distinguished by darker, shorter and heavier bill, white in wings and fast, slightly undulating flight. **HH** Locally common resident. In pairs or small family parties in forest and various adjacent wooded habitats. May congregate in large feeding flocks outside breeding season. Arboreal. **Voice** Vocal. A series of high-pitched, scolding whistles varying in pitch and tempo *pyi-pyi-pyi-pyi-PYI-PYI-PYI-…*; in display ending with drawn-out *pieeu*. [CD8:13a]

Piping Hornbill *Bycanistes fistulator* 50–60 cm

B. f. fistulator*.** Medium-sized, compact forest hornbill. Wings black with ***broad white tips to secondaries; outer tail feathers broadly tipped white. Upper breast black; rest of underparts white. ***Flight slightly undulating with fast, shallow wingbeats interspersed by glides*.** **Adult male** Grooved bill dusky with pale base and tip. **Adult female** Bill smaller. **Juvenile** Bill entirely dusky and smaller still. Similar-sized Pied Hornbill distinguished by more slender, cream-coloured bill without obvious casque, broad, entirely black wings, and flapping flight. Black-and-white-casqued Hornbill much larger, with much larger casque, more white in wings and tail (latter also differently patterned), and less white on underparts. **HH** Locally common resident, SW Senegal (Casamance). In pairs or family trios in canopy of lowland and gallery forest. May congregate in feeding flocks outside breeding season. Arboreal. Wings make rushing sound in flight. **Voice** Vocal. Distinctive, loud, harsh, nasal laughing. Also shrill piping *peep-peep-peep*. [CD8:17]

Black-and-white-casqued Hornbill *Bycanistes subcylindricus* 65–78 cm

B. s. subcylindricus*.** Large black-and-white forest hornbill with ***blackish bill and two-toned casque*. *Central tail feathers black, rest white with broad black band on centre. Wings black with ***secondaries and inner primaries white*** with black base (outer primaries all black); greater coverts broadly tipped white. Underparts to upper belly black; lower belly, thighs and undertail-coverts white. **Adult male** High casque with pale base. **Adult female** Smaller, with smaller bill and casque. **Juvenile** Smaller, entirely dusky bill without casque. Beware confusion with Piping Hornbill, which is much smaller, with smaller, differently shaped casque, less white in wings, and much more white on underparts. **HH** No certain records. A few claims of singles, W Gambia (possibly confused with Piping Hornbill?). Normally in pairs or small family parties in canopy of mature lowland forest. Favours edges and large clearings; also in adjacent woodland. Flight strong and direct, with loud, swishing wingbeats interspersed by glides. **Voice** Calls loud and nasal, including abrupt *keh!* or *heh!* and a slow series of sarcastic laughing *hah-hah-hah-hah-hah-…* [CD8:18]

Yellow-casqued Hornbill *Ceratogymna elata* 70–90 cm

Very large*, *black forest hornbill with white outer tail feathers. Heavy wingbeats make distinctive swishing and buzzing sound. **Adult male** Bill and lower part of casque blackish; upper part of casque creamy-horn. Bare skin around eye and inflatable bare throat wattles pale blue. **Adult female** Smaller, with rufous head and neck; neck feathers with white bases; bill and casque smaller and all creamy-horn. **Juvenile** Similar to adult female but bill smaller, without casque; no wattles. Young male presumably acquires all-black head at end of first year. **HH** Scarce resident, SW Senegal (Casamance). In pairs or small family parties in canopy of mature lowland forest; often at edges. Penetrates gallery forest and forest patches in savanna. Several may gather at fruiting trees. Seasonal movements related to search of fruiting trees recorded. Red List category: Near Threatened. **Voice** Far-carrying, resonant, plaintive nasal trumpeting. [CD8:22]

♂
♀
juv
African Pied Hornbill
♂
♀
juv
Piping Hornbill
♂
♀
juv
♂
Black-and-white-casqued Hornbill
♀
♂
Yellow-casqued Hornbill

BARBETS AND TINKERBIRDS Stocky birds with large heads and heavy bills. Mostly arboreal and frugivorous, nesting in self-excavated tree holes. Calls characteristic, typically consisting of a monotonously repeated single note.

Yellow-fronted Tinkerbird *Pogoniulus chrysoconus* — 11.5 cm

P. c. chrysoconus. Small barbet with ***golden-yellow forecrown***, black-and-white streaked upperparts and head-sides, and ***lemon-yellow underparts***. Rump and uppertail-coverts pale yellow. **Juvenile** has black forecrown. **HH** Common resident in wooded grassland, savanna and bush. Often in mixed-species flocks. **Voice** One of the most characteristic savanna sounds: a monotonous series of *poops* repeated for long periods without pause. Also series of hoarse *kwèp* notes and *kwrrrr kwrrrr...* trills in variable series. [CD8:32]

Yellow-rumped Tinkerbird *Pogoniulus bilineatus* — 10 cm

P. b. leucolaimus. Small barbet with black upperparts, lemon-yellow rump, ***two conspicuous white facial stripes*** and white throat. Breast pale greyish becoming pale yellowish on belly. **Juvenile** duller, with some yellow-tipped crown and back feathers; bill with pale base. **HH** Very local, Gambia; more widespread and locally common, S Senegal. In various forest and woodland types. Forages in canopy or mid-level. **Voice** A monotonous, far-carrying, rhythmic series of 3–6 (sometimes more) *kok* or *poop* notes, uttered with short pause between each series and endlessly repeated, even during the heat of the day. Also series of *krrw* notes and rattling *kkkkkk*. [CD8:31]

Red-rumped Tinkerbird *Pogoniulus atroflavus* — 13 cm

The only tinkerbird with ***bright red rump***. Has ***three yellow facial stripes***. Underparts yellow, washed olive-grey from breast down. **Juvenile** duller, with olive-tipped crown and back feathers, paler underparts and pale base to bill. **HH** Claims from SW Senegal (Casamance) unproven and highly unlikely. Known range from Guinea east. In lowland forest and clearings. Forages singly at all levels, but usually high. **Voice** Rhythmic series of *poop* calls (occasionally distorted into a nasal *klunk*), resembling those of Yellow-rumped Tinkerbird, but slower (*c.*1/sec), lower pitched and in much longer series. Also series of trills *kirrr kirrr kirrr...* or *krukukkk...* [CD8:29]

Hairy-breasted Barbet *Tricholaema hirsuta* — 16.5 cm

T. h. hirsuta. Stout barbet with black head and throat, and ***bold white supercilium and moustachial stripe***. Upperparts black spotted yellow; underparts greenish-yellow, streaked black on breast, spotted on belly. **Juvenile** has pale bill with black base. **HH** Claims from SW Senegal (Casamance), unproven. Known range from E Sierra Leone east. Frequents forest and forest edge. Singly or in pairs; occasionally gathering in numbers at fruiting trees or joining mixed-species flocks. **Voice** Deep *hoops*, uttered in continuous, rhythmic series at variable speed (most often *c.*1-2/sec, but also much faster). Also accelerating *hoop hoop-hoophoopoopoopoopop*. [CD8:34–35]

Vieillot's Barbet *Lybius vieilloti* — 15 cm

L. v. rubescens **(& *buchanani*?)**. Diagnostic combination of ***red face***, pale yellow underparts and rump, and largely drab greyish-brown upperparts and tail. Red spots from throat to central belly. **Juvenile** duller, with less red and bill paler at base and tip. ***L. v. buchanani*** (possibly N Senegal) somewhat paler, but races poorly defined. **HH** Common resident in savanna woodland and thorn scrub. In pairs or small groups. **Voice** Duet a distinctive rhythmic series of far-carrying, melodious *poop* calls, preceded by chattering or purring notes *kekkekkekkkk-urrrrr poop poop poop-eh-poop epoop poop-eh...* Lone bird may utter single *poop* note; also single *krwèh*. [CD8:38]

Bearded Barbet *Lybius dubius* — 25 cm

Large black-and-red barbet with yellowish bill and legs, ***distinctive 'beard'*** and ***black breast-band***. White patch on lower back conspicuous in flight. **Female** has small black spots on white flanks. **Juvenile** duller, with dull black breast and dusky base to bill. **HH** Common resident in open woodland, wooded grassland and cleared areas with suitable trees (esp. large fig trees). Singly, in pairs or small groups. **Voice** Harsh, low, grating notes (such as *khroaa*), uttered singly or in series. [CD8:44]

Yellow-fronted Tinkerbird

Yellow-rumped Tinkerbird

Red-rumped Tinkerbird

Hairy-breasted Barbet

Vieillot's Barbet

Bearded Barbet

HONEYGUIDES Drab coloured and generally inconspicuous birds. Outer tail feathers white and generally dark tipped, often conspicuous in flight. Feed on wax, insects and fruit and are brood parasites. Hatchlings possess bill hooks with which they pierce the eggs or kill the young of hosts. Hosts and eggs of some species unknown.

Cassin's Honeybird *Prodotiscus insignis* — 11–12 cm

P. i. flavodorsalis. Flycatcher- or warbler-like with ***all-white outer tail*** and ***fine pointed bill***. Yellowish-green above with rump feathers tipped bright yellow; grey below; narrow white eye-ring. White feathers from flanks to rump-sides fluffed out in display, but normally concealed. Tail frequently spread in undulating flight, with white outer feathers conspicuous. **Juvenile** duller, more dusky above, with pale bill and legs. **HH** Status unknown. One claim, Senegal (Missirah, Jun 1974). Nearest known population, Guinea. In canopy or mid-strata of evergreen and gallery forest, frequenting edges, clearings and second growth. Restless, gleaning insects from leaves and branches like warbler and hawking insects from perch like flycatcher. Parasitises flycatchers, warblers, sunbirds and probably white-eyes. **Voice** Distinctive buzzy *tsrrr-tsrrr-...*, uttered in flight.

Wahlberg's Honeybird *Prodotiscus regulus* — 12–13 cm

Resembles Cassin's Honeybird but browner, with ***outer tail feathers broadly tipped dark*** (all white in juvenile). Grey-brown above, grey below becoming whitish on belly. Tail frequently spread in undulating flight, revealing inverted T pattern. **Juvenile** has outer tail feathers entirely white; also paler above and yellower below; gape orangey. **HH** Afrotropical vagrant, Gambia (Kulari Swamp, Sep 2006). Habits as Cassin's Honeybird, but occurs in woodland and bush (not evergreen forest) and forages lower and in smaller trees. Parasitises hole-nesters and species with globular nests. Distribution in W Africa poorly known. **Voice** Buzzy *tsrrr-tsrrr-...*, uttered in flight; also high-pitched *tsip* notes. Song a long, buzzy *tsrrrrrrrrrrrrrrrrrrrrrr...* [CD8:50]

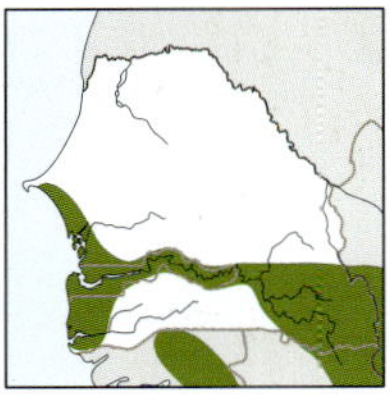

Lesser Honeyguide *Indicator minor* — 14–15 cm

I. m. senegalensis. Inconspicuous, with stubby bill, grey head, usually with dusky submoustachial stripe and sometimes white on lores. Upperparts olive-green with blackish feather centres and dull greenish-yellow fringes producing rather indistinct streaking. Underparts grey becoming whitish on throat and belly; flanks have some dusky streaks. Bill dark; base of lower mandible pinkish. **Juvenile** is greener above and greyer below; loral and submoustachial marks faint or lacking; throat streaked dusky. **HH** Uncommon and local in various types of woodland. Gleans foliage and branches; catches insects in flight. Frequents beehives. Displays include bounding, circling flight often with fanned tail and 'drumming' sound. Parasitises hole-nesters or species with deep nests, incl. bee-eaters, kingfishers, barbets, woodpeckers, swallows, starlings and others. **Voice** Song, uttered from traditional perches, a far-carrying, deliberate, rhythmic series of 10–30 identical notes *wrip wrip wrip wrip...* introduced by faint *pee-yew* or *tyeew*; rhythm variable. [CD8:56]

Spotted Honeyguide *Indicator maculatus* — 16.5–17.5 cm

I. m. maculatus. Distinctive but unobtrusive honeyguide with ***olive underparts spotted creamy-green*** on breast and upper belly, paler and streaked on throat and undertail-coverts. **Juvenile** has speckled forehead; spotting below more extensive and distinct. **HH** Scarce and local. Usually in evergreen and gallery forest. At all levels, from canopy to undergrowth, even on ground. Inconspicuous. Hosts unknown, Buff-spotted Woodpecker suspected. **Voice** Purring *brrrrrrrr...* Also an extended raptor-like mewing *peeeoo-peeeoo* uttered at dawn. [CD8:53]

Greater Honeyguide *Indicator indicator* — 18.0–19.5 cm

Stout with tail pattern typical of genus. **Adult male** Distinctive head pattern with ***whitish ear patch, black throat*** and ***pink bill***. Upperparts dark grey-brown with white edges to wing-coverts and white streaks on rump and uppertail-coverts. Breast greyish-white, becoming white on rest of underparts. **Adult female** Duller than male and lacking its head pattern; rather bulbul-like. Bill dark. **Juvenile** Top of head and upperparts brown with darker mask, ***throat and breast pale lemon to orange-yellow***, bill dark. **HH** Fairly common in various wooded habitats. May guide humans to bee's nests by fluttering from tree to tree, uttering chattering call. Flight undulating, conspicuously revealing white outer tail feathers. Occasionally joins mixed-species flocks. Parasitises hole-nesters or species with deep nests, incl. bee-eaters, kingfishers, hoopoes, wood-hoopoes, woodpeckers, swallows, starlings and others. **Voice** Distinctive, far-carrying series of *WHIT-birr WHIT-birr...* ending with single *WHIT*. 'Guiding' call a loud, fast chattering. [CD8:55]

Cassin's Honeybird
Wahlberg's Honeybird
juv
Lesser Honeyguide
ad
Spotted Honeyguide
juv
♀
♂
Greater Honeyguide

WOODPECKERS (Plctes 90–91) Small to large birds with strong, straight, pointed bills. Characteristically cling to trees, using stiff tail as prop. Large.y insectivorous. Often in pairs. Sexes differ in head pattern, males usually with (more) red on head. Nest in self-excavated holes in trees or termitaria. Flight strong and bounding, with brief bursts of flapping interspersed by glides on completely folded wings. All are resident.

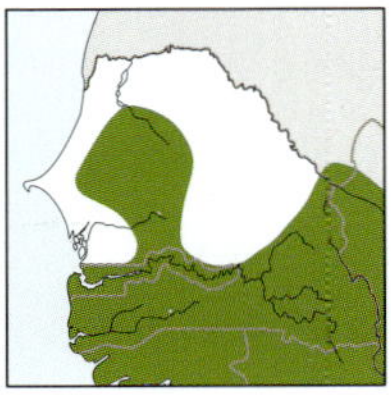

Cardinal Woodpecker *Dendropicos fuscescens* 14–15 cm

D. f. lafresnayi. Small woodpecker with green upperparts, faintly barred mantle and streaked underparts. Head-sides and throat whitish with some faint dark streaking and narrow malar stripe; rest of underparts pale yellowish lightly streaked blackish; some orange on uppertail-coverts. **Adult male** Forehead and forecrown brownish; hindcrown and nape red. **Adult female** Forehead to nape brownish-black. **Juvenile** Dull version of adult, with red crown (brighter in male) and black nape. **HH** Fairly common. Usually in pairs, in various types of woodland. Occasionally in mixed-species flocks. **Voice** Rapid series of harsh, rattling notes *kree-kree-kree-kree...* Other rapidly uttered notes include *kwik-ik-ik-ik...* [CD8:75]

Fine-spotted Woodpecker *Campethera punctuligera* 21 cm

C. p. punctuligera. Medium-sized woodpecker with green upperparts and ***pale yellow underparts finely spotted black***, most densely on breast (tiny spots may be hard to see in field). Head-sides and throat white. Tail golden-yellow barred brown. **Adult male** Forehead to nape bright red; ***red moustachial stripe***. **Adult female** Forecrown black speckled white; indistinct, speckled moustachial stripe. **Juvenile** Forecrown plain black; moustachial stripe blacker than in adult female. **HH** Fairly common. Singly, in pairs or small family groups in open woodland, foraging on trunks and branches, regularly descending to base of tree or ground. Joins mixed-species flocks. **Voice** A ringing series, often starting with a few *kip* calls: *kip-kip-kip-kieeh-kieeh-kieeh*, often uttered simultaneously by pair. [CD8:63]

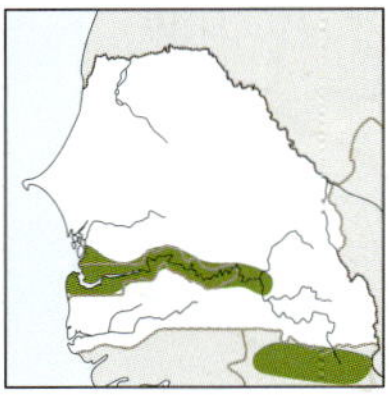

Golden-tailed Woodpecker *Campethera abingoni* 20 cm

C. a. chrysura. Medium-sized woodpecker with dull green upperparts ***finely spotted and barred yellowish-white***, and pale yellowish underparts ***boldly streaked blackish***, most densely on breast. Head-sides and throat white streaked blackish. Tail deep yellow with broad brownish bars and golden-yellow shafts. **Adult male** Crown and moustachial stripe dusky-red; nape red. **Adult female** Forecrown and moustachial stripe black speckled white. **Juvenile** As adult female but duller and more heavily streaked below. **HH** Scarce (possibly overlooked). Singly or in pairs in gallery forest and woodland. Rather unobtrusive. Sometimes joins mixed-species flocks. **Voice** Single, drawn-out *k-heeeeew*. Drums softly and infrequently. [CD8:66]

Grey Woodpecker *Dendropicos goertae* *c.*20 cm

D. g. goertae. Fairly large woodpecker with diagnostic ***grey head and underparts***, golden-olive upperparts, ***red rump***, and orangey streak on belly. **Adult male** Hindcrown and nape red. **Adult female** Head plain grey, lacking any red. **Juvenile** As adult but duller; both sexes have some red on crown. **HH** Fairly common. In various types of woodland, gallery forest and forest edge. Avoids closed forest. Usually in pairs or small family parties. Forages at all levels. **Voice** Shrill, loud *krreet-krreet-krreet*, descending in volume and sometimes preceded by *kik-kik-kik-*; fast and hard *kwik-wik-wik-wik; kee-krrirrt-krrirrt...* and variations. [CD8:80]

♀
♂
Cardinal Woodpecker
juv
♂
♀
Fine-spotted Woodpecker
♀
♀
♂
♂
Golden-tailed Woodpecker
Grey Woodpecker

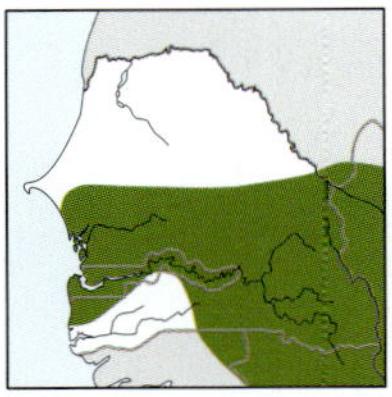

Brown-backed Woodpecker *Dendropicos obsoletus* 13–14 cm

D. o. obsoletus. Small, ***mainly greyish-brown*** woodpecker with ***plain back and white-spotted wings and tail***. Grey-brown ear-coverts surrounded by white. Underparts faintly streaked brownish. **Adult male** Hindcrown bright red. **Adult female** Entire crown grey-brown. **Juvenile** As adult; both sexes have some red on nape. **HH** Fairly common. Usually in pairs in woodland. Often joins mixed-species flocks. **Voice** Rapid *ki-ki-keeew-keeew-keeew* and *krreet-krreet-krreet*. [CD8:82]

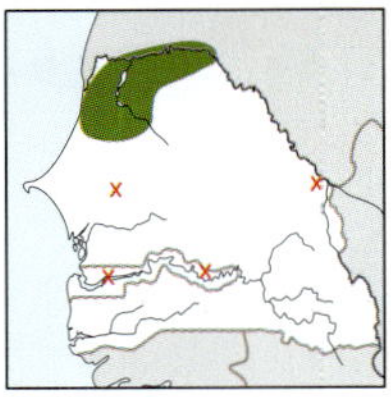

Little Grey Woodpecker *Dendropicos elachus* 11–13 cm

Very small, rather 'scruffy' looking, ***pale greyish-brown*** woodpecker with conspicuous ***red rump*** and white-barred back and wings. Greyish-brown ear-coverts surrounded by white line. Underparts whitish with pale brown spotting and barring. Head markings fade. **Adult male** has red hindcrown and nape. **Adult female** lacks red. **Juvenile** As adult female, but even duller. **HH** Uncommon to scarce resident, N Senegal. A few old records from Gambia. Relatively little known. Occurs in arid areas with small, scattered trees. **Voice** Sharp, very rapid rattling *kree-kree-kree-kree-...* similar to Cardinal Woodpecker, but harder and faster. [CD8:72]

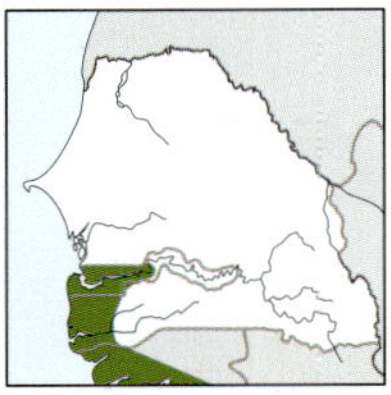

Buff-spotted Woodpecker *Campethera nivosa* 14–16 cm

C. n. nivosa. Small, ***dark olive*** woodpecker, with ***underparts densely spotted yellowish-white*** (may appear barred). Head-sides and throat densely streaked olive-brown on dirty white. Tail blackish. **Adult male** Red patch on nape. **Adult female** No red nape patch. **Juvenile** Similar to adult female, with browner, more barred breast and greyer top of head. **HH** Uncommon and local. Singly or in pairs, quietly searching branches and vines at lower levels of lowland and gallery forest. Frequently joins mixed-species flocks. Nest usually excavated in ant nests and termitaria. **Voice** Mostly silent. Occasionally a soft *peeer*. [CD8:70]

Little Green Woodpecker *Campethera maculosa* *c.* 16 cm

Medium-sized forest woodpecker with ***pale yellowish-green underparts boldly barred dark***. Head-sides and throat buffish with fine dark specks. Upperparts plain dark golden-olive. Tail black. **Adult male** Forehead and crown dusky-red, nape red. **Adult female** Crown black finely barred white. **Juvenile** As adult female but upperparts greener, underparts paler and more irregularly barred. **HH** Scarce, S Senegal (Casamance); rare (vagrant?), Gambia (Tendaba). Singly or in pairs in lowland forest, clearings and edges. **Voice** Drawn-out, plaintive *huweeeeh*, and shorter, harsh *whee* and hard *kewik*. [CD8:67]

WRYNECKS *Woodpecker relatives with cryptic plumage, elongated shape and soft tail, which is not used as brace when clinging to trees. Sexes similar. Feed mainly on ants.*

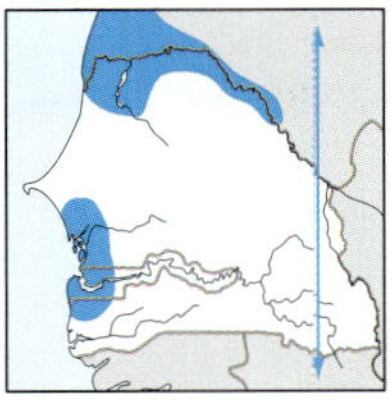

Eurasian Wryneck *Jynx torquilla* 16–18 cm

J. t. torquilla. Cryptically patterned, bark-like plumage. Dark stripes from eye onto neck-sides, from centre of crown onto back and on edge of scapulars. Densely barred underparts buffish-cream on throat, becoming whitish on lower breast and belly. Bill rather short and pointed. **HH** Uncommon to scarce Palearctic migrant (Nov–Apr). Singly, in open habitats. Unobtrusive and easily overlooked. Often feeds on ground. Perches on branches; also clings to tree like woodpecker. Shallow bounding flight, with short series of wingbeats interspersed by glides on closed wings. **Voice** Usually silent in winter quarters; occasionally a series of loud *kièh-kièh-kièh-...*, recalling small falcon. [CD3:33]

♀
♂
Brown-backed Woodpecker
♀
♂
Little Grey Woodpecker
♀
♂
Buff-spotted Woodpecker
♀
♂
Little Green Woodpecker
Eurasian Wryneck

LARKS (Plates 92–93) Terrestrial birds with cryptic plumage pattern and coloration, often varying within a species to match dominant shade of soil. Fresh and worn plumage often quite different: with wear, pale feather tips and fringes abrade, leaving only dark centres, resulting in more uniform plumage. Sexes similar, except in sparrow larks. Juveniles similar to adults (or adult females) but feathers of upperparts often have pale fringes, giving a spotted or scaly appearance; this plumage moulted immediately after fledging. Walk or run on the ground.

Singing Bush Lark *Mirafra cantillans* 13 cm

M. c. chadensis. Small, sandy-coloured lark with ***pale, stubby bill*** and ***white outer tail feathers*** (conspicuous on landing). Streaked dusky above, whitish below with faintly streaked breast washed buff; creamy-white supercilium. Flappet Lark is darker and more streaked above, with rufous on tail and darker, less stubby bill. **HH** Locally fairly common to common wet-season breeder, N Senegal; scarce and local, mainly dry-season visitor, Gambia. In dry open grassland with scattered bushes. **Voice** Rapid medley of short phrases of varied notes; same phrase often repeated several times, from ground or low perch, or in fluttering display flight. [CD8:88a]

Kordofan Lark *Mirafra cordofanica* 14 cm

Small, pale sandy-rufous lark with distinctive ***tricoloured tail pattern*** (rufous, black and white). Stout whitish bill. When fresh, feathers of upperparts fringed buff with narrow blackish subterminal crescents. Underparts whitish, with pale cinnamon-rufous wash and faint dusky cinnamon speckles on breast. Singing Bush Lark has less cinnamon plumage with shorter, sandy grey-brown, dark brown and white tail, lacking black and rufous. **HH** Rare intra-African migrant, N Senegal. In open arid areas with red sandy soil (matching plumage) and low, scattered scrub. **Voice** Series of short, varied phrases with trills, chirps, whistled notes and imitations (phrases are not repeated, unlike e.g. Singing Bush Lark); from ground, bush tops or in high display flight. [CD8:88b]

Flappet Lark *Mirafra rufocinnamomea* 13–14 cm

M. r. buckleyi. Small lark ***with diagnostic wing-clapping display flight***. Cinnamon-brown streaked blackish above; pale buffish supercilium. Throat whitish or buff; rest of underparts orange-buff, indistinctly streaked blackish and rusty on breast. Bill bicoloured. In flight, shows rufous wing patch. **HH** Status unclear. Probably scarce resident of Gambia River drainage system in both countries. Singly or in pairs, in various grassy habitats. **Voice** Most characteristic sound a dry, rattling *prrrrrrp*, produced by rapid wing-flaps during undulating display flight. [CD8:93]

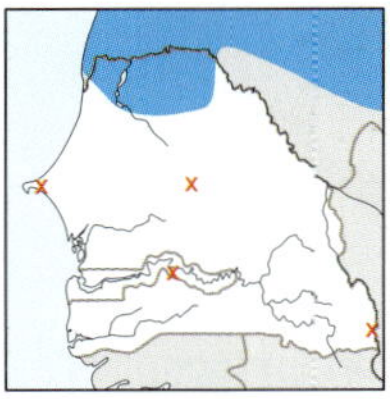

Greater Short-toed Lark *Calandrella brachydactyla* 14–16 cm

C. b. brachydactyla/rubiginosa. Pale lark with ***streaked upperparts, broad pale supercilium*** and ***long tertials almost completely covering primaries***. ***Underparts whitish and unstreaked, with variable narrow, dark crescent-shaped patch on breast-sides***; breast variably washed pale buff and occasionally with some faint streaks. Tail feathers blackish, broadly edged cinnamon on central pair, mainly white on outer pair. **HH** Locally common Palearctic winter visitor (Sep–Apr), N Senegal; vagrant, Gambia. In open arid and semi-arid country. Often in flocks. **Voice** Hard, dry *prrt, prrt-trrt*. Song in wintering grounds a weak jumble of harsh and buzzy notes. [CD3:42]

Sun Lark *Galerida modesta* 14 cm

G. m. modesta. Small, sandy-rufous lark with ***prominent supercilium surrounding streaked cap***, and ***heavily streaked upperparts and breast***. Tail blackish edged pale rufous. Crown feathers may be raised to form short crest. **HH** Uncommon to locally fairly common, Senegal; scarce, Gambia (first record in 1994). Singly or in pairs in open grassy habitats, often with rocky areas or lateritic soils. In small flocks outside breeding season. **Voice** Short, fast, grating warble; from ground or bush, or in flight. More varied and sustained song in hovering display flight. [CD8:97]

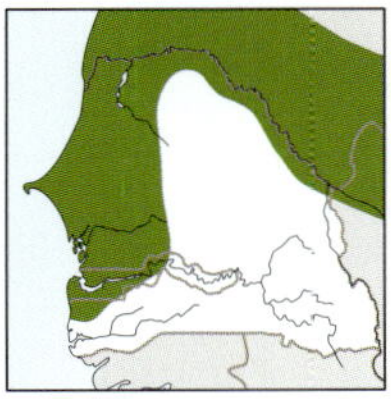

Crested Lark *Galerida cristata* 17 cm

G. c. senegallensis. Rather stocky with ***distinct spiky crest***, relatively short tail, and slightly decurved, longish bill. Pale grey-brown above, off-white below with streaked breast. **HH** Common along entire coast, becoming less common inland. In pairs or family parties in various open habitats. **Voice** Liquid *doo-leeoo* and *diui*. Song varied and sustained, with clear whistles and throaty twitters. Also fast and clear *tee-titee-titiu*. From perch or ground and in flight. [CD3:46]

Singing Bush Lark

Kordofan Lark

Flappet Lark

Greater Short-toed Lark

Sun Lark

Crested Lark

Rufous-rumped Lark *Pinarocorys erythropygia* 18 cm

Distinctive, large, slender lark with ***well-marked head pattern***, ***rufous rump*** contrasting with dark upperparts, long rufous-edged tail and heavily marked breast. **HH** No certain records. Claimed in Gambia from unusual habitat (Bund Road, 1961). Singly, in open savanna woodland and farmland, preferring recently burnt areas. Frequently perches in trees. **Voice** Short, often repeated series of clear, far-carrying whistles with some variation, in high display flight with dangling legs. [CD8:94]

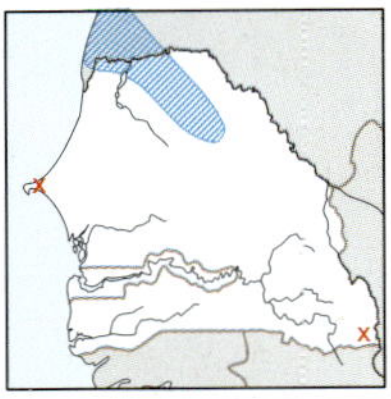

Greater Hoopoe Lark *Alaemon alaudipes* 19–22 cm

A. a. alaudipes. Large, elongated, pale sandy lark with ***long, black, decurved bill and long grey-white legs***. In flight, ***conspicuous black-and-white wing pattern*** diagnostic. **Adult female** Similar, but smaller. **Juvenile** Upperparts browner, feathers have buff tips and fringes, and dusky subterminal bars. **HH** Scarce non-breeding migrant from Sahara, N Senegal. **Voice** Piping, accelerating and ascending series of pure drawn-out whistles, occasionally with short buzzy trill. [CD3:37]

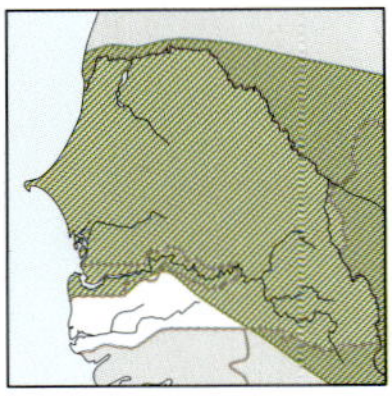

Chestnut-backed Sparrow Lark *Eremopterix leucotis* 12 cm

E. l. melanocephalus. Small dark lark with distinctive male plumage. **Adult male** ***Black head with large white patch on ear-coverts and small white half-collar on nape***. ***Upperparts rich chestnut***. ***Underparts and underwing black*** with whitish flanks. Bill pale horn. **Adult female** Much duller and variable, with dusky mottled head and pale buff underparts, mottled brownish on breast. **Juvenile** Similar to adult female, but feathers of head and upperparts broadly tipped white. **HH** Common dry-season breeding visitor. In pairs in various open areas with bare soil and short grass or scrubs; partial to recently burnt ground. Often in small flocks when not breeding. Erratic. **Voice** Rapid phrase with harsh, grating notes, from ground or low perch and in flight. [CD8:98]

Black-crowned Sparrow Lark *Eremopterix nigriceps* 11–12 cm

E. n. albifrons. Small, compact lark with distinctive male plumage and, in both sexes, ***black underwing-coverts*** and black outer tail-feathers. **Adult male** Bold pied head pattern with ***white forehead***. Upperparts pale grey-brown; underparts black. **Adult female** Pale sandy-cinnamon above, faintly streaked on crown; patch around eye and neck-sides whitish. Below whitish with pale cinnamon breast-band. **Juvenile** Similar to adult female, but feathers of head and upperparts broadly tipped pale buff with faint narrow dark subterminal bar; breast diffusely spotted dusky. **HH** Uncommon to locally common resident in dry open areas with sparse vegetation, N Senegal; vagrant, Gambia (Kaur, Jan 2006). Appears to be spreading southwards. In pairs or, when not breeding, in small flocks. **Voice** Abrupt, buzzing *eezp* or *jeep*; also a quiet *djib*. Song, usually given in flight, a pleasant, simple, rapid phrase of high, clear whistles *chiWEE-chreep* or *chichiWEE-chireep*, and variations. [CD8:99]

Temminck's Horned Lark *Eremophila bilopha* 14–15 cm

Pale sandy-coloured lark with ***distinctive, contrasting adult face pattern***. **Adult male** Crown, nape and upperparts pale orange-sandy or sandy-cinnamon; ***narrow black band on forecrown ending in delicate 'horns' on sides***; forehead, head-sides and underparts white with black patch from lores to cheeks and black breast-band; breast-sides and flanks washed orange-sandy. Tail blackish with white edges and pale orange-sandy centre. **Adult female** Black areas often duller and slightly reduced; horns shorter. **HH** Palearctic vagrant, Gambia (Fajara, Western Division, Feb 2002). **Voice** Calls *tsip*, *seeuu* and *sweeeup*. [CD3:52]

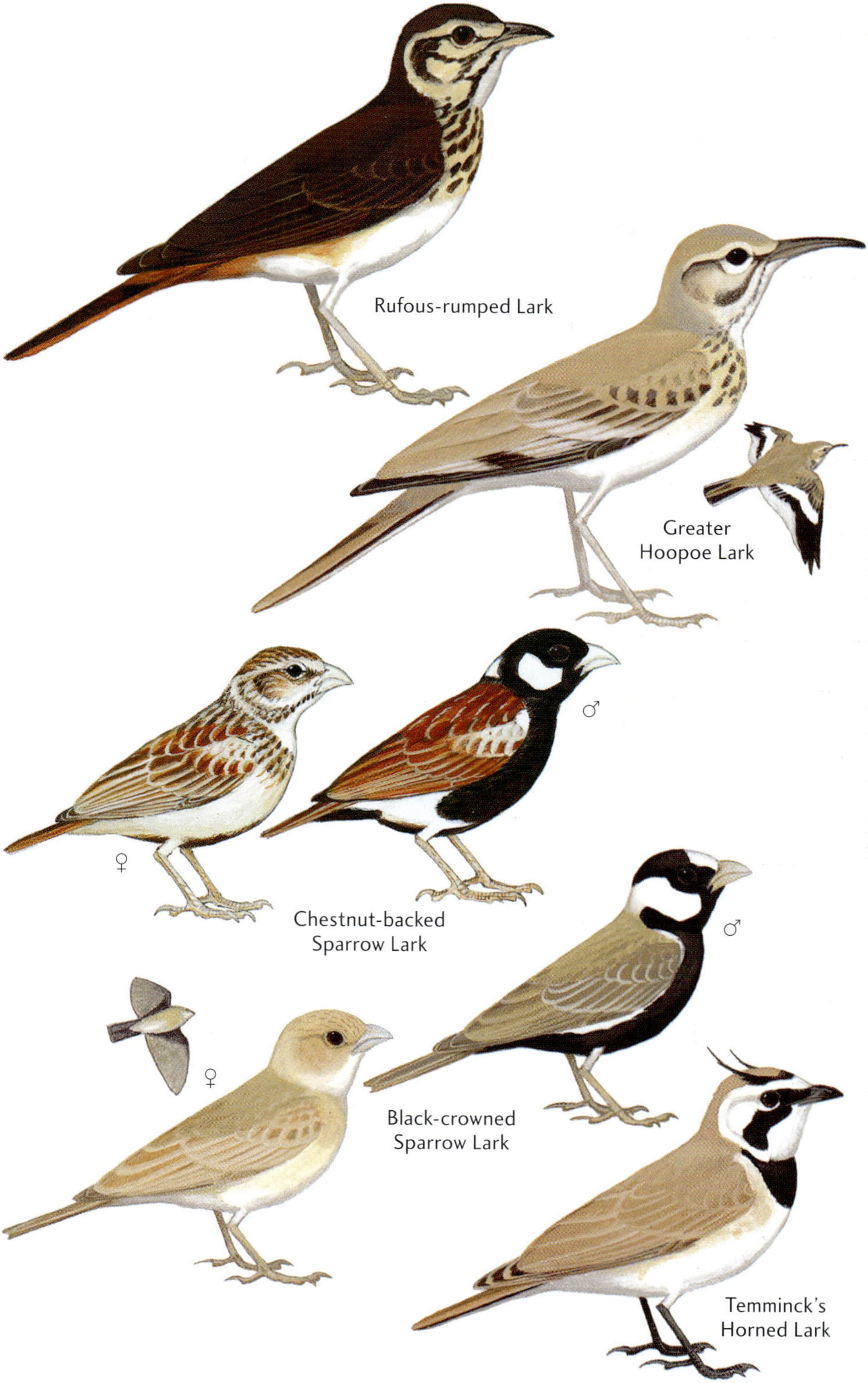
Rufous-rumped Lark
Greater Hoopoe Lark
♂
♀
Chestnut-backed Sparrow Lark
♂
♀
Black-crowned Sparrow Lark
Temminck's Horned Lark

PLATE 94: SWALLOWS I

SWALLOWS (Plates 94–96) *Distinctive, highly specialised aerial insectivores with slender body, short neck and pointed wings. Bill very short, but gape broad; legs short. In some, tail deeply forked with long streamers, often with white 'windows' (white patches on inner webs of feathers). Sexes similar; in long-tailed species, females have shorter outer tail feathers. Juveniles similar but duller. Regularly perch on wires and bare branches. Most gregarious when not breeding (some also when breeding), often forming mixed groups with other hirundines. Many are migratory.*

Fanti Saw-wing *Psalidoprocne obscura* — 17 cm

Adult Small, ***all-blackish*** swallow with ***very long, strongly forked tail*** (tail fork male 5.0–7.5 cm, female 3.0–6.5 cm). Greenish gloss visible in favourable circumstances. **Juvenile** Browner; tail shorter. **HH** Fairly common and possibly increasing but local and partially migratory, with numbers augmented in wet season. In pairs or small flocks in forest clearings, along tracks and edges, in farmbush, woodland and savanna. Flight slow and fluttering. Pairs breed solitarily, in burrow excavated in earth bank or similar site. **Voice** Mainly silent. Call a soft *sheep*. [CD9:3]

MARTINS *Small to medium-sized hirundines with square or slightly forked tails. Most have brown upperparts and brown or white underparts.*

Plain Martin *Riparia paludicola* — 11–12 cm

R. p. paludibula. Small martin with ***brownish throat and breast***. Head, upperparts, tail and underwing brown. Throat and breast greyish-brown becoming white on belly and undertail-coverts. Tail notched, almost square. **HH** Scarce to rare non-breeding visitor. Usually in small flocks, over or near rivers and lakes, but outside breeding season also away from water. Occasionally singly. Flight weak and fluttering. Associates with other hirundines. Dry-season breeder, nesting in burrows dug in sand banks, generally in small colonies. **Voice** Calls include rasping *chtrrr* and harsh *steeh*. Song a soft twitter. [CD3:53]

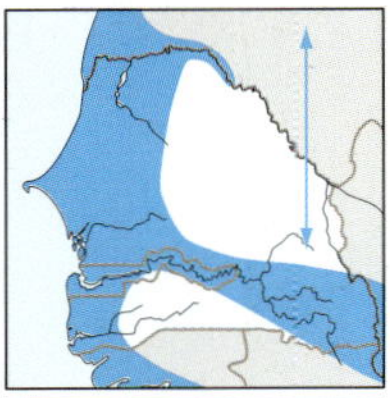

Common Sand Martin *Riparia riparia* — 12–13 cm

R. r. riparia. Small martin with ***pure white underparts*** and ***well-defined brown breast-band***. Brown above; tail slightly forked. **HH** Common Palearctic migrant (Sep–May). Over various open habitats. Gregarious; often in large flocks and associating with other hirundines. **Voice** Dry, rasping *chrrp*. [CD3:54]

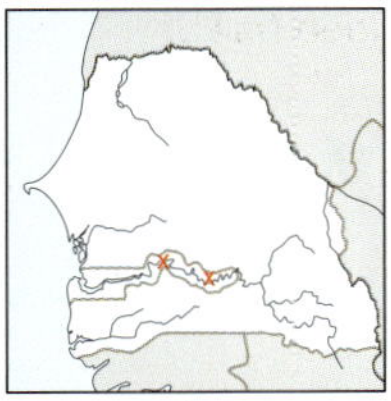

Banded Martin *Riparia cincta* — 17 cm

R. c. cincta (?). Conspicuously ***large*** martin with ***well-defined breast-band*** and square tail. Resembles giant Common Sand Martin. Brown above, white below, including underwing-coverts; short white supercilium. **HH** Afrotropical vagrant, Gambia. Singly or in small flocks. Forages low over grassland, with slow and erratic, tern-like flight. Associates with other hirundines. **Voice** Mostly silent. Call *chrip*. [CD9:6]

Rock Martin *Ptyonoprogne fuligula* — 12–13 cm

P. f. bansoensis. Small, compact, ***all-brown*** hirundine (dark brown above, paler below) with ***two small white patches in almost square tail*** (conspicuous when tail spread). **HH** Rare, probably non-breeding visitor, SE Senegal; uncertain record, Gambia. Normally associated with inselbergs, rocky outcrops, cliffs and gorges, also villages and towns. Flight slow, with much gliding. Breeds in pairs or small colonies; nest a mud cup attached to rocky surface, bridge or similar place. **Voice** Mostly silent. Calls include *wik* and a high-pitched *sree*. Song a soft twitter. [CD3:56]

Crag Martin *Ptyonoprogne rupestris* — 14–15 cm

Stocky, brown hirundine with white patches in slightly notched tail and ***blackish-brown underwing-coverts contrasting strongly with grey-brown flight feathers***. Grey-brown above; pale buffish below, brownish on lower abdomen. Rock Martin has similar tail pattern, but less contrasting underwing-coverts. **HH** Rare Palearctic winter visitor, N Senegal; vagrant, Gambia (one record, Central River Division, Nov). Normally associated with cliffs and gorges, but on migration also over low, open areas. Flight slow, with frequent gliding. **Voice** Mostly silent. Calls include a soft *chrp* or *prrt*. [CD3:57]

ad
Fanti Saw-wing
ad
Plain Martin
ad
Common Sand Martin
ad
Banded Martin
ad
Rock Martin
ad
Crag Martin

PLATE 95: SWALLOWS II

SWALLOWS *Swallows of the genus* Cecropis *have glossy blue upperparts and long tail streamers as those in* Hirundo *with which they are often merged, but the rump is contrastingly rufous. Sexes similar, but females have shorter tail streamers. Juveniles are duller and have even shorter tails.*

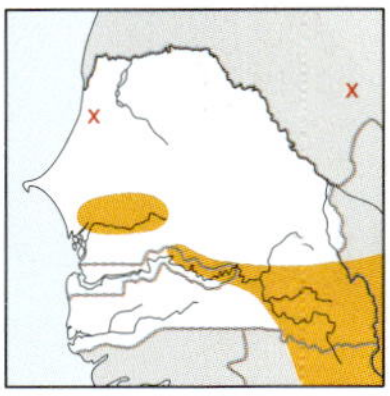

Lesser Striped Swallow *Cecropis abyssinica* — 15–19 cm

C. a. puella. The only rufous-rumped swallow with ***rufous head*** and ***heavily streaked underparts***. Tail has long outer feathers and small white patches. **HH** Uncommon wet-season breeding visitor, Senegal; rare, Gambia. Singly, in pairs or small flocks in various open wooded habitats; also forest edges; occasionally mangroves. Forages low or at mid-height, often over water. Pairs usually breed solitarily; mud nest bowl shaped with tunnel entrance plastered to overhanging surface of natural or artificial sites. **Voice** Wheezy *cheeew*. Song varied and nasal, including liquid, squeaky, buzzy and gurgling notes. [CD9:10]

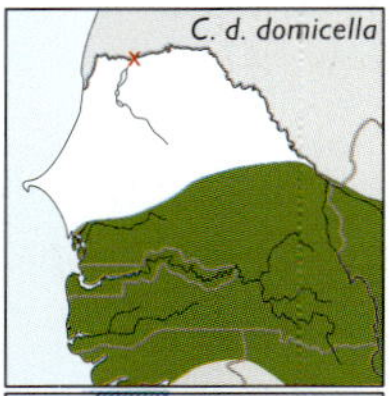

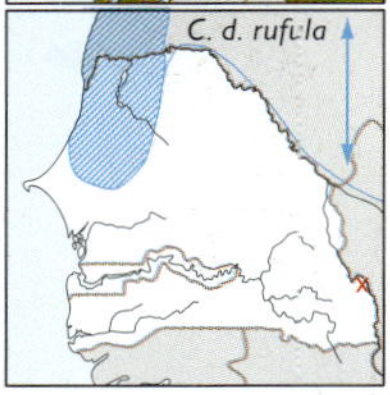

Red-rumped Swallow *Cecropis daurica* — 16–17 cm

Medium-sized rufous-rumped swallow with ***dark blue cap***, ***rufous collar***, long, often inwardly curving tail streamers, and ***diagnostic clear-cut black undertail-coverts***. No white in tail. Resident and partial intra-African migrant race ***domicella*** (West African Swallow) has plain orange-rufous rump, creamy-white underparts, and glossy dark blue crown, bordered by rufous forehead and narrow supercilium. Palearctic ***rufula*** has pale forehead (dark blue of crown not reaching bill), ***two-toned rump*** (lower rump paler), and pale creamy-buff underparts with faint dark streaks. Rufous and buff parts of all forms pale with wear. **HH** *C. d. domicella* is fairly common south of 14°N, scarce north to c.15°N. *C. d. rufula* is uncommon passage migrant, N Senegal (Oct–Nov & Feb–May). Singly, in pairs or small flocks in various types of savanna. Pairs breed solitarily or in small, loose colonies; mud nest bowl shaped with tunnel entrance plastered against variety of natural or artificial sites. **Voice** Soft *djuit* or *chreet*. Song a varied, quiet twittering with nasal notes and rasping trills. [CD3:55]

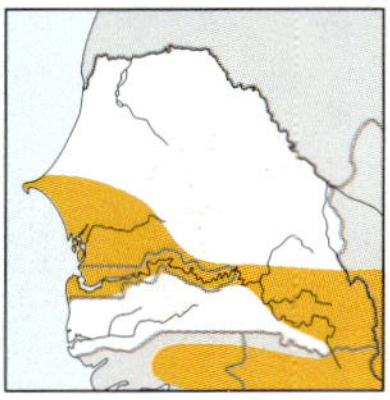

Rufous-chested Swallow *Cecropis semirufa* — 18–21 cm

C. s. gordoni. Large, dark blue-and-rufous swallow with ***dark blue hood***, incomplete collar, orange-rufous underparts and ***very long tail streamers***. Top of head to below eye and onto ear-coverts glossy dark blue connecting, on hindneck, with similarly coloured upperparts. Has rufous underwing- and undertail-coverts, and white patches in tail. **HH** Uncommon to fairly common wet-season breeding visitor. In pairs or small flocks in farmbush, degraded savanna, grassland, villages and towns. Flight slow and buoyant on broad wings, with much gliding. Forages at all heights. Pairs breed solitarily; mud nest is bowl shaped with tunnel entrance plastered to overhanging rock, human habitation, hollow tree or similar place. **Voice** Gurgling *trlrrrrr* or *chip-chip-chleeeeurrrr*; also soft *dee-uuuh*, short *chip* and high-pitched *weet-weet*. [CD9:8]

Mosque Swallow *Cecropis senegalensis* — 21–23 cm

C. s. senegalensis. Resembles Rufous-chested Swallow but is even larger and heavier, with pale head-sides and rufous nuchal collar surrounding dark blue cap, shorter tail streamers, no white in tail and usually whiter underwing-coverts. **HH** Uncommon to locally fairly common; more numerous in wet season, when breeding. Singly, in pairs or small flocks in various open habitats; frequently around villages and towns. Flight slow, with much gliding; sometimes recalling small falcon. Typically forages high over treetops. Pairs breed solitarily; mud nest bowl shaped with tunnel entrance plastered against overhanging surface of natural or artificial sites. **Voice** Nasal *nyaa*; a distinctive, piping, reedy note, recalling sound of tiny trumpet, and various guttural notes. Song consists of various rather slow, short series of nasal, whining and creaking sounds. [CD9:9]

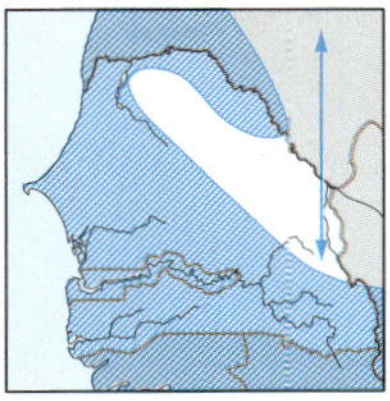

Common House Martin *Delichon urbicum* — 13–15 cm

D. u. urbicum. Small, rather compact martin with ***conspicuous white rump***. **Adult breeding** (from Apr) Head, upperparts and tail glossy blue-black; rump and entire underparts white; underwing with pale buff-grey underwing-coverts and dark silvery flight feathers. Tail forked. **Adult non-breeding** (from Aug) White of rump, throat and flanks often mottled pale buff-brown. **Immature** Duller and browner. **HH** Fairly common but irregular Palearctic migrant (Oct–Jun); most numerous on passage. In small or large flocks over various open habitats. Flight fluttering with much gliding. Forages at all heights, often very high. **Voice** Hard *prrt prrpt*. [CD3:59]

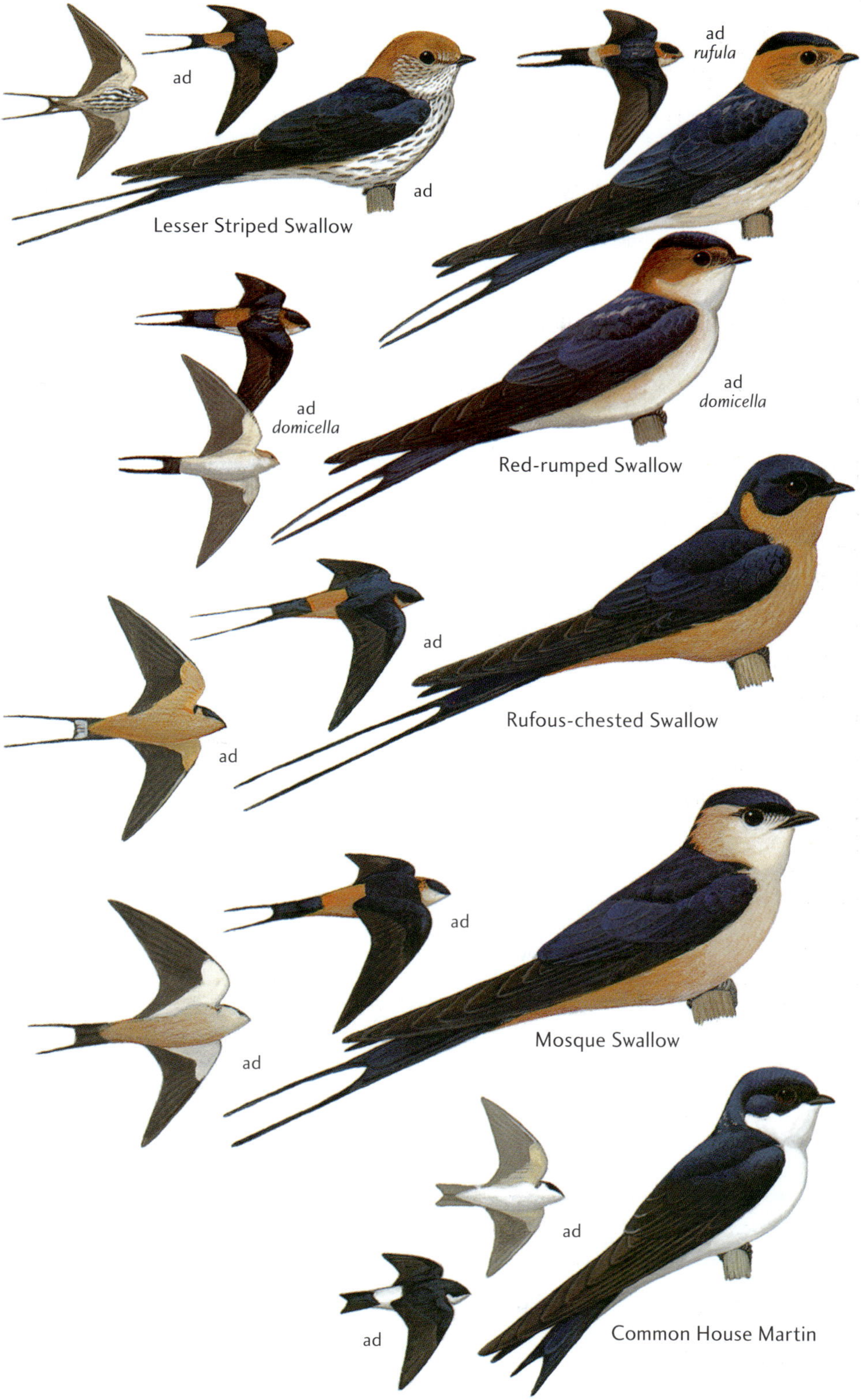

ad
ad
Lesser Striped Swallow
ad
rufula
ad
domicella
ad
domicella
Red-rumped Swallow
ad
ad
Rufous-chested Swallow
ad
ad
Mosque Swallow
ad
ad
Common House Martin

Grey-rumped Swallow *Pseudhirundo griseopyga* 14 cm

P. g. melbina. Small, slim swallow with deeply forked tail and ***greyish rump contrasting with dark blue upperparts***. Top of head to nape brown-grey; underparts white. **HH** Rare non-breeding Afrotropical visitor (Nov–Feb). Patchily distributed; movements inadequately known. Singly, in pairs or small flocks in grassland, wooded savanna and similar open areas. Usually forages low. **Voice** Mostly silent. Soft, nasal *chwèèp*. [CD9:7]

Swallows of the genus Hirundo *have entirely glossy blue upperparts and most have long tail streamers. Construct mud nests plastered under overhanging surface or on ledge.*

Pied-winged Swallow *Hirundo leucosoma* 12 cm

The only hirundine with ***long white patches on upper wing***. Glossy steel-blue above, pure white below; forked tail with white patches. **HH** Uncommon resident; local movements reported. Singly or in pairs, occasionally in small flocks, over wooded savanna and farmbush. Flight fast with much banking and turning. Forages low, often flying between trees and bushes, like saw-wings. Associates with other hirundines. Pairs breed solitarily. Mud nest an open cup, plastered to surface of artificial site. **Voice** Mostly silent. Call a low *chut*. [CD9:19]

Wire-tailed Swallow *Hirundo smithii* 14 cm

H. s. smithii. Slender, blue-and-white swallow with ***rufous cap*** bordered by black lores and blue-black ear-coverts, ***and very thin, wire-like, tail streamers***. Underparts pure white with ***blue-black patch on breast-sides and lower flanks***. **HH** Generally uncommon resident. In pairs or small flocks in open habitats, usually near water. Forages low or at mid-height. Occasionally with other hirundines. Pairs breed solitarily; mud nest an open cup, plastered to overhanging surface of natural or artificial site. **Voice** Mostly silent. Call a quiet *chit*. Song a soft twittering. [CD9:18]

Ethiopian Swallow *Hirundo aethiopica* 13 cm

H. a. aethiopica. Small blue-and-white swallow with ***rufous forehead, buff-white throat and no breast-band***. May recall Wire-tailed Swallow, but has rufous restricted to forehead, tail streamers shorter and less wire-like, and no patches on lower flanks. **HH** Afrotropical vagrant, S Senegal (bred Niokolo Koba). Spreading westwards in W Africa. Normally over various open habitats, often near water; also in villages and towns. Forages low or at mid-height. Occasionally with other hirundines. Pairs breed solitarily; mud nest an open cup, plastered to overhanging surface of natural or artificial site. **Voice** Soft *chit* or *cheep*. Song a melodious twittering. [CD9:21]

Red-chested Swallow *Hirundo lucida* 15 cm

H. l. lucida. **Adult** Very similar to Barn Swallow but has shorter outer tail feathers, ***more white in tail***, dark rufous of throat extending to upper breast, narrower breast-band (narrowest in centre; sometimes broken), and ***whiter underparts***. In flight, more compact silhouette with broad-based, more triangular wings. **Juvenile** separated from juvenile Barn Swallow by narrower breast-band and more white in tail. **HH** Common resident and partial intra-African migrant; movements inadequately known. Singly, in pairs or small flocks over open habitats and villages and towns. Flight fast and agile. Forages low or at mid-height. Often with other hirundines. Pairs usually breed solitarily or in small loose colonies. Mud nest an open cup, plastered to overhanging surface of natural or artificial site, often under bridges. **Voice** Similar to Barn Swallow. [CD9:24]

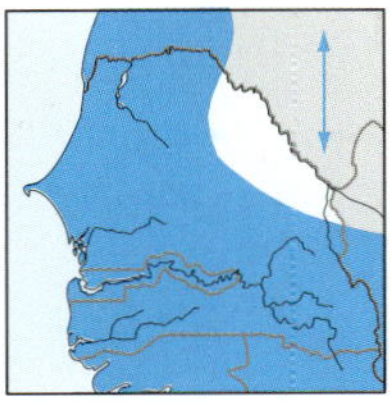

Barn Swallow *Hirundo rustica* 15–19 cm

H. r. rustica. **Adult breeding** Head, upperparts and tail glossy dark blue (duller when worn); ***forehead and throat dark rufous bordered by dark blue breast-band***. Tail with small white patches on inner webs and very long outer feathers. Lower breast to undertail-coverts whitish. **Adult non-breeding** Duller; rufous on throat may become very pale (even whitish); no tail streamers. **Immature** Duller and browner; forehead and throat paler; tail streamers shorter. **HH** Common Palearctic migrant (mainly Sep–early May). In small or large flocks in all habitats, also in villages and towns. Flight fast and agile with much banking and turning. Forages low or at mid-height. **Voice** Vigorous *whit-whit*; also a sharp *siflit*. Song, occasionally uttered in winter quarters, a fast, varied twitter including clear notes, rasping sounds and gurgling rattles. [CD3:58]

ad
ad
Grey-rumped Swallow
ad
ad
Pied-winged Swallow
ad
Ethiopian Swallow
ad
Wire-tailed Swallow
ad br
ad
Barn Swallow
ad non-br
Red-chested Swallow

WAGTAILS *Small, slender, terrestrial insectivores with long, frequently pumped tails.*

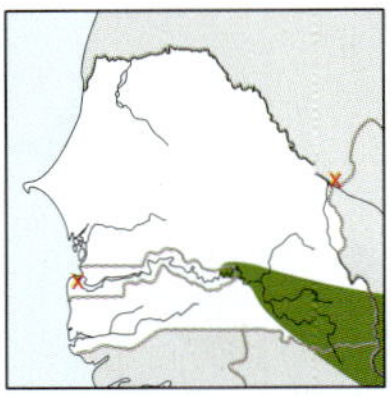

African Pied Wagtail *Motacilla aguimp* 19 cm

M. a. vidua. **Male breeding** ***Black-and-white*** wagtail with ***large white wing patch*** and ***broad black breast-band***. Head and upperparts black with long, broad white supercilium and white patch on neck-sides. **Male non-breeding/adult female** Similar, but black less intense. **Juvenile** Much duller, with black replaced by dark grey-brown. **HH** Uncommon, SE Senegal; scarce, Gambia. Singly, in pairs or family parties near water and habitations. Tame. **Voice** Clear, whistled *tluwsee*. Song a varied series of quite melodious, short phrases of clear piping and whistling notes. [CD9:27]

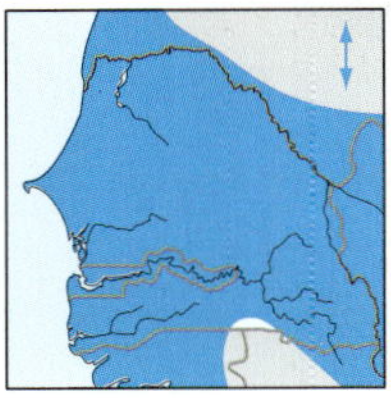

White Wagtail *Motacilla alba* 18 cm

M. a. alba. Grey, black and white wagtail with ***white face***. **Male breeding** (from Feb) Forehead, forecrown and head-sides white; hindcrown and nape black, sharply contrasting with grey upperparts; wing feathers black broadly fringed white. Tail black with white outer feathers. Throat to upper breast black; rest of underparts white. **Female breeding** Similar, but black on rear of head diffusely merging into grey upperparts. **Male non-breeding** Throat white (sometimes tinged cream) bordered by black breast-band; black of crown duller, mixed with grey. **Female non-breeding/immature** Usually entire top of head (incl. forehead) grey; head-sides variably tinged greyish and cream. **HH** Common Palearctic migrant (Sep–Apr). In variety of habitats, incl. fields, gardens, villages, edges of lakes and rivers, dams. **Voice** Sharp *chissik*; alarm *chik*. [CD3:62]

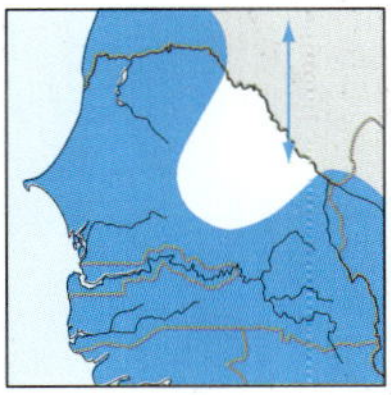

Yellow Wagtail *Motacilla flava* 16–17 cm

Commonest and most widespread wagtail, with comparatively short tail and, in all plumages, generally ***some yellow on underparts***. Several races occur, differing mainly in head pattern of males in breeding plumage. Non-breeding plumages usually indistinguishable and similar to female *flava* but duller; immatures are even more washed out. Illustrated forms are all in breeding plumage. **Adult male *flava*** Blue-grey head, white supercilium and wholly yellow underparts. Moult to breeding plumage starts Jan–Feb, completed Mar–Apr. **Adult female *flava*** Duller, with less distinct head pattern and paler underparts, often becoming whitish on throat and breast. **Adult male *iberiae*** Head darker grey, supercilium narrower and throat white. ***M. f. cinereocapilla*** similar but supercilium absent or restricted to streak behind the eye. ***M. f. thunbergi*** Top of head slate-grey, head-sides blackish, and supercilium indistinct, often absent. ***M. f. flavissima*** Head yellow-green. **HH** Common Palearctic migrant (mainly Sep–May). Singly and in small to large flocks in variety of habitats, generally open areas, incl. short grassland, lawns, airfields, marshes, dams, ricefields and farmland, but also forest clearings and tracks. **Voice** Rather loud *pseew*. [CD3:60]

Grey Wagtail *Motacilla cinerea* 18–19 cm

M. c. cinerea. Slimmer, longer tailed than Yellow Wagtail with ***grey upperparts, contrasting blackish wings and, in all plumages, yellow on rump and vent***. Very long tail and broad white band on upper- and underwing conspicuous in deeply undulating flight. Legs brownish–pink. Yellow Wagtail has olive-tinged upperparts and relatively longer, blackish legs. **Male breeding** (from Feb) Throat black bordered by white moustachial stripe, rest of underparts yellow. **Female breeding** Similar, but throat whitish variably mottled black. **Male non-breeding** Similar, but throat white; yellow on underparts paler, brightest on breast and vent. **Female non-breeding** Very similar to male non-breeding, but yellow on underparts paler. **Immature** As adult female non-breeding, but upper breast buffier. **HH** Rare Palearctic migrant (Oct–Mar). Various habitats, normally near water. Often pumps whole rear body and tail. **Voice** Disyllabic *chzizik*, similar to White Wagtail's but shorter, more metallic. [CD3:61]

Citrine Wagtail *Motacilla citreola* 17–18 cm

M. c. citreola/werae. Resembles Yellow Wagtail but has, in all plumages, ***grey upperparts and two conspicuous white or whitish wingbars*** (narrowing with wear). Tail and legs slightly longer. **Male breeding** Bright lemon-yellow head bordered by black nuchal band. **Male non-breeding/adult female** Crown greyish washed yellow; broad yellow supercilium extending around greyish ear-coverts; no black on nape. Yellow coloration duller; undertail-coverts whitish. **First-winter** Characteristic head pattern with broad whitish ear-covert surround, palish lores (often tinged buff) and narrow, weak, dusky malar stripe. Yellowish tinge to head gradually appears in Dec–Jan. **HH** Palearctic vagrant, Senegal (Jan 1999). **Voice** Sharp, rasping *tsrreep*.

juv
ad
African Pied Wagtail
♂ br
ad non-br
White Wagtail
Yellow Wagtail
♀
♂ br
flavissima
♂ br
iberiae
♂ br
flava
♂ br
thunbergi
♂ br
cinereocapilla
Grey Wagtail
♂ br
♂ non-br
ad
♂ br
1st-winter
Citrine Wagtail

PIPITS *Terrestrial insectivores with rather cryptic, variably streaked brown plumage. Slightly resemble larks, but slimmer. Large species conspicuously long legged. Sexes similar. Non-breeding plumage usually similar to breeding; pale tips and fringes to wing feathers broader when fresh. For identification, note extent of streaking on upper- and underparts, colour of outer tail feathers, and calls.*

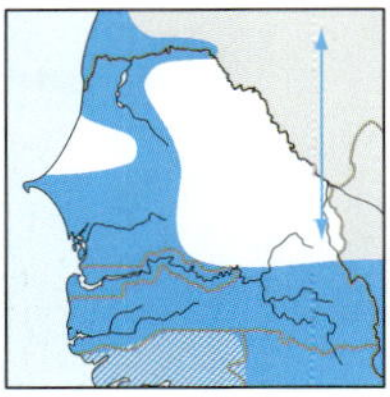

Tree Pipit *Anthus trivialis* 15 cm

A. t. trivialis. Slim, small pipit with olive-brown upperparts streaked blackish, warm buff supercilium and ***plain rump***. Underparts whitish washed creamy-buff on throat, breast and flanks, with blackish malar stripe and ***coarse blackish streaking on breast*** extending thinly on flanks. White outer tail feathers. **HH** Uncommon to locally fairly common Palearctic migrant (Sep–Apr). Singly, in twos or small flocks in various open wooded habitats. Forages on ground, often moving stealthily and slowly wagging tail when pausing. Frequently perches in trees or on bushes. **Voice** Call a single *tzeep* in flight. [CD3:66]

Red-throated Pipit *Anthus cervinus* 15 cm

Small pipit with ***boldly streaked upper- and underparts***. In non-breeding and first-winter plumages always separable from Tree Pipit by heavy blackish streaking on breast ***extending in two broad lines along entire flanks*** and ***streaked*** (not plain) ***rump***; voice also different. **Adult breeding** ***Diagnostic rusty-red face, throat and often also breast***. In female, face and throat usually less bright. **Adult non-breeding** Colour of face and throat variable but always duller. **Immature** Lacks any rufous; blackish malar stripe usually distinct and ending in ill-defined, wedge-shaped patch on lower throat-sides. **HH** Uncommon Palearctic migrant (Sep–May). In various types of grassland and cultivation. **Voice** Call a distinctive, drawn-out, high-pitched, thin *pseeeeh*, slightly trailing off towards end and usually uttered singly; longer, less rasping and more piercing than call of Tree Pipit. [CD3:67]

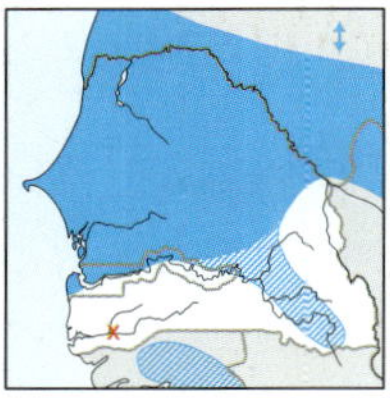

Tawny Pipit *Anthus campestris* 16–17 cm

A. c. campestris. Easily identified by ***pale sandy-brown upperparts with contrasting dark centres to median wing-coverts***. Head sandy with creamy supercilium, dark lores and narrow dark malar stripe. Wing feathers blackish-brown, broadly tipped and edged sandy-buff when fresh; outer tail feathers white. Underparts plain cream-white. **Immature** similar; but sometimes with traces of streaked juvenile plumage on upperparts and breast. **HH** Uncommon to fairly common Palearctic migrant (Oct–Apr), Senegal; uncommon, Gambia. Singly in open habitats, incl. short dry grassland and fields. Walks rather horizontally and creeps through short grass like wagtail; also wags tail. **Voice** Sparrow-like *chlip* and *cheeup*. [CD3:64]

Long-billed Pipit *Anthus similis* 17–19 cm

A. s. asbenaicus (?). ***Large, grey-buff pipit*** with ***diffusely streaked upperparts***, prominent buffish supercilium; narrow dark malar stripe and ***very lightly streaked breast***. Wing feathers brown, broadly tipped and edged sandy-buff when fresh. Tail blackish-brown with buffish-white outer feathers. **HH** Scarce Afrotropical migrant/wanderer, N Senegal (Djoudj, Nov–Mar). Nearest population in C & E Mali. **Voice** Rather soft *chee* or *djeep*. Song a simple series of 2–3 detached, sparrow-like notes repeated at intervals *chriu shree chewee* or *tsreep shree shruw*; from ground, tree or large rock, or in slow, fluttering display flight. Plain-backed Pipit has similar vocalisations, but sings much less often and song usually consists of only 1–2 notes repeated at intervals. [CD9:28]

Plain-backed Pipit *Anthus leucophrys* 17 cm

A. l. ansorgei. Large pipit with ***plain, unstreaked upperparts***. **Adult** Medium grey-brown above, with prominent whitish supercilium, rather indistinct dark moustachial and malar stripes, and dark brown wing feathers broadly fringed buffish. Tail blackish with buff outer feathers. Underparts pale buff streaked dark on breast. **Juvenile** Browner, with heavier breast streaking and narrow whitish fringes to upperpart feathers. **HH** Uncommon to locally fairly common resident. Singly or in pairs in short grassland, cultivation, ricefields and burnt ground. Forages on ground, walking rather upright; wags tail. Occasionally perches on low vantage point. **Voice** Soft *chee-chee* on take-off. Song a rather slow and monotonous series of single or alternating *swree* and *chirrup* notes, given from ground or low perch. [CD9:30]

ad
Tree Pipit
ad br
ad non-br
1st-winter
Red-throated Pipit
ad
Tawny Pipit
ad
Long-billed Pipit
ad
Plain-backed Pipit

PLATE 99: LONGCLAW AND CUCKOOSHRIKES

LONGCLAWS *Robust ground-dwellers, with relatively short tail, strong legs and long claw on hind toe.*

Yellow-throated Longclaw *Macronyx croceus* 20–22 cm

Adult Brown above streaked blackish, with broad yellow supercilium; ***bright yellow below with black necklace***. ***White tail corners conspicuous in flight***. Female often somewhat duller. Immature duller still, with ill-defined necklace. **Juvenile** Buff below with necklace of dark spots. **HH** Uncommon resident. Singly or in pairs in variety of grassy habitats. Forages on ground, but often perches on vantage point. Flight distinctive: low and slow, bursts of stiff fluttering alternated with glides, with spread tail. **Voice** Far-carrying, melodious and repeated whistle *teeuwheee* or *twee-eu*, uttered by both sexes and by male in display flight, sometimes followed by *tiri-tiri-ti* or variation. [CD9:36]

CUCKOOSHRIKES *Unobtrusive arboreal birds. Mainly singly or in pairs, quietly gleaning foliage in search of insects, usually in canopy.*

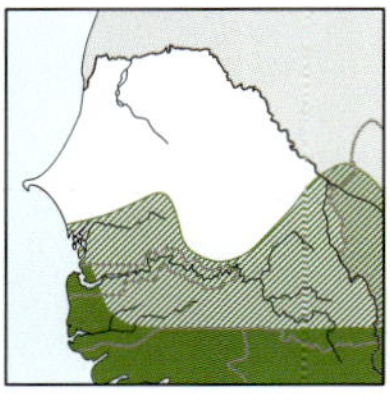

Red-shouldered Cuckooshrike *Campephaga phoenicea* 20 cm

Adult male Entirely glossy blue-black with ***large, conspicuous scarlet shoulder patch*** (sometimes orange or golden-yellow). Bill black with swollen yellow or pinkish gape wattles. **Adult female** Greyish olive-brown above, white barred black below. Wing feathers and outer tail feathers edged yellow; rump and uppertail-coverts barred black. **Juvenile** Similar to adult female but upperparts barred black; underparts more heavily barred, bars spade-shaped on belly; underside of tail feathers only narrowly tipped yellow. **HH** Uncommon to fairly common wet-season breeder, moving north with rains. Singly or in pairs in forest patches, gallery forest, woodland and thickets in savanna. Usually in upper and mid-storeys, occasionally lower, quietly foraging in foliage. Occasionally in mixed-species flocks. **Voice** Commonest call a short *tsuk*; also a single clear loud whistle *heeew* and a double *huu-tseew*. Song a jumble of high-pitched and scratchy notes, somewhat reminiscent of sunbird and often uttered in duet. [CD9:39]

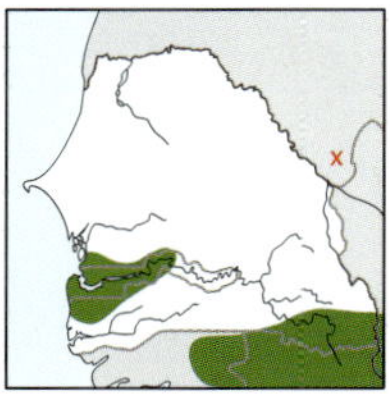

White-breasted Cuckooshrike *Coracina pectoralis* 25 cm

Large, grey-and-white cuckooshrike with relatively stout bill. **Adult male** Head, upperparts, throat and upper breast pale grey, sharply delimited from pure white rest of underparts. Lores black; white eye-ring. **Adult female** Lores paler and upper throat whitish bordered paler grey. **Juvenile** Similar but barred black-and-white above; spotted blackish below. **HH** Scarce, Senegal; rare to uncommon and decreasing, Gambia. Singly or in pairs in well-wooded savanna. Regularly joins mixed-species flocks. **Voice** Rather thin, high-pitched *seeeu*, easily passing unnoticed. Various other calls, incl. weak *tsip-tsip* and *tsitsitsi*, louder *wreewree-tiuptiup*, vigorous and abrupt *tsiu*, and rolling *wrrllu-wrrllu-...* [CD9:43]

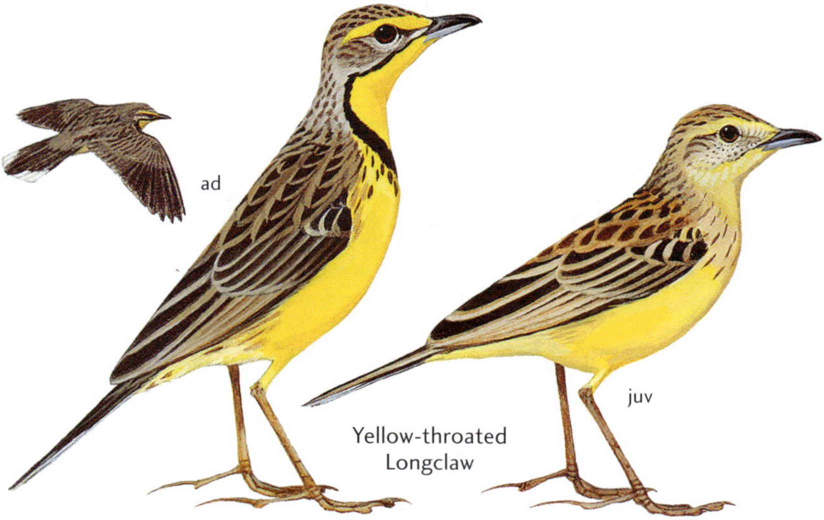
ad
juv
Yellow-throated Longclaw

♀
♂
Red-shouldered Cuckooshrike
♂
orange-shouldered variant
♀
♂
White-breasted Cuckooshrike

PLATE 100: BULBULS I

BULBULS (Plates 100–101) *Small to medium-sized, mostly arboreal birds. Sexes similar, males usually slightly larger; juveniles/immatures similar to, but generally duller than, adults. Family is well represented in Africa; all species but one are endemic. Most are inconspicuously coloured in various shades and combinations of dull green, grey, brown and yellow, and are associated with forest habitat. Frugivorous and insectivorous; occasionally flycatch. Mainly sedentary. Vocalisations are an important aid to identification.*

Slender-billed Greenbul *Andropadus gracilirostris* 18 cm

A. g. gracilirostris. A conspicuously ***two-toned***, rather slim, ***canopy-dwelling bulbul***. Olive-brown above, brownish-grey below. **HH** Locally common, SW Senegal (Casamance). Singly or in pairs, largely restricted to canopy of various forest types and cultivation with large trees. Occasionally joins mixed-species flocks. **Voice** Clear, drawn-out whistle *tseeeeu*, frequently uttered. Also a seasonal song of 4–5 whistled notes *whee-ti-twheew-ti-twhee* or *whuut-hEET whuut-hEET…*, with slight variations. [CD9:54]

Little Greenbul *Andropadus virens* 16.5 cm

A. v. erythropterus. Skulking, wholly dirty olive-green bulbul with distinctive, easily learnt song. Absence of any contrast in plumage is a distinguishing feature. No eye-ring. **Juvenile** Similar to juvenile Yellow-whiskered Greenbul, but has darker legs and rusty wash to wings. **HH** Common. In variety of habitats, incl. forest edge, rank scrub, abandoned cultivation, thicket, etc. Sings all day and nearly year-round, from deep cover, usually 2–3 m above ground. **Voice** Song a few subdued chuckling notes followed by a rapidly ascending, pleasant bubbling warble, increasing in volume and abruptly ending on a clear, high, rising note. Call a dry *kuk-kuk-kuk…* [CD9:49]

Yellow-whiskered Greenbul *Andropadus latirostris* 17 cm

A. l. congener. Olive-green forest bulbul with ***bright yellow malar stripe*** and distinctive ***orange-brown or yellow-brown legs***. Tail dark grey-brown (appearing almost black). Bill blackish. Yellow 'whiskers' often puffed out. **Juvenile** No or only rudimentary whiskers, but ***pinkish legs*** key to identity. Bill blackish with dark orange-yellow gape and base. **HH** Uncommon and local, SW Senegal (Casamance). Singly, at all levels of various forest types and dense scrub. Rather shy. **Voice** Song a monotonous series of c.12 *chruk* notes, slightly increasing in volume. Call a repeated *chuk* and rapid rattling *ditditditdit…* (alarm). [CD9:55]

White-throated Greenbul *Phyllastrephus albigularis* 17 cm

Rather small bulbul. ***White throat contrasts with grey head and olive-grey breast***; pale yellow belly. **HH** Scarce and local, SW Senegal (Casamance). In pairs or family groups in primary and secondary forest. Prefers edges with dense undergrowth and lianas. Usually forages very low, but also in middle and upper levels. Joins mixed-species flocks, but often in small monospecific groups. Rather unobtrusive. **Voice** Song a variable, rapid series of clear, scolding notes, usually rising, then dying away, often preceded by loud, rolled *turrr*. More often heard contact call a low, soft *trrr-trrr, trrr-trrr*. [CD9:75]

Grey-headed Bristlebill *Bleda canicapillus* 20.5–22.0 cm

B. c. morelorum. Shy and skulking, but large and vocal bulbul with ***slate-grey head***, pale grey loral spot, and ***three outer tail feathers broadly tipped yellow*** (conspicuous in flight). **HH** Locally fairly common. In pairs or small parties in forest undergrowth (mostly at edges and clearings), gallery forest and woodland thickets. Joins mixed-species flocks and attends ant columns. **Voice** Varied. Song far carrying with cheerful, ringing quality, typically a loud initial note, a brief pause, then a series on descending scale. Other notes include loud *CHEEup* and *kyuw*, often in long series. [CD9:80]

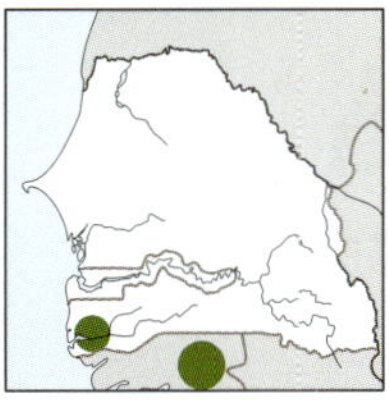

Red-tailed Greenbul *Criniger calurus* 20 cm

C. c. verreauxi. Fairly large, handsome bulbul with ***pure puffy white 'beard' contrasting strongly with grey head and bright yellow breast and belly***. Upperparts and flanks dull olive-green. Tail olive-green (not red in this race!) slightly tinged brownish. Blue-grey orbital ring. **HH** Scarce, SW Senegal (Casamance). Vocal, occurring in pairs or small parties in lower and mid-levels of closed forest. Frequent member of mixed-species flocks. **Voice** Song a cheerful rising *chup-chup-chwirulup*. Also a three-note *tsik-tyu-tyip* and rapid series *tyu-tyutyutyu-tyip-tyip-tyip…* Alarm *tsik*. [CD9:84]

Slender-billed Greenbul
juv
ad
Little Greenbul
juv
ad
Yellow-whiskered Greenbul
White-throated Greenbul
Grey-headed Bristlebill
Red-tailed Greenbul

Simple Leaflove *Chlorocichla simplex* 21 cm

Large bulbul with ***white throat*** and ***broken eye-ring*** contrasting with rest of plumage. Dark brown above, olive-brown below, becoming whitish in centre of belly. Superficially similar in general colour as Common Bulbul, but latter lacks white throat and white crescents around eye. **HH** Single record, SW Senegal (Casamance). Singly or in pairs in farmbush, forest regrowth and thickets. Rather shy and skulking, but often emerges in the open. **Voice** Song a frequently uttered, subdued nasal chattering. Call a scolding *wherr* and short, clipped *kwit!* [CD9:63]

Yellow-throated Leaflove *Chlorocichla flavicollis* 22.5 cm

C. f. flavicollis. Large bulbul with bright yellow throat contrasting with dark olive-brown head and olive-grey underparts. Upperparts and tail dark olive-brown. **HH** Fairly common. Singly, in pairs or noisy family parties, mainly in lower and mid-levels of dense woodland and thickets; also low swamp forest. Fairly shy, but vocal. **Voice** Call a loud, nasal *chow* and *kyip*. Song a string of *chows* interspersed with shorter notes; in quality reminiscent of Swamp Palm Bulbul. [CD9:64]

Swamp Palm Bulbul *Thescelocichla leucopleura* 23 cm

Easily identified, large and vocal bulbul with dark olive-brown upperparts, ***broad white tips to outer tail feathers***, and ***creamy-white underparts*** washed grey-buff on upper breast. **HH** Uncommon, SW Senegal (Casamance); vagrant, Gambia (Abuko, 1968). Associated with palm trees, particularly *Raphia*, in lowland and gallery forest. In small, mobile and noisy groups. **Voice** Raucous nasal cackling like tape-recorded conversation being played too fast. Also other loud, nasal, scolding calls. [CD9:65]

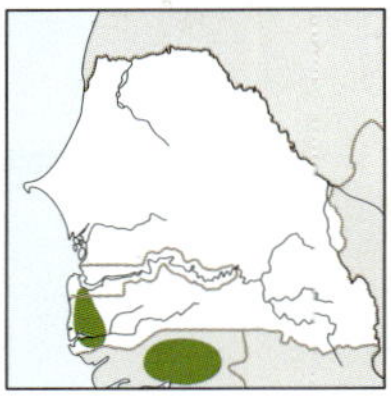

Leaflove *Pyrrhurus scandens* 22 cm

P. s. scandens. Large, handsome, distinctively coloured bulbul with pale grey head, rufous-edged flight feathers, and pale chestnut tail. Throat white; breast and flanks greyish-buff becoming creamy-buff on lower breast and belly. **HH** Scarce, SW Senegal; rare, Gambia. In small, noisy groups in forest fringing rivers and swampy places within forest; also forest patches in savanna. At all levels, though mainly in canopy and middle stratum, working its way through dense liana tangles. Joins mixed-species flocks. Easily located and identified by its incessant call. **Voice** Song starts with some subdued, nasal notes, then bursts into loud, resonant, pleasant conversational cackling, including *kyop-kee-kyop-kyop-ke-kyop--...*, uttered for long periods by family groups. (Caution is needed when attempting to identify this species on voice alone: Swamp Palm Bulbul and Yellow-throated Leaflove sound quite similar.) [CD9:66]

Common Bulbul *Pycnonotus barbatus* 18–20 cm

P. b. inornatus. Africa's ubiquitous, most conspicuous and familiar bird. Mainly earth-brown, with blackish-brown head and throat, pale brown breast and whitish belly. Peaked crown gives head distinctive shape. **HH** Common. In almost all habitats, incl. gardens and other man-made environments, but absent from closed forest (though penetrates edges). Pugnacious and noisy. Usually in pairs or small (family?) groups. **Voice** Song a cheerful phrase of 3–6 notes, e.g. *chuk chuk twirulup* or *chuk twee tu twuri* or *kwik kweek kwuyu*, etc.; most adequately rendered by popular *quick quick, doctor, quick!* Calls include a variety of chattering and ringing notes. [CD3:71]

Western Nicator *Nicator chloris* 21–24 cm

Large. Olive-green above ***with bold yellow spots on wings***. Underparts pale grey becoming greyish-white on upper throat and centre of belly; undertail-coverts yellow. White supraloral spot; head-sides ('cheeks') olive-yellow; narrow but distinct yellow eye-ring. Bill black, heavy and hooked, recalling bush-shrike. **HH** Scarce, SW Senegal (Casamance); rare or possibly extinct, Gambia. Skulking but inquisitive species of tangles, dense foliage and thick cover, in forest and its gallery extensions. Usually singly; sometimes joins mixed-species flocks. Mostly at mid-level, but occasionally lower, even on ground (like a thrush). **Voice** Powerful, melodious and very varied song comprising an explosive crescendo of notes, clear whistles, guttural rattles, etc. Occasionally mimics other species. Calls a loud, abrupt *tok!*, often in long, accelerating series; may recall *kuk* of Yellowbill, but latter 'drier' and less powerful. [CD14:1a] **Note** Recently proposed to be placed in a separate family, Nicatoridae.

Simple Leaflove
Yellow-throated Leaflove
Swamp Palm Bulbul
Leaflove
Common Bulbul
Western Nicator

THUSHES AND ALLIES (Plates 102–106) Large and diverse family (Turdidae), which consists of small to medium-sized songbirds with usually strong legs, mostly square or rounded tails, and a broad range of vocalisations; some are notable songsters.

White-tailed Alethe *Alethe diademata* — 18 cm

Adult Dark olive-chestnut above with ***rusty-orange central crown-stripe***; head-sides grey. Tail blackish with ***outer three feathers broadly tipped white***. Underparts white washed grey on breast-sides and flanks. Crown feathers can be raised when excited, forming short crest. **Juvenile** Spotted orange-rufous above; underparts orange-rufous scalloped blackish on throat and breast; whitish tail corners. With age, spots gradually disappear from upperparts, lower underparts become whitish, white feathers appear on throat and breast. **HH** Locally common resident, SW Senegal (Casamance) in forest, gallery forest and forest patches in savanna. Mainly on or near forest floor. Frequently attends ant columns. **Voice** Three sweet ascending whistles, *huu hee hueee*; third note sometimes absent. Also a subdued song with mixture of whistles and imitations. Calls: whistled *huu*, often in long series; also harsh notes. [CD10:3b]

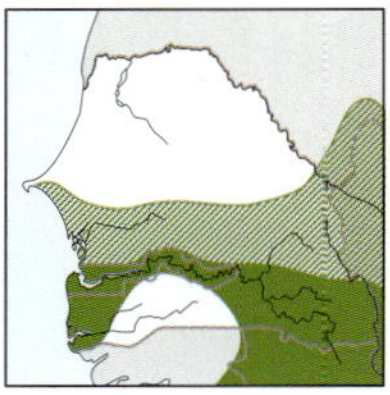

Snowy-crowned Robin Chat *Cossypha niveicapilla* — 20.5–22.0 cm

C. n. niveicapilla. Relatively large with ***white crown*** and slate-grey upperparts separated from head by narrow ***rufous hind-collar***. Forehead and head-sides black. Rump and tail orange-rufous with black central rectrices. **Juvenile** Blackish spotted rufous above, rusty-buff scalloped blackish below. **HH** Locally fairly common to uncommon resident, inhabiting thickets in savanna, forest edges and clearings, overgrown farmland, gallery forest and gardens. Singly or in pairs. Forages mainly on or near ground. **Voice** Vocal. Song rich, melodious and sustained, with characteristic fast delivery and including many imitations; some motifs repeated a few times, often with slight variations and thus somewhat reminiscent of African Thrush. Calls include *heeee* (contact) and guttural *krrr* (alarm). [CD10:2b]

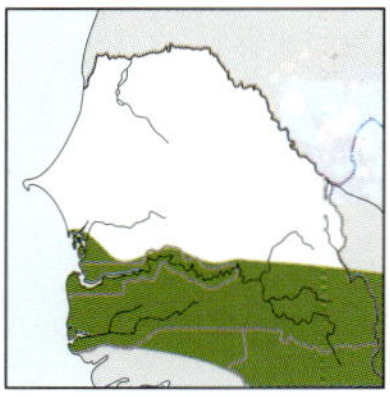

White-crowned Robin Chat *Cossypha albicapilla* — 26 cm

C. a. albicapilla. ***Very large and long-tailed robin chat with white crown*** (black scaling to crown disappears with wear; blackish feather bases give crown a somewhat greyish appearance). Head-sides (***to chin***) and upperparts black ***without rufous hind-collar***. **Juvenile** Paler; some rufous spots on wing coverts. **HH** Locally fairly common to common resident in thickets in savanna, riverine scrub, gallery forest and large overgrown gardens south of 14°N. Singly or in pairs. Forages mainly on ground. Skulking but not shy. **Voice** Song varied, sustained and very fast, including scratchy notes and trills, also imitations. Call a penetrating *sweeuee*. [CD10:3a]

African Thrush *Turdus pelios* — 23 cm

T. p. chiguancoides. Typical thrush, ashy-brown above, paler below, with ***conspicuous yellow bill***. Throat variably streaked dusky; underparts pale ashy-brown on breast becoming whitish on belly and washed pale orange-buff on flanks. **Juvenile** Similar, but with rufous wash over face and breast, pale tips to coverts, and crescentic markings on breast and flanks. **HH** Fairly common to locally common in south; rare, N Senegal. Some movements recorded; more common in rains. Singly or in pairs in variety of wooded habitats, incl. cultivation with some trees, gardens and wooded savanna. Forages on ground. **Voice** Song far carrying, melodious and variable, with series of short repetitions of simple motifs, e.g. *toolee toolee toolee weetyuuwee weetyuuwee teewit teewit teewit swrreep swrreep leepoo leepoo leepoo churp churp churp...* etc. Call a hard *chuk*, often repeated in dry series *chukukukukuk...* [CD10:40]

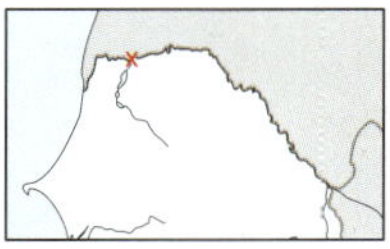

Song Thrush *Turdus philomelos* — 23 cm

T. p. philomelos. Upperparts warm brown; boldly spotted underparts cream-white washed buff on breast and flanks. Legs pinkish. In flight has orange-buff underwing-coverts. Flight fast and direct. **HH** Palearctic vagrant, N Senegal (one trapped, Richard-Toll, Nov 1958). **Voice** Short, dry *tsik* in flight. [CD3:99]

European Robin *Erithacus rubecula* — 14 cm

Adult Small, compact and rounded with ***orange-red face and breast***. Olive-brown above; lower underparts dirty buffish becoming whitish in centre of belly. **HH** Palearctic vagrant (ship-assisted?), coastal Gambia (Mar 2008; race unknown). Forages on ground, hopping with upright posture. **Voice** Calls include a dry *tik* often in fast series *tik-tikikikikkk*. [CD3:77]

juv
ad
White-tailed Alethe
juv
ad
Snowy-crowned Robin Chat
juv
ad
White-crowned Robin Chat
juv
ad
African Thrush
Song Thrush
European Robin

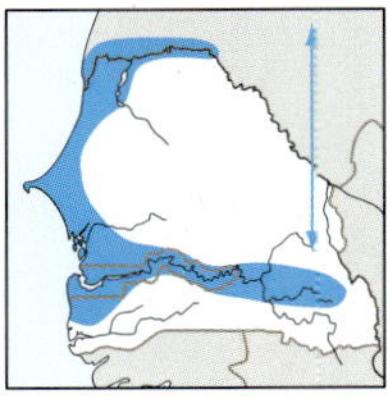

Common Nightingale *Luscinia megarhynchos* 16.5 cm

L. m. megarhynchos. Plain, warm brown above with ***open face*** and ***rufous tail***, pale grey-brown below becoming whitish on throat and belly. Whitish eye-ring. **HH** Locally fairly common to common Palearctic visitor and passage migrant (late Aug–early May) in scrub, farmbush, overgrown clearings, gardens, etc. Usually in dense cover. **Voice** High *hueet*, hard *tak*, low *karrr*. Song loud, rich and melodious, including a variety of clear whistles, vigorous chuckles and piping notes, which are often repeated; characteristic are a rapid, accelerating series of full *tchiok* notes and in-drawn whistles *heeeet-heeeet-heeeet-*...; often uttered in winter quarters. [CD3:78]

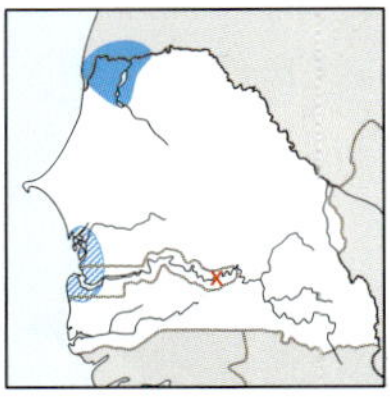

Bluethroat *Luscinia svecica* 14 cm

Unobtrusive with dark brown upperparts, whitish supercilium and ***diagnostic rufous sides to basal half of tail*** (conspicuous in flight). **Male breeding** (from Feb) Unmistakable. ***Throat and upper breast bright blue*** with white (***cyanecula***) or rufous (**nominate**) central spot. **Male non-breeding** Throat buff-white bordered by dappled black breast-band, the latter variably suffused blue. **Female** As non-breeding male, but usually lacks blue suffusion in band surrounding throat or rufous on breast. **HH** *L. s. svecica* is uncommon to locally fairly common Palearctic visitor, N Senegal (mainly mid Sep–Apr); rare to uncommon, Gambia. *L. s. cyanecula* rare, N Senegal. Usually in scrub near water, ricefields, and in damp places. Forages on or near ground. **Voice** Hard *tak* and *hueet*. [CD3:79]

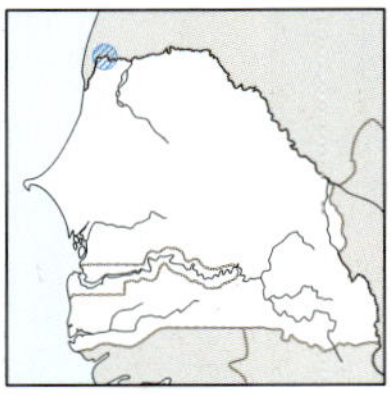

Black Redstart *Phoenicurus ochruros* 14 cm

P. o. gibraltariensis. **Adult male** Dark grey above; white wing panel; bright rufous tail; black face and breast. **Adult female** Entirely slate-grey (greyer than female Common Redstart); tail as male. **HH** Rare Palearctic visitor, N Senegal (Djoudj, Dec–Apr). Habits similar to Common Redstart. **Voice** Call *tsip*, often followed by *tak-tak-tak*. [CD3:81]

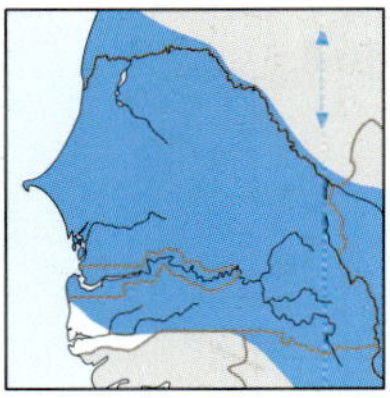

Common Redstart *Phoenicurus phoenicurus* 14 cm

P. p. phoenicurus. Dainty, slim chat with upright stance and, in all plumages, ***bright rufous tail, which is almost constantly shivering***. **Male breeding** (from Jan) Blue-grey above with black face bordered above by white streak, and orange-rufous below. In Oct–Dec, pale feather tips partially obscure bright colours, which are gradually revealed through wear. **Female** Pale grey-brown above with narrow pale eye-ring, mainly buffish below, often tinged pale orange on breast and flanks. **HH** Fairly common Palearctic migrant (Sep–early May) to various wooded habitats. Catches insects in the air or on ground, sallying from concealed perch. A male trapped in N Senegal (Feb 1993) had pale wing panel typical of SW Asian *P. p. samamisicus*. **Voice** Call *huweet* (*tuk-tuk*). [CD3:82]

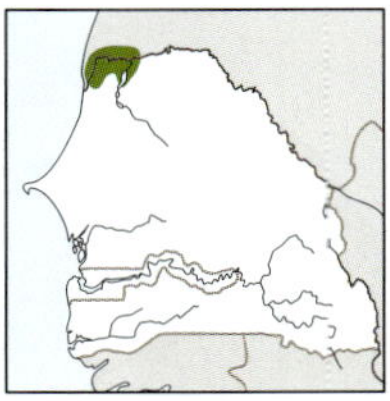

African Stonechat *Saxicola torquatus* 12.5 cm

S. t. moptanus. Small chat with relatively large, round head and short tail. **Adult male** Black head and upperparts; white neck-sides, rump, uppertail-coverts and underparts; chestnut breast patch. In flight, white wing patch and rump conspicuous. **Adult female** Streaked dark brown above, white rump and uppertail-coverts; ***indistinct pale supercilium***. Compare Whinchat. **HH** Uncommon resident, Senegal delta. Flicks wings and tail. **Voice** Call (*wheet*) *trek-trek*. Song: short clear, twittering and scratchy phrases. [CD3:84]

European Stonechat *Saxicola (torquatus) rubicola* 12.5 cm

Male non-breeding Browner above than African Stonechat; rump rusty-brown mottled white. **Adult female** Faint or no supercilium, darker below than African Stonechat with rusty-brown rump. **HH** Possible Palearctic vagrant, N Senegal (recorded coastal Mauritania). **Voice** [CD3:84]

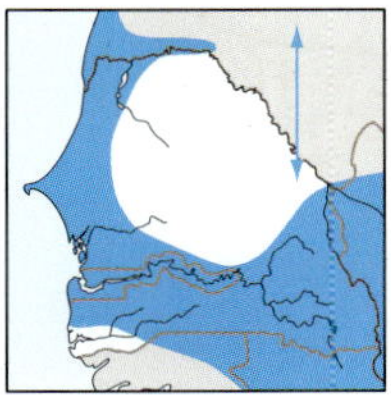

Whinchat *Saxicola rubetra* 12.5 cm

Small, streaky buffish-brown chat with upright stance and ***broad buff or white supercilium***. **Male breeding** (from Jan/Feb) White supercilium, blackish head-sides, small white patch on edge of wing, and ***white sides to base of tail*** (conspicuous in flight). Underparts orange-buff becoming buffish-white on belly. **Female** similar but paler. **Adult non-breeding** Very buffish overall; head-sides brownish. **HH** Uncommon to locally fairly common Palearctic visitor and passage migrant (Sep–May). Uses prominent, low perches in open areas, incl. farmland, degraded savanna, wooded grassland, etc. **Voice** Call (*whu*) *tek-tek*. Usually silent. [CD3:86]

Common Nightingale
♂ cyanecula
♀ svecica
♂ svecica
Bluethroat
♀
♂
Black Redstart
♂ br
♀
♂ non-br
Common Redstart
♀
♂
African Stonechat
ad non-br
Whinchat
European Stonechat
♀
♂
♂ br

WHEATEARS *Small, ground-dwelling insectivores. Black-and-white tail pattern an important identification mark. Breeding plumage acquired after partial moult in winter quarters before return migration. Usually silent; most common call a hard* chak.

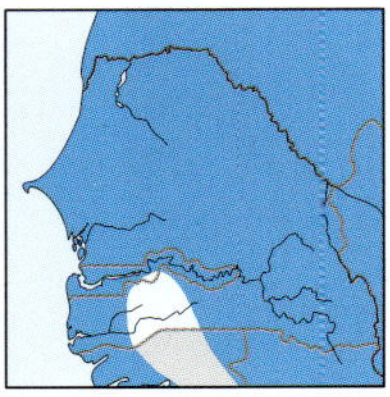

Northern Wheatear *Oenanthe oenanthe* 15–16 cm

In all plumages, identified by ***white lower rump and tail with characteristic black inverted T*** (terminal band of even width on distal third). **Male *oenanthe* breeding** (from Feb/Mar) ***Top of head to back blue-grey*** (diagnostic), black mask bordered by white supercilium, black wings, and sandy-buff underparts (richer on breast, becoming whitish on belly). Greenland race ***leucorhoa*** on average larger, brown-grey above, deeper (rather orange-buff), more extensively and uniformly coloured below. Distinctive N African ***seebohmi*** (Seebohm's Wheatear) has face and throat black (often connected to black wing at shoulder); underwing-coverts black. **Male non-breeding** Dull grey-brown above with variable darkish mask (sometimes very indistinct); wing feathers black broadly fringed orange-buff. In *seebohmi* black throat partially or wholly obscured by pale grey feather fringes. **Female breeding** Duller and browner above than male breeding, with variably distinct mask (sometimes lacking); supercilium buffish. Most female *seebohmi* have throat variably mottled blackish (but some plain throated). **Female non-breeding/first-winter** Similar to male non-breeding but lack mask; wing feather centres dark brown. **HH** Palearctic winter visitor (Sep–Apr). Nominate and *leucorhoa* are common N Senegal, uncommon further south (Senegambia main wintering area for *leucorhoa*); *O. o. seebohmi* is scarce, N Senegal. Singly in various open habitats. Forages mainly by running short distances, then pecking at prey. Bobs head, flicks wings and tail. Defends individual winter territory. **Voice** Hard *chak* (alarm), less often *wheet-chak*. Song consists of variable short phrases including whistles, scratchy notes and imitations. [CD3:89-90]

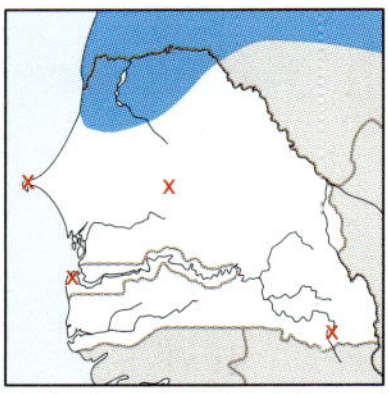

Isabelline Wheatear *Oenanthe isabellina* 16–17 cm

Robust, long-legged, plain sandy-coloured wheatear. T-patterned tail with ***broad*** black terminal band of even width covering *c.* half of tail. Wing feathers brown broadly fringed sandy-brown (thus barely contrasting with upperparts when fresh), with ***contrasting blackish alula.*** Some adult female/first-winter Northern very similar, but distinguished by darker centres to wing-coverts. **HH** Uncommon Palearctic winter visitor (Oct–Mar), N Senegal. Very rare, Gambia and SE Senegal. Usually singly in open country with scattered bushes. Runs rather than hops. **Voice** Hard *chak* (alarm) and *wheew*. Song, uttered in winter quarters, consists of variable short phrases including whistles, scratchy notes and imitations; similar to Northern Wheatear's. [CD3:95]

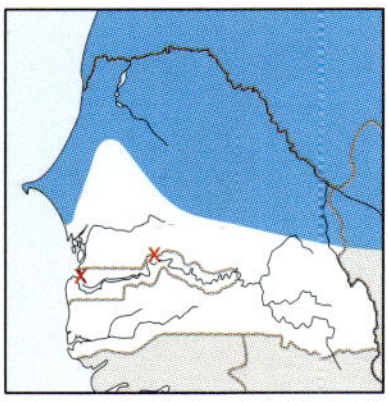

Black-eared Wheatear *Oenanthe hispanica* 14.5 cm

O. h. hispanica. Rather slim and long-tailed wheatear. T-patterned tail ***with uneven, sometimes broken, terminal bar.*** Has white-throated and black-throated forms in both sexes (former more common). **Male breeding** Rich sandy-buff; ***black mask not connected to black wings.*** Breast warm sandy-buff becoming whitish on rest of underparts. **Male non-breeding** Similar, but basic pattern overlaid by buffish feather tips; greater coverts and inner secondaries broadly tipped and fringed rich buff, forming pale wing panel. **Female non-breeding/first-winter** Sandy-brown; indistinct mask; dark brown wings broadly fringed buffish. **HH** Uncommon Palearctic migrant (Sep–Apr), N Senegal; rare, Gambia. Singly in *Acacia* savanna and dry farmland, often on shrubs. **Voice** Hard *chak*. Song, sometimes given in winter quarters, a rapid jumble of scratchy notes and whistles. [CD3:91]

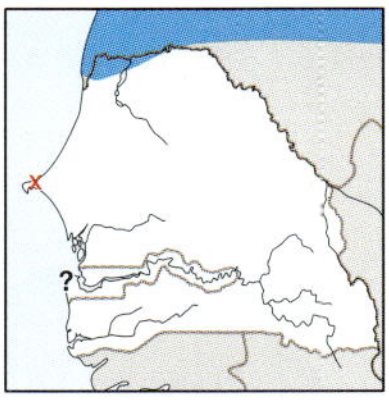

Desert Wheatear *Oenanthe deserti* 14–15 cm

O. d. homochroa. Sandy-coloured wheatear with ***diagnostic black tail in all plumages.*** **Male breeding** ***Black face and throat connected to black wings at shoulder.*** Scapulars whitish. **Male non-breeding** Slightly greyer above; black areas fringed white or pale sandy. **Female non-breeding** Deep sandy-buff, sometimes with greyish tinge; no black on face. Isabelline Wheatear differs from female by noticeably larger size and paler wing. **HH** Rare N African winter visitor (Nov–Feb), N Senegal. Unsubstantiated records from Gambia. Usually singly in arid habitats with some vegetation. **Voice** Calls include *hweee* and hard *tuk*. Song, often given in winter quarters, a characteristic, plaintive, descending *swee-you* or *deedjrruu*, frequently repeated. [CD3:94]

Heuglin's Wheatear *Oenanthe heuglini* 14–15 cm (not illustrated)

Adult Long-legged wheatear with dark brown upperparts and ***rufous-buff underparts.*** Tail T-patterned with ***broad*** black terminal band (about half of tail). Narrow white supercilium bordered by dark eye-stripe. Sexes similar. **Juvenile** Similar, but spotted dark rufous-buff above and scalloped dusky below. **HH** Vagrant, Senegal (Djoudj NP, Jan 2007). Migrates between Sahel and savanna zones elsewhere in W Africa. Often wags rear body. **Voice** Sharp *chak*.

♂ br
oenanthe
♂ br
seebohmi
♂ non-br
seebohmi
♀
oenanthe
♀
seebohmi
Northern Wheatear
♂ non-br
oenanthe
♂ non-br
♀ non-br
♂ non-br
Isabelline
Wheatear
Black-eared Wheatear
♂ br
♂ br
Desert
Wheatear
♀ non-br
♂ non-br

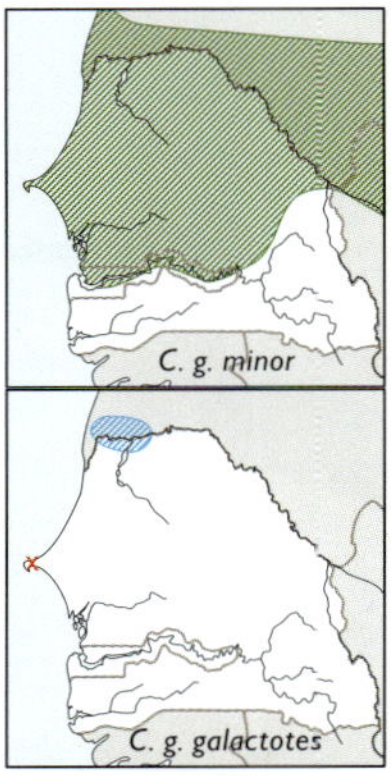

Rufous Scrub Robin *Cercotrichas galactotes* 15 cm

Distinctive. **Adult *minor*** Plain sandy rufous-brown above with rufous rump and fan-shaped rufous tail boldly tipped black and white. Creamy-white supercilium bordered below by narrow blackish eye-stripe. Underparts plain sandy-white. **Nominate** has broader subterminal black band to tail. **Juvenile** Similar. **HH** *C. g. minor* (African Scrub Robin) is a locally common resident, mainly from central Senegal south (but also recorded in north). Palearctic *C. g. galactotes* is an uncommon to locally common visitor (Nov–May), mainly N Senegal (also recorded near Dakar; distribution inadequately known). Dry-country species, occurring in open grassland with shrubs, degraded savanna, farmland, etc. Mainly on ground or in low bushes, constantly fanning and cocking tail. **Voice** Song loud and sustained, with short whistled phrases, regularly repeated; from perch or in butterfly-like display flight. Calls a hard *tek tek*, a hoarse *tseeeip*, a low *chrrr* and *deeu*. [CD3:80]

Black Scrub Robin *Cercotrichas podobe* 18 cm

C. p. podobe. Unmistakable. **Adult** Slender, long-legged, ***black*** chat with white-tipped undertail-coverts and ***long fan-shaped tail boldly tipped white***. Rufous in wing visible in flight. **Juvenile** Duller, more sooty-brown; undertail-coverts have indistinct pale tips. **HH** Not uncommon resident in arid savanna, thorn scrub with clumps of palm trees and dry brush, Senegal. Mainly on ground or in low bushes, fanning and cocking tail. Sings frequently and almost year-round. Not shy. **Voice** Far carrying song with fairly short, varied phrases, resembling song of Rufous Scrub Robin. Also a sustained babble of sweet, fluty and scratchy notes. [CD10:16]

Northern Anteater Chat *Myrmecocichla aethiops* 19 cm

M. a. aethiops. Long-legged, blackish-brown chat with large white wing patches visible in flight. **Juvenile** is warmer brown; throat and breast feathers edged buff; gape yellow during first 3 months. **HH** Locally frequent to uncommon resident. In pairs or small parties in open grassland and farmland with scattered bushes, trees and, usually, termitaria. Mainly on ground, but readily uses perches. Tame and conspicuous. **Voice** Song, from perch, a varied mixture of clear whistles and short, hard trills. Call a penetrating *tseeu* and *heeh*. [CD10:24]

Sooty Chat *Myrmecocichla nigra* 16 cm

Adult male Entirely glossy black with conspicuous ***white shoulder patches***. **Adult female** Entirely blackish-brown. **Juvenile** Similar to adult female. **HH** A few records from SE Senegal (two near Kédougou, Mar 1972; a few pairs with males singing c.70 km NNE of Tambacounda, Feb 1983). Singly or in pairs in open grassland with termitaria and few or no bushes; also farmland, by roads and on recently burnt areas. Slowly moves tail up and down. Fairly tame. **Voice** Song consists of clear whistles occasionally interspersed with short, low, hard trill; from perch or in short flight. Sometimes with imitations. Call: various whistles. [CD10:25]

White-fronted Black Chat *Myrmecocichla albifrons* 15 cm

M. a. frontalis. Small, slim chat. **Adult male** Entirely black with ***white patch on forehead***. Flight feathers whitish below. **Adult female** Similar, but lacks white forehead. **Juvenile** Browner and slightly mottled. **HH** Uncommon to locally fairly common resident in open savanna woodland; also cultivated areas and clearings. Partial to recently burnt ground. Mostly perches in small trees and bushes, from where it pounces on prey on ground. Frequently flicks tail. Singly or in pairs. **Voice** Song: short phrases with sharp and rolling notes, including frequently repeated *uwheetirr*; sometimes with mimicry. Call a high-pitched, penetrating *heet*. [CD10:26]

galactotes
minor
Rufous Scrub Robin
Black Scrub Robin
♂
♀
Northern Anteater Chat
♂
Sooty Chat
♀
♂
White-fronted Black Chat

Familiar Chat *Cercomela familiaris* 14 cm

C. f. falkensteini. Rather nondescript, ***plain brown*** chat with ***rufous rump and outer tail***. Blackish centre of tail and terminal band form inverted T. Narrow pale eye-ring; ear-coverts tinged rusty. Underparts pale greyish becoming whitish on central belly. **Juvenile** Spotted buff above, barred dusky below, giving scaly appearance. **HH** Local and scarce to uncommon, extreme SE Senegal. Singly or in pairs in open, wooded habitats, especially around rocky outcrops, inselbergs, bare stony ground and erosion gullies. On landing, often raises and lowers tail and flicks wings. **Voice** Song a soft series of whistled and chattering notes. Call a high *whee* (*chak-chak*); also *cher-cher*. [CD10:20]

Blackstart *Cercomela melanura* 14 cm

C. m. ultima. Plain brownish-grey above with ***black uppertail-coverts and tail***. **HH** Single claim from Gambia (3–4 photographed near Fatoto, Feb 1974). Nearest populations in NE Mali (Hombori). Normally frequents rocky areas in arid country. Constantly flicks tail open, often simultaneously drooping and flicking wings half-open. **Voice** Call a harsh note; alarm a high-pitched *hiih*. [CD10:22]

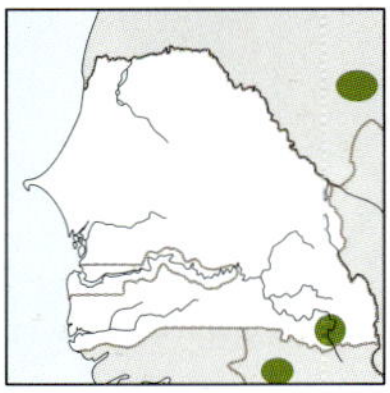

Mocking Cliff Chat *Thamnolaea cinnamomeiventris* 20 cm

T. c. bambarae. Fairly large chat with characteristic coloration. **Adult male** ***Black above*** with ***small white shoulder patch*** and rufous rump. Below, breast black, lower underparts rufous. **Adult female** Duller and lacking white on shoulder, black areas more sooty, rufous areas browner, less bright. **Juvenile male/female** Like respective adults but duller; male has less white on shoulder. **HH** Local, SE Senegal. Usually in pairs on rocky outcrops, inselbergs, escarpments and cliffs in savanna. Often slowly raises and lowers tail, fanning it on upswing. **Voice** Song melodious, rich and far carrying, with clear, vigorous and varied whistles or rapid imitations. Call a penetrating *seeeo* and *seeu-seeu*. [CD10:28]

Blue Rock Thrush *Monticola solitarius* 20 cm

M. s. solitarius. Larger, with longer bill and tail than Common Rock Thrush. **Adult male** Wholly dark blue (appearing blackish at distance). In fresh plumage (Sep–Dec) blue colour obscured by pale brownish feather tips. **Adult female** Upperparts and ***tail dark brown***; underparts buff-brown, throat and breast scaled, rest barred blackish-brown. **Immature** As adult female, but slightly paler brown; upperparts spotted buffish; underparts appear more spotted; barring on lower underparts paler and less sharply defined. **HH** Rare to scarce Palearctic migrant (Oct–early Apr). Year-round at Cap de Naze, S Senegal; breeding suspected. Mainly in rocky habitats. **Voice** Mainly silent. Calls include *tak-tak* (contact), *chuk-chuk* (alarm) and high-pitched *tsee*. [CD3:97]

Common Rock Thrush *Monticola saxatilis* 18.5 cm

Chat-like, rather long-billed thrush with ***short, dark-centred, rufous-orange tail*** in all plumages. **Male breeding** (from Apr) Head and upperparts greyish-blue with white patch on back, underparts orange-rufous. **Male non-breeding** Upperparts dark brown and mottled; underparts rufous and crescent-marked. **Adult female** As non-breeding male but paler, without white on back. **HH** Rare to scarce Palearctic migrant (Nov–Mar). In various habitats, incl. rocky plateaux, savanna, recently burnt bush, cultivated areas. Solitary. **Voice** Short *chak*. Mostly silent in winter quarters. [CD3:96]

Familiar Chat
Blackstart
♀
Mocking Cliff Chat
♂
♀
Blue Rock Thrush
♂
♀
♂ br
Common Rock Thrush
♂ non-br

WARBLERS (Sylviidae, Plates 107–112) *Large, varied family of mostly small songbirds. Sexes similar, except in* Sylvia*; juveniles unspotted. Mainly insectivorous. Broad range of vocalisations. Occur in all habitats. Identification should usually be based on a combination of plumage characters, habitat, behaviour, voice and distribution.*

ACROCEPHALUS WARBLERS *Dull-coloured species with mainly brown upperparts, paler underparts, rounded tails and prominent bills. Occur in reedbeds, marshes and rank vegetation; Palearctic migrants also in less typical habitats, far from water.*

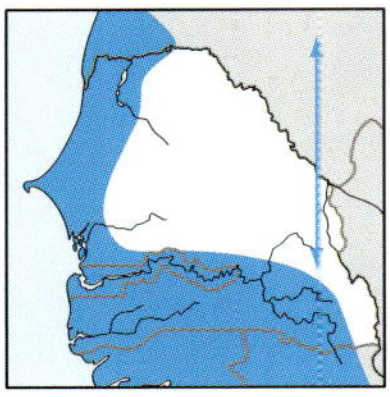

Sedge Warbler *Acrocephalus schoenobaenus* 13 cm

Easily distinguished from other small *Acrocephalus*, except Aquatic Warbler, by ***conspicuous creamy-buff supercilium*** bordered by dark eye-stripe, and ***streaked upperparts***. Rump unstreaked rusty-brown contrasting with dark brown upperparts and tail. Legs greyish. **First-winter** may have ill-defined creamy crown-stripe and faintly speckled breast. **HH** Fairly common Palearctic visitor (Oct–Apr) in reedbeds, rank vegetation by rivers and lakes, inundated grassland and similar habitats. Inquisitive and attracted to imitation of call. **Voice** Song a fast series of extremely varied rasping and musical notes, most repeated several times, including clear trills and imitations. Call *tuk*. Alarm *trrr*. [CD4:6]

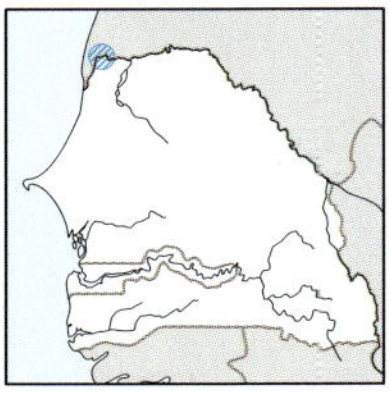

Aquatic Warbler *Acrocephalus paludicola* 13 cm

Boldly marked *Acrocephalus* with creamy-buff supercilium and sharply defined crown-stripe; lores usually pale. Upperparts (incl. rump) yellowish-buff heavily streaked blackish-brown (becoming greyish-brown with wear), mantle bordered by two broad buffish longitudinal stripes ('tramlines'); tail spiky. Underparts buffish-white; breast-sides and flanks washed buff and finely streaked. Legs pale pinkish. **First-winter** bright straw coloured with no or very little streaking on underparts. **HH** Scarce Palearctic winter visitor, NW Senegal (Djoudj), in extensive, water-logged open marshes with high grasses and very few or no bushes or trees. Little known and rarely recorded due to skulking behaviour. Red List category: Vulnerable. **Voice** Song (regularly uttered from Jan–Feb) includes call alternated with short series of rapidly repeated notes e.g. *krrr hweeweeweeweewee, krrr diidiidiidiidii, krrr dudududududu kirrr...* etc. Call a sharp *krrr krrr*. [CD4:5]

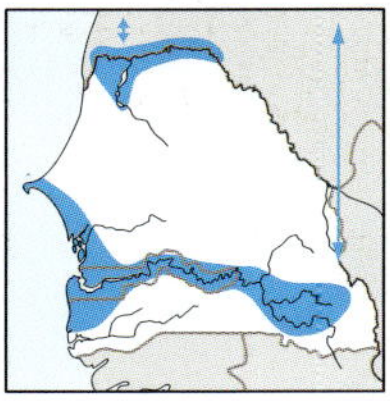

European Reed Warbler *Acrocephalus scirpaceus* 12.5–13.0 cm

A. s. scirpaceus. Plain brown above with ***contrasting rusty-tinged rump***. Rather flat forehead and crown in combination with long thin bill produce tapering profile. Pale supercilium faint and short; narrow whitish eye-ring. Underparts whitish with rusty-brown wash on breast-sides and flanks. Note primary projection. **HH** Common Palearctic visitor (Oct–Apr) in dense dry herbage, riverside thickets with tall grass, scrub, cassava plantations and gardens. Skulking but not shy; attracted to imitations of its call. **Voice** Song a series of slow, grating and nasal syllables repeated 2–5 times e.g. *kerr-kerr-kerr kirrik-kirrik-kirrik peet-peet-peet trer-trer-trer tjetje-tjetje...*, occasionally including imitations. In winter quarters song is usually more subdued. Call *krrrr* and *chrrrt*. Alarm a sharp *tek, tek* and *krrrt*. [CD4:7]

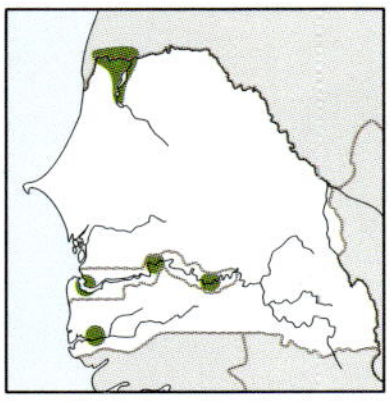

African Reed Warbler *Acrocephalus (scirpaceus) baeticatus* 12.0–12.5 cm

Very similar to European Reed Warbler and virtually indistinguishable in field but slightly smaller with shorter, rounder wing not extending beyond rump. Plumage slightly brighter coloured (***cinnamomeus***), or a shade darker with slight grey tinge (***guiersi***). Juvenile as adult. **HH** Locally fairly common resident, N Senegal (*guiersi*); scarce to rare, Gambia and S Senegal. In reedbeds, rank grass on lakeshores and similar wet or moist habitats. Status inadequately known through similarity with European Reed Warbler. **Voice** As European Reed Warbler. [CD10:52]

Marsh Warbler *Acrocephalus palustris* 12.5–13.0 cm

Very similar to European Reed Warbler and extremely difficult to separate when not singing. Typically more uniform olive grey-brown above (lacking warm brown tones), rump distinctly less rusty, wings slightly longer with narrow pale tips to 7–8 primaries visible, legs usually paler. **HH** Palearctic vagrant, N Senegal (Djoudj, Jan 1994). **Voice** Calls: sharp *tek*, low *tuk*, grating *krrr* (alarm). Song extremely rich and varied, including many imitations and liquid trills. [CD10:53]

Sedge Warbler

Aquatic Warbler

European Reed Warbler

African Reed Warbler

Marsh Warbler

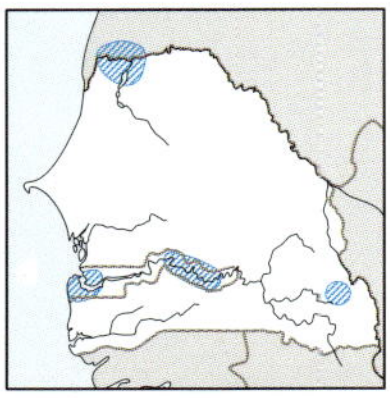

Great Reed Warbler *Acrocephalus arundinaceus* 19–20 cm

A. a. arundinaceus. Like very large and bulky European Reed Warbler, with much stouter bill, angular head and more prominent supercilium. **HH** Uncommon to scarce Palearctic visitor (Oct–May) in variety of habitats, incl. scrub, wooded grassland, dense herbage, cassava plantations and swamps. **Voice** Song a rather slow series of far-carrying, low, raucous, grating notes interspersed with high squeaks, repeated 2–3 times e.g. *krr-krr karra-karra-karra-keet-keet, kruruk-kruruk kreek-kreek*… In winter quarters utters both subdued and full song. Call a hard *krek*. [CD4:9]

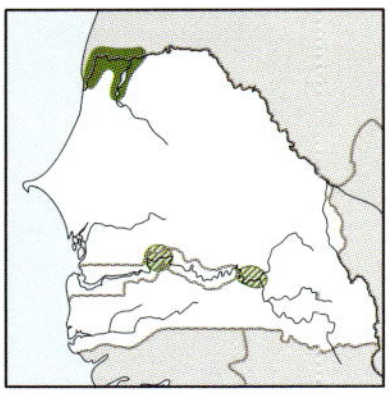

Greater Swamp Warbler *Acrocephalus rufescens* 18 cm

A. r. senegalensis. Robust, with dark brown upperparts, faint greyish supercilium, and greyish-white underparts washed tawny on flanks. **Juvenile** has warmer plumage tones. Great Reed Warbler is heavier and has warmer brown upperparts, obvious and pale supercilium, stouter bill and longer primary projection. **HH** Uncommon to locally fairly common resident in reedbeds, marshes, moist herbage and sugarcane plantations. Solitary and secretive but inquisitive. **Voice** Song: loud croaking, churring and gurgling notes. Call a harsh *chrr*, low *kreeok* and *chok!* [CD10:55]

LOCUSTELLA *WARBLERS Dull-coloured Palearctic migrants, with rounded, almost graduated tails and very long undertail-coverts. Skulk low in dense vegetation.*

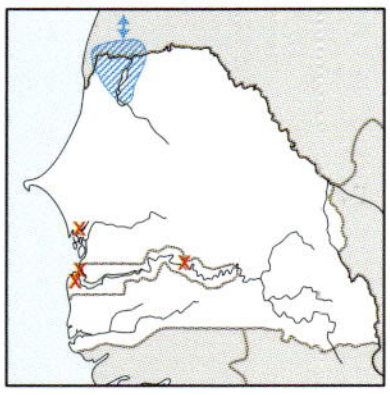

Grasshopper Warbler *Locustella naevia* 12.5 cm

L. n. naevia. Secretive, darkish warbler with diffusely streaked olive-brown upperparts, faint supercilium and graduated, rounded tail. Dirty whitish below, often washed brownish on breast with some diffuse streaking on breast, rear flanks and undertail-coverts. **HH** Scarce Palearctic visitor, Senegal (mainly on passage, Sep–Oct & Mar–May); rare, Gambia. Frequents dense rank herbage and regrowth. Unobtrusive, keeping low to ground, walking or running mouse-like through dense vegetation. Flight flitting, low and only over short distances. Bobs tail, thereby showing streaky undertail-coverts. **Voice** Call a short *twit* or *pit*. Song (exceptionally heard in winter quarters) a monotonous, insect-like reeling, uttered on constant high pitch. [CD4:3b]

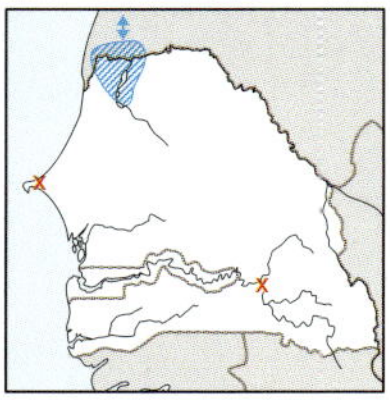

Savi's Warbler *Locustella luscinioides* 14 cm

L. l. luscinioides. Resembles dark European Reed Warbler (Plate 107), but larger and with broad, rounded tail and curved primaries (straight in all *Acrocephalus*). Plain warm brown above; narrow pale supercilium. Whitish below becoming buffish-brown on breast and flanks; ***undertail-coverts very long, cinnamon-buff, often with paler tips***. **HH** Scarce Palearctic visitor, Senegal (in north recorded Dec–Mar). In dense scrub and herbage near or in water. **Voice** Call a sharp *tswik*. Song (exceptionally uttered in winter quarters) an insect-like reel similar to that of Grasshopper Warbler but faster, lower pitched and more buzzing. [CD4:4]

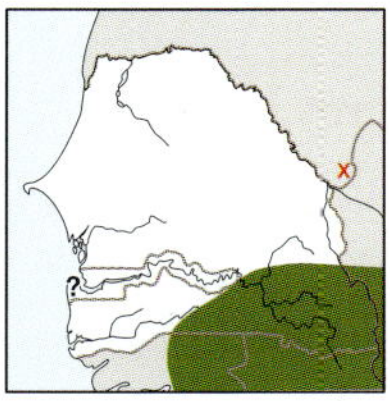

African Moustached Warbler *Melocichla mentalis* 19–20 cm

M. m. mentalis. ***Large, bulky warbler*** of uncertain affinities with ***fairly long, broad, rounded tail***. Plain earth-brown above with rufous-brown forehead, whitish supercilium and ***yellowish eye***. ***Black malar stripe*** contrasts with white throat; rest of underparts buff. Tail blackish-brown. **Juvenile** lacks rufous on forehead; wing-coverts tipped buff; underparts somewhat mottled; eye dark. **HH** Fairly common, SE Senegal; one unsubstantiated record, coastal Gambia (Dec 1973). In rank herbage in savanna. Usually hidden in tall grass or scrub but song attracts attention. Sometimes quite confiding, singing in the open. **Voice** Song a distinctive, cheerful and vigorous *tup-tup-twiddle-diddle-dee*. Call a hoarse, querulous *pièh-pièh-pièh-*… [CD11:38]

Great Reed Warbler

Greater Swamp Warbler

Grasshopper Warbler

Savi's Warbler

African Moustached Warbler

HIPPOLAIS *WARBLERS* *Plain coloured and arboreal, with prominent, rather broad, flattened bills and square tails.*

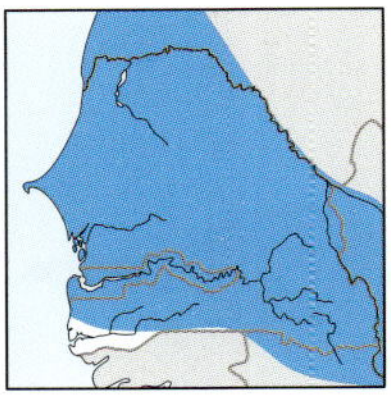

Western Olivaceous Warbler *Hippolais (pallida) opaca* 12–14 cm

Strikingly pale and rather elongated with long bill and flat crown. Upperparts dull pale grey to greyish-brown; rump slightly warmer coloured; short pale supercilium faint and typically not extending behind dark eye. Underparts whitish washed pale buff on breast-sides and flanks. Bill strong with convex ('bulging') sides. **HH** Common Palearctic visitor (Oct–Apr; records in all months). In wooded savanna, scrub, overgrown plantations and gardens. Solitary and restless. **Voice** Song a rapid, rather hoarse warble, including harsh, often repeated notes; somewhat reminiscent of European and African Reed Warblers. Call *chek!* and *churr-churr.* [CD4:11]

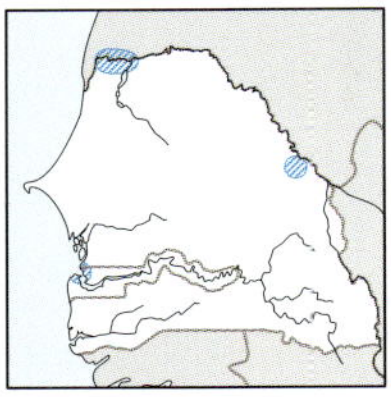

Eastern Olivaceous Warbler *Hippolais pallida* 12–13.5 cm

H. p. reiseri (Saharan Olivaceous Warbler). Similar to Western Olivaceous Warbler and often impossible to distinguish in field. Somewhat smaller and slightly paler and sandier, with pale wing panel (lacking in *H. opaca*). Bill finer with straight or slightly concave sides. **HH** Scarce N African migrant (Oct–Mar); status and wintering range inadequately known due to identification difficulties. Behaviour as Western Olivaceous but frequently pumps tail downwards (*H. opaca* holds tail still). Often hides within foliage. **Voice** Song a rapid, harsh warble, with short motifs being repeated before moving on. Calls include a hard *chek!* or *tak!* and *churr-churr.* [CD4:11]

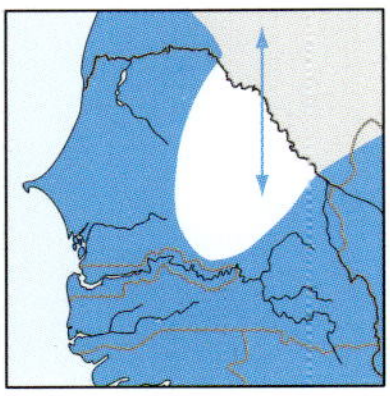

Melodious Warbler *Hippolais polyglotta* 13 cm

Olive-green above, yellow below. ***Pale lores*** and faint supercilium confer open-faced expression. In autumn, prior to moult, may appear very washed out with pale grey-green upperparts and creamy underparts. Very similar to Icterine Warbler but ***wings shorter*** (primary projection *c.* half of tertial length), ***wing panel lacking or indistinct***, crown rounder. Bill mainly orange-pink. Legs brownish (grey in Icterine). **HH** Fairly common Palearctic visitor (Oct–early May) in various wooded and scrubby habitats. **Voice** Song a rapid, sustained and varied babble, sometimes including imitations. Call a chattering, sparrow-like *tchèèèrr* or *tret-tret-…* [CD4:12]

Icterine Warbler *Hippolais icterina* 13.5 cm

Very similar to Melodious Warbler but has ***peaked*** (not rounded) ***crown, longer wings*** (primary projection almost equal tertial length) and, following moult in winter quarters, ***pale wing panel*** (formed by yellow fringes to inner secondaries and tertials). Bill relatively long (emphasised by sloping forehead) and mainly orangey. **First-winter** Duller, with greyish-green upperparts and pale underparts washed yellow. **HH** Palearctic vagrant, Senegal (Feb 2006) and Gambia (Oct 2005). Frequents open woodland and scrub. Raises crown feathers when excited. **Voice** Song a varied and sustained jumble of harsh, melodious and discordant, often repeated notes, including a distinctive nasal and creaky *GEE-a* (recalling sound made by squeezing rubber toy) and imitations. In winter quarters often a quieter and less varied subsong. Call *tek* and *deedeweet.* [CD10:60]

EREMOMELAS *Small and active warblers, occurring in pairs or small parties and often joining mixed-species flocks.*

Senegal Eremomela *Eremomela pusilla* 10 cm

Common savanna species with greyish head, green upperparts, white throat and upper breast, and bright lemon-yellow lower underparts. **Juvenile** duller. **HH** Common resident. In small parties, actively foraging and flitting between trees in wooded savanna. Frequently in mixed-species flocks. Breeds cooperatively. **Voice** Cheerful little trill, constantly uttered when foraging. Dawn song a rhythmic, monotonous *whirp-whirp-whirp-…,* delivered from perch, prior to and during breeding season; may be uttered continuously for up to 30 minutes. [CD11:44]

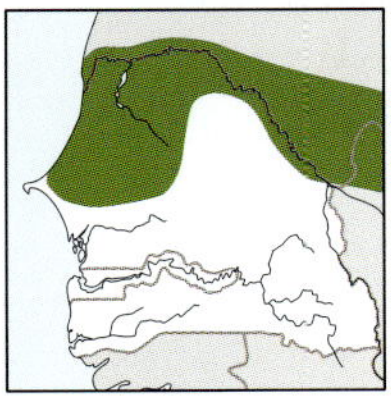

Yellow-bellied Eremomela *Eremomela icteropygialis* 10 cm

E. i. alexanderi. Head and upperparts pale brownish-grey; faint supercilium, throat and breast greyish-white; rest of underparts lemon-yellow. **Juvenile** duller; wing-coverts and tertials tipped and edged buff. **HH** Fairly common resident, N Senegal; locally south of 15°N. In pairs or small parties in arid *Acacia* scrub. **Voice** Short, clear song, similar to Northern Crombec's. [CD11:39]

Western Olivaceous Warbler

Eastern Olivaceous Warbler

Melodious Warbler

Icterine Warbler

Senegal Eremomela

Yellow-bellied Eremomela

PHYLLOSCOPUS *WARBLERS Very active, arboreal birds with largely greenish upperparts, pale underparts, a pale supercilium and a dusky eye-stripe.*

Willow Warbler *Phylloscopus trochilus* 11.0–12.5 cm

P. t. trochilus. Olive-green upperparts with yellowish supercilium; underparts pale yellowish on throat and breast, becoming whitish on belly; pale brownish legs. Adults in fresh plumage (after moult Jul–Aug) and first-winters are yellower overall, with greener upperparts and evenly yellow underparts. **HH** Common Palearctic visitor (mainly Sep–May) to various wooded habitats. Singly or in small parties; sometimes in mixed-species flocks. Restless, flicking wings and tail. **Voice** Soft, plaintive *hooeet*. Song sweet, short and slightly melancholy, starting with a few faint notes, growing louder, then fading and ending in a short flourish *se-se-see-see-swee-swee-sweet-sweet-sweeut-sweetoo*; occasionally uttered in winter quarters, esp. before northward migration. [CD4:31]

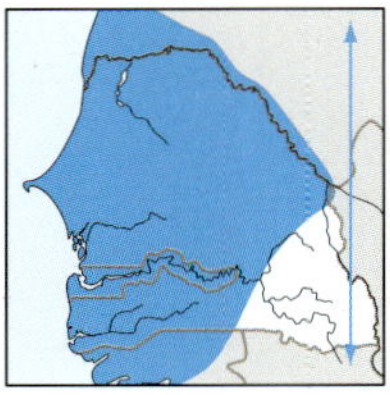

Common Chiffchaff *Phylloscopus collybita* 10–12 cm

Resembles Willow Warbler, but is less yellow and has more uniform, darker plumage, ***dark legs***, shorter primary projection (about half of tertial length), and more compact shape. Upperparts dull olive-brown to olive-green; underparts dull buffish-white with variable amount of yellow. Races vary slightly in plumage tones: **nominate** mainly brownish-olive above and dull buffish-yellow below; ***abietinus*** colder, greyer above, whiter, less yellow, below. **HH** Common Palearctic visitor (Sep–Apr; both races c.50%) to open, dry and wooded habitats. Often pumps tail up and down when foraging. **Voice** Call *hweet*, more monosyllabic than Willow Warbler's. Song a rhythmic *chiff-chaff-chiff-chiff-chaff-...* introduced with low *krr-krr*; uttered mainly from Dec. [CD4:28]

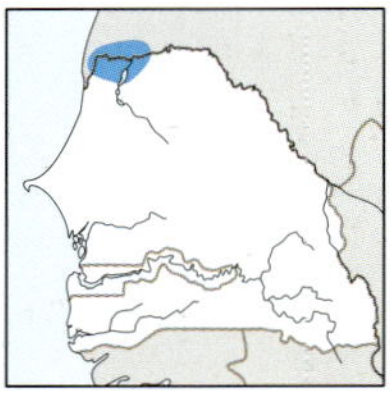

Iberian Chiffchaff *Phylloscopus (collybita) ibericus* 10–12 cm

Extremely similar to Common Chiffchaff (differences very slight and variable) and not safely separable unless voice heard. Upperparts greener with faint or no brownish tinge; head-sides washed yellow and usually lacking buff-brown tinge (rather dusky buff in Common Chiffchaff); supercilium usually more distinct and more lemon-yellow; underparts on average cleaner lemon-yellow and white with, typically, vivid lemon-yellow (occasionally pale yellow) undertail-coverts. Bill and legs paler. **HH** Palearctic migrant reported from N Senegal. Distribution inadequately known. Habits as Common Chiffchaff. **Voice** Slightly nasal, downslurred *fiuu* or *peeuw*. Immatures also utter a high-pitched *heep*. Song rhythmic and ending in short trill *tup-tup-tup-weet-weet-tsu-tchutututu*. [CD4:30]

Wood Warbler *Phylloscopus sibilatrix* 11.0–12.5 cm

Relatively bright and well-marked. Upperparts yellow-green with well-defined yellow supercilium; head-sides to throat and upper breast lemon-yellow; rest of underparts ***pure white***. Long wings (reaching beyond tail base) may produce short-tailed appearance (primary projection ≥ tertial length). Adults in fresh plumage (following moult Jun–Sep) and first-winters have less bright upperparts and paler yellow on underparts. **HH** Rare Palearctic passage migrant. Active, but quieter than Willow Warbler, and does not flick wings and tail. **Voice** Song an accelerating series of *sip* notes ending in shivering trill, often preceded by clear, piping *piu-piu-piu-piu*. Call *piu*. Sings and calls irregularly in winter quarters, but increasingly prior to migration. [CD4:27]

Western Bonelli's Warbler *Phylloscopus bonelli* 10.5–12.0 cm

Pale. Upperparts pale greyish-brown with narrow yellowish-green edges to wing and tail feathers (***forming green panel on closed wing***, contrasting with dark-centred tertials), and ***yellowish-green rump*** (often obscured by wings). Underparts whitish with buffish suffusion. Dark eye obvious within ***bland face***; faint pale supercilium. **HH** Fairly common to uncommon Palearctic visitor (Sep–Apr) to dry scrub and bushy savanna. **Voice** Call *hoo-eet*. Song a short, shivering trill, reminiscent of finale of Wood Warbler's song, but slower, lower pitched and lacking acceleration. [CD4:26]

Yellow-browed Warbler *Phylloscopus inornatus* 9.0–10.5 cm

Greyish-green above with ***long, prominent, pale yellow supercilium*** and ***double yellowish wingbar***. Tertials edged whitish or pale yellow. Underparts off-white; legs pinkish-brown. **HH** Palearctic vagrant, N Senegal (Dec 2003) and Gambia (Dec 2005). Restless and often flicking wings. **Voice** Sharp *sweest*, frequently uttered.

Willow Warbler
yellow individual
grey individual
Common Chiffchaff
yellow individual
grey individual
Iberian Chiffchaff
Wood Warbler
Western Bonelli's
Warbler
Yellow-browed
Warbler

SYLVIA WARBLERS *(Plates 111–112) Palearctic scrub and woodland warblers. Have two characteristic vocalisations: a sharp tak and a harsh churr. Song a sweet pleasant warble.*

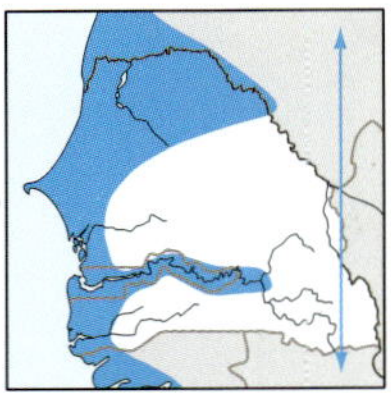

Blackcap *Sylvia atricapilla* 13.5–15.0 cm

S. a. atricapilla. ***Plain grey*** with diagnostic ***black*** (male) ***or rufous*** (female) ***cap***. **HH** Uncommon to locally fairly common Palearctic passage migrant and visitor (Oct–early May) in various wooded habitats. **Voice** Most frequently heard calls are a hard *tek* and *chrrr*. Song a rich warble starting with a hurried subdued jumble followed by clear whistles and ending in a loud flourish. In winter quarters mostly a subdued, less varied, but more sustained version. [CD4:15]

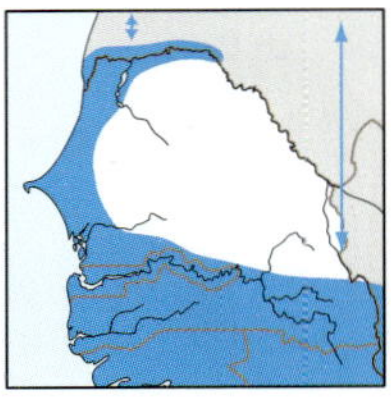

Garden Warbler *Sylvia borin* 13.0–14.5 cm

S. b. borin. ***Featureless***, plain grey-brown and buff with gentle, open face, rather short, stubby bill, rounded crown and greyish legs. Lack of any distinctive features is in itself diagnostic. **HH** Uncommon to locally fairly common Palearctic passage migrant and visitor (main passage Sep–Nov & Apr–early May) in various wooded and bushy habitats. **Voice** Calls a sharp *tek* and a low, harsh *chrrr*. Song a sustained, melodious, even warble of largely mellow notes; in winter quarters often subdued. [CD4:14]

Barred Warbler *Sylvia nisoria* 15.5–17.0 cm

Robust, with relatively long tail. **Adult** Crescentic barring on breast-sides and flanks; tail tipped and edged white (often evident as white corners in flight). **First-winter** Brown to grey-brown above with diagnostic ***paler tips and fringes to wing-coverts, tertials and uppertail-coverts***; faint barring on undertail-coverts. **HH** Palearctic vagrant, N Senegal (Djoudj, Feb 1990 & Apr 1993) and Gambia (Ginak Island, Feb 1999) in dry bush and woodland. Skulking, shy and rather slow. Frequently raises crown and flicks tail. Flight strong and usually heavy. **Voice** A harsh *chak* and a rattling *chrrrrt-t-t-t-...* [CD10:61]

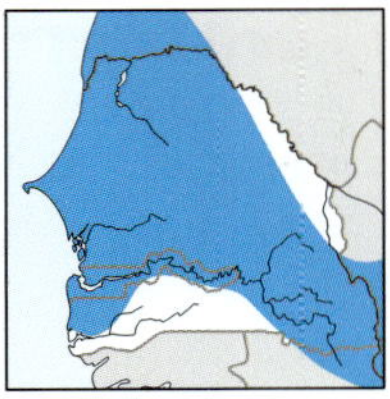

Common Whitethroat *Sylvia communis* 13–15 cm

S. c. communis. Medium-sized, rather slim warbler with ***rusty wing panel*** and ***white outer tail feathers***. **Adult male** Head grey, throat pure white, rest of underparts whitish with buffish-pink wash to breast. **Adult female** Head browner, underparts buffish. **First-winter** Very much as adult female but duller; outer tail feathers fringed buffish-white. **HH** Locally common to scarce Palearctic visitor (Sep–Apr). In dry scrub, wooded savanna, gardens. **Voice** Calls include sharp *tak*, grating *charrr* and nasal *wed wed wed*. Song a brief, hurried, scratchy warble. [CD4:16]

Lesser Whitethroat *Sylvia curruca* 12.5–13.5 cm

S. c. curruca. Greyish upperparts, white throat and dark ear-coverts (forming usually distinct mask, barely noticeable in some). Tail edged white; breast and flanks tinged pinkish-buff. Legs dark grey. Adult female Subalpine Warbler has buffish underparts and pinkish legs. First-winter Sardinian and Rüppell's Warblers have pale brownish or red-brown legs. **HH** Rare Palearctic visitor, N Senegal (Oct–Mar) and Gambia (two records, Jan & Feb 1996). In dry scrub and *Acacia* woodland. Skulking; often low. **Voice** Call a sharp, frequently uttered *tak* or *chek*, also *charr*. Song a brief muffled warble followed by fast, far-carrying rattle *djedjedjedjedjedje*. [CD4:17]

♀
Blackcap
♂
Garden Warbler
ad
Barred Warbler
1st-winter
♀
Common Whitethroat
♂
Lesser Whitethroat

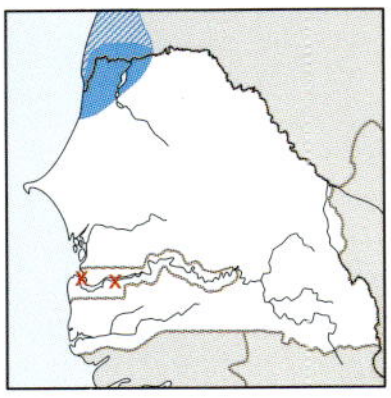

Spectacled Warbler *Sylvia conspicillata* 12–13 cm

S. c. conspicillata. Small and delicate ***with striking rusty wing panel***. Resembles miniature Common Whitethroat. Narrow white eye-ring (broader above eye). Head rounded with rather steep forehead; wings relatively short. **Adult male** Grey head; black area from bill to eye; white throat. **Adult female** Browner; blackish tail contrasts with rest of upperparts. **HH** Palearctic visitor (Nov–Apr), Senegal (occasionally frequent in NW) and coastal Gambia (rare). Mainly singly or in pairs in dry scrub near coast, moving restlessly through and between low bushes, cocking tail and moving it sideways. Frequently hops on ground. **Voice** Call a distinctive harsh rattle *trrrrrrr*; also *tek*. Song a typical, short, fast, high-pitched *Sylvia* chatter. [CD4:21]

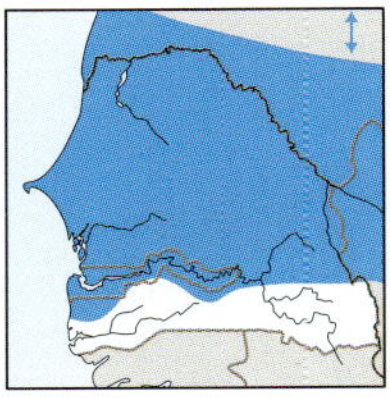

Subalpine Warbler *Sylvia cantillans* 12–13 cm

Adult male *inornata/iberiae* (Western Subalpine Warbler) Easily identified by blue-grey upperparts, fairly uniform rusty-orange underparts (slightly warmer-coloured in *inornata*), with only central belly white, ***distinct white moustachial stripe and red orbital ring***. ***S. (c.) subalpina*** (Moltoni's Warbler) paler and more buffish-pink below (not rusty-orange). **Adult female** Paler and duller. Pale grey-brown upperparts; pinkish-buff to buffish-white underparts; paler orbital ring. **Immature** Browner, less grey. Buffish-brown above; buff below; pale fringes to wing feathers; pale orbital ring. Female/first-winter plumages sometimes difficult to separate from those of other *Sylvia*. **HH** Common Palearctic visitor (Sep–Apr; *iberiae* and *inornata*). Wintering range of *subalpina*, breeding on Mediterranean islands, unclear (at present only definitely recorded in SE Mauritania, NE Nigeria and N Cameroon). In dry scrub; also gardens, mangroves. Frequently cocks tail. **Voice** Call a hard but quiet *tek* (*iberiae* and *inornata*) or rattling *trrrt* (*subalpina*). Song (occasionally uttered in winter quarters) consists of regularly spaced phrases of a fast, clear and scratchy warble. [CD4:20] **Note** If split from *S. cantillans*, scientific name of Western Subalpine Warbler becomes *S. inornata*.

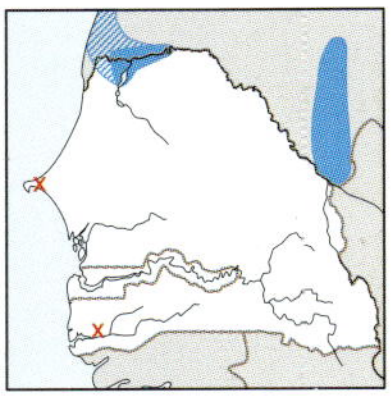

Sardinian Warbler *Sylvia melanocephala* 13–14 cm

S. m. melanocephala. **Adult male *Black hood contrasting with pure white throat***. Conspicuous red orbital ring. Upperparts grey; ***underparts whitish with greyish flanks***. **Adult female** Hood dusky-grey, upperparts brown, flanks dull brown. **First-winter male/female** As respective adult but males have browner upperparts and females a brown hood; tail corners buffish; orbital ring brownish. **HH** Palearctic visitor (Nov–Mar), Senegal (uncommon in north; reported once from south). In bush country and wooded savanna. Skulking but not particularly shy; inquisitive. Frequently cocks tail. **Voice** Call a frequently uttered, harsh rattle *chretetikitikitik...*, fast, harsh *treet-treet* or *chreet* notes, and hard *tsek*. Song an even, pleasing warble, incorporating harsh calls. [CD4:19]

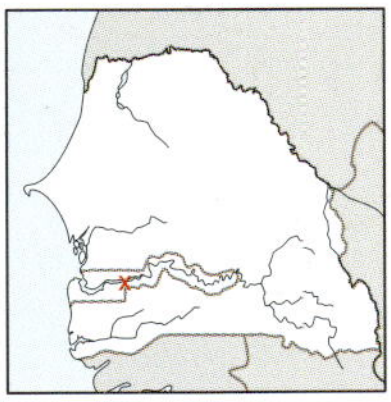

Rüppell's Warbler *Sylvia ruppeli* 12.5–13.5 cm

In all plumages, has grey upperparts and contrasting pale fringes to dark tertials and greater coverts. **Adult male** Crown, head-sides to upper breast black with ***bold white submoustachial stripe***. Conspicuous red orbital ring. **Adult female** Variable. Some as male but duller and browner, with mottled black crown and throat, less contrasting submoustachial stripe, and buffish-white underparts. Others have greyish head and whitish throat lacking dark mottling. **First-winter male** Black on head less uniform; upperparts browner. **First-winter female** Head grey; throat pale; no dark mottling; wings browner. Poorly marked females and first-winters could be confused with female/first-winter Sardinian Warbler, but latter have browner upperparts and less conspicuous fringes to wing feathers. **HH** Palearctic vagrant, Gambia (Tendaba, Nov 2006). Skulking but not shy. **Voice** Call a sparrow-like chatter; also sharp *tak*. Song a series of short, fast, dry, chattering phrases. [CD10:62]

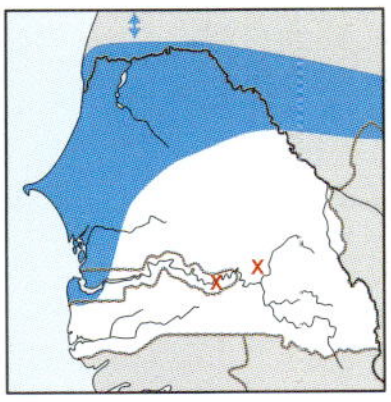

Western Orphean Warbler *Sylvia hortensis* 15–16 cm

Rather ***large, stout*** warbler with ***black or dusky face***, white throat and ***white outer tail feathers***. **Adult male** Blackish hood; pale eyes. **Adult female** Grey-brown crown. First-winter similar, but paler; crown brown as upperparts; outer tail feathers tinged buffish; eye usually dark. **HH** Uncommon to scarce Palearctic visitor (Sep–May), N Senegal and Gambia, in dry scrub and wooded savanna. Probably under-recorded. Not shy, though often remains within cover, quietly gleaning insects in tops of trees and bushes. In flight may recall chat. **Voice** Call a sharp *tak*. Alarm a rattling *trrrr*. [CD4:13]

♀
♂
Spectacled Warbler
♂
♀
Subalpine Warbler
1st-winter
♀
♂
Sardinian Warbler
1st-winter
♀
Rüppell's Warbler
♂
♀
♂
Western Orphean Warbler

CROMBECS Very small size and extremely short tail create distinctive jizz. Often in mixed-species flocks.

Northern Crombec *Sylvietta brachyura* 9 cm

S. b. brachyura. Very small, dumpy and unobtrusive. Grey above, tawny below, with buffish supercilium and dusky eye-stripe. **Juvenile** has wing-coverts tipped tawny. **HH** Fairly common to common resident. Singly, in pairs or in small family parties in wooded grassland. Actively inspects branches, occasionally hanging upside-down. Often joins mixed-species flocks of insectivores. **Voice** Song a short, sweet, clear warble. Contact call a soft, repeated *krrr*; also a soft, high-pitched *tsweet-krrr*. [CD11:50]

Green Crombec *Sylvietta virens* 9 cm

S. v. flaviventris. Top of head dark brown with faint pale supercilium and pale brownish face; upperparts olive-green. Throat and upper breast pale brown merging with yellow on lower breast and upper belly; lower belly whitish. **Juvenile** is mostly yellow below. **HH** Rare and very local, coastal Gambia; probably more widespread, SW Senegal (Casamance). In pairs or small family groups in second growth, dense shrub and tangles along tracks and clearings; mostly at mid-levels. **Voice** Song a short, sweet, rather high-pitched, clear whistle, descending in scale. Call a dry *prrt, prrt*. [CD11:48]

Lemon-bellied Crombec *Sylvietta denti* 8 cm

S. d. hardyi. Tiny and inconspicuous, with olive-green upperparts and lemon-yellow underparts. Crown olive-grey; head-sides and throat greyish-white with some pale rufous on ear-coverts and lower throat. Breast tinged olive. **Juvenile** has wing-coverts and tertials tipped greenish-white. **HH** No certain records. Claimed from Gambia but habitat unusual. Normally singly or in pairs within forest canopy, occasionally venturing lower at edges. Frequents mixed flocks of small insectivores. **Voice** Song a rapid series of identical, high-pitched notes (usually up to seven), *tswee-tswee-tswee-tswee-...* easily passing unnoticed, or of two notes on different pitch *tsee-tsu-tsee-tsu-tsee-tsu-...* [CD11:49]

CAMAROPTERAS Small, with rather long and straight bills; colour of thighs contrasting with belly. Skulk in dense vegetation, mostly low down. Have distinctive mewing calls. Pouch-like nest sewn into leaves of a shrub. Placed in family Cisticolidae.

Grey-backed Camaroptera *Camaroptera brachyura* 11.5 cm

C. b. brevicaudata. Small grey warbler with contrasting ***yellowish-green wings***. Central belly whitish; thighs orange-buff. Assumes non-breeding plumage in dry season (Oct–May) when upperparts become ashy-brown and underparts paler. **Juvenile** is dull olive-green above, pale lemon-yellow below. **HH** Common resident. Frequents dense shrubbery in various habitats. Tail often held cocked at sharp angle over back. One of the most widespread and common warblers. **Voice** Call a distinctive mewing note. Song variable, but always a rhythmic repetition (variable in speed) of a single note, e.g. *churrup-churrup-churrup-..., churp-churp-churp-..., chrup chrup chrup..., kechup-kechup-...*, etc. [CD11:32]

Olive-green Camaroptera *Camaroptera chloronota* 11 cm

C. c. kelsalli. Skulking, but song loud and distinctive. Dark olive-green above, uniformly slate-grey below. Lores and ear-coverts tinged rufous; thighs cinnamon. **Juvenile** has upperparts more yellowish-green; underparts pale lemon-yellow washed grey on flanks. **HH** Unproven claims from SW Senegal (Casamance; specimens from Niokolo Koba 1956 & 1967 were misidentified Grey-backed Camaropteras) and Gambia (two coastal sight records, Jan 1984 & Oct 1986). Occurs within dense undergrowth at forest edges. **Voice** Song a remarkable, loud and sustained series of a single note uttered at same pitch and without interruption for up to several minutes before ceasing abruptly. [CD11:35]

Northern Crombec
ad
Green Crombec
juv
Lemon-bellied Crombec
juv
Grey-backed Camaroptera
ad
Olive-green Camaroptera

CISTICOLAS (Plates 113–114) Large genus of similar-looking species, occurring in grassy habitats. Tail strongly graduated. Males usually larger than females. Juveniles are more rufous and often have yellow on underparts. Usually breed during rains, which is best season to observe them, as males are singing and displaying. Occur singly or in pairs, after breeding in small family groups. In the dry season some may form small parties, sometimes with other cisticolas. Among the hardest species to identify on plumage characters alone, but vocalisations and aerial displays, if any, diagnostic. For correct identification, note size (small, medium or large), upperpart pattern (streaked or plain), head pattern (colour of crown and head-sides, presence or absence of streaking and supercilium), tail colour and pattern. Then check against known range. All species in Senegambia are residents.

Medium-sized Plain upperparts (plumage warmer coloured in non-breeding):

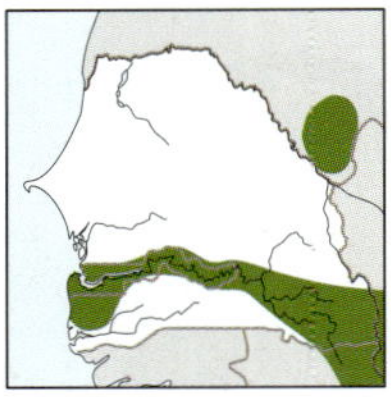

Red-faced Cisticola *Cisticola erythrops* 12–14 cm

C. e. erythrops. ***Forehead and head-sides rufous*** (diagnostic); crown and upperparts dark olivaceous-brown, without contrasting rufous wing panel. **HH** Fairly common in various grassy and bushy habitats; often associated with swamps, watercourses and drainage ditches. Forages low, skulking in rank herbage. **Voice** Rhythmic succession of loud, varied notes, e.g. *ch-ch-ch trweet-trweet-trweet WEET-WEET-WEET plik-up plik-up plik-up WEET WEET WEET…*; from low perch and usually in duet. Alarm *cheap cheap cheap*. [CD10:90]

Singing Cisticola *Cisticola cantans* 11.5–14.0 cm

C. c. swanzii. ***Chestnut crown and wing panel*** contrasting with brown upperparts; head-sides pale brown washed rufous. **HH** Common. Frequents thick bushes and rank undergrowth; generally in drier areas than Red-faced Cisticola. **Voice** Single, repeated, variable note, such as *krwleep krwleep krwleep…* or *p-lip p-lip…* and *kwiplip kwiplip…* etc. Occasionally duets, female uttering low grating notes. Calls include thin *tsit tsit* and drawn-out *cheerr cheerr*. [CD10:91]

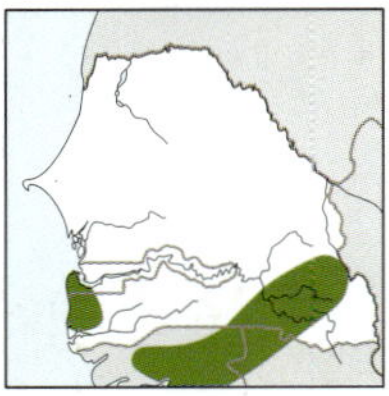

Whistling Cisticola *Cisticola lateralis* 12.5–14 cm

C. l. lateralis. **Adult breeding** Greyish-brown to dull rufous-brown above; flight feathers edged rufous. Underparts whitish washed greyish on breast-sides, becoming buff or grey-buff on flanks and undertail-coverts. **Adult non-breeding** Rufous-brown above, rich buff below becoming whitish on throat and belly. **HH** Locally fairly common in savanna woodland, overgrown farmland with scattered trees, bushy forest edge and large forest clearings with low scrub and tall trees. **Voice** Short, vigorous and melodious whistled phrase, from prominent perch. Call *thup*, often repeated. Alarm harsh and rasping. [CD10:87]

Small Plain upperparts not contrasting with top of head; typically sing from treetop or other high perch:

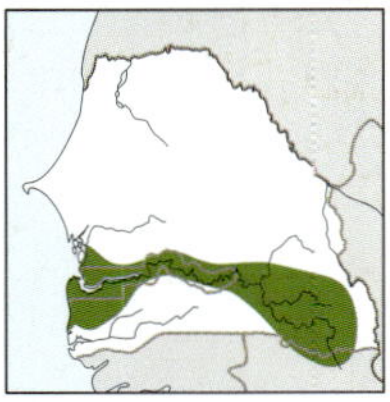

Short-winged Cisticola *Cisticola brachypterus* c.10 cm

C. b. brachypterus. **Adult breeding** Upperparts dull brown. Rufous fringes to wing feathers contrast slightly with duller mantle. **Adult non-breeding** Upperparts warmer brown with faint dusky streaks. **HH** Fairly common in various types of wooded grassland, incl. grassy edges of farmland with scattered trees, scrub at forest edge, second growth and grassy clearings in woodland. **Voice** Continuously repeated short, rapid series of 2–3 high-pitched notes descending in pitch e.g. *see-se-swu see-su-swu…* and variations. [CD10:81]

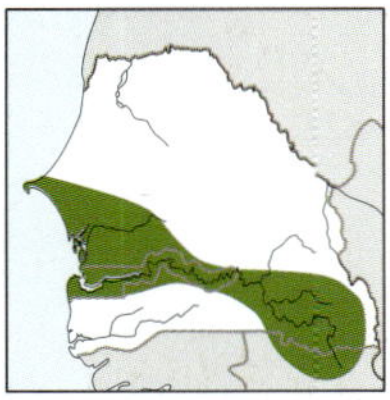

Rufous Cisticola *Cisticola rufus* 10 cm

Upperparts plain rusty-brown, more rufous on rump. Underparts white washed pale rusty-brown. **HH** Fairly common in wooded grassland. **Voice** Song 2–3 thin descending notes e.g. *see see hu, see see hu, …*. Also rapid, hesitant series *tsutitititsutititu…* [CD10:82]

Medium-sized Plain or streaked upperparts contrasting with rufous top of head:

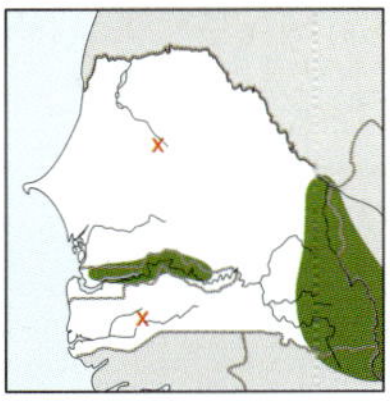

Dorst's Cisticola *Cisticola guinea* 13 cm

Plain reddish cap contrasts with grey-brown upperparts. Upperparts plain in breeding, diffusely streaked dark brown in non-breeding plumage. Underparts mainly whitish faintly washed buff, esp. on flanks and thighs, with buff vent. **HH** Locally fairly common in open grassy areas with scattered trees or bushes. **Voice** Drawn-out trill of variable speed, often preceded by short vibrant note and followed by simple motif, repeated several times, e.g. *chirrrrrrr tsu-wheet tsu-wheet tsu wheet…*, given from atop stem, bush or small tree. [CD10:80]

Red-faced Cisticola
Singing Cisticola
ad non-br
ad br
Whistling Cisticola
ad br
ad non-br
Short-winged Cisticola
Rufous Cisticola
ad br
ad non-br
Dorst's Cisticola

Medium-sized and large *Streaked upperparts:*

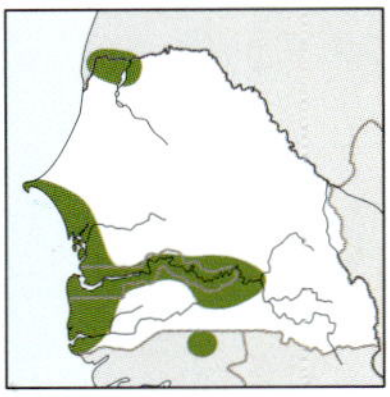

Winding Cisticola *Cisticola galactotes* 12–15 cm

C. g. amphilectus. Medium-sized, streaked cisticola occurring in moist habitats. **Adult breeding** Forehead dull rust-brown; russet-edged flight feathers. **Adult non-breeding** Warmer coloured; tail longer. **HH** Fairly common in rank grass and sedge of marshes, temporary water pans and similar wet places. **Voice** Monotonously repeated, single, drawn-out rasping note *zrrrreeeeeee*, from low perch in herbage. [CD10:95]

Croaking Cisticola *Cisticola natalensis* 12.5–16.0 cm

C. n. strangei. Large, bulky cisticola with stout bill (with strongly curved culmen). **Adult breeding** Upperparts dark earth-brown streaked dusky; almost plain when worn. Tail ash-brown tipped white. Bill black. Female noticeably smaller. **Adult non-breeding** Upperparts buffish-brown boldly streaked black. Tail longer and tipped buff. Bill more horn-coloured. **HH** Scarce, Senegal; rare, Gambia. In rank grass with scattered low bushes and edges of cultivation. **Voice** Loud *klink klunk* in display flight at medium height; slower *kluuu klink!* and *kzeee klunk!* when perched. Occasionally uttered by birds in non-breeding plumage. Croaking alarm call. [CD10:75]

Small *Streaked upperparts (warmer coloured in non-breeding); aerial displays; open grassy areas:*

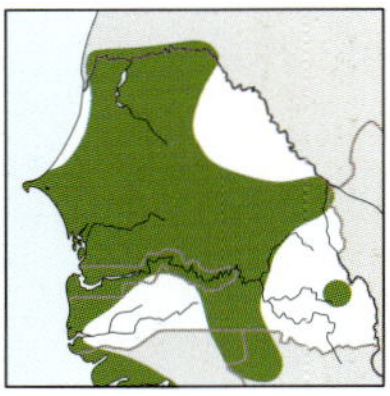

Zitting Cisticola *Cisticola juncidis* 10 cm

C. j. uropygialis. **Male breeding** Top of head brownish-buff heavily streaked black; head-sides buffish; nape pale, vaguely streaked; rump dull, mostly plain rufous-brown. Underparts mainly buffish, becoming whitish on throat and central belly. **Male non-breeding/female** More buff; streaking less heavy. Overall less rufous than Black-backed Cisticola, with brownish-buff (not rufous-brown) crown, which is always streaked, buffish supercilium, rather dull rufous-brown (not orange-rufous) rump and no orange-rufous on lower flanks. **HH** Common in wet and dry grassy areas without trees in savanna, edges of farmland and similar habitats. Unobtrusive outside breeding season, foraging mostly low in rank grass or on ground. **Voice** Monotonous repetition of single harsh *tsip*, *zit* or *tslik*, from perch or, more often, in high, undulating display flight, one note at each dip; without wing-snapping. Alarm call a rapid *tik-tik-tik.-*... [CD4:36]

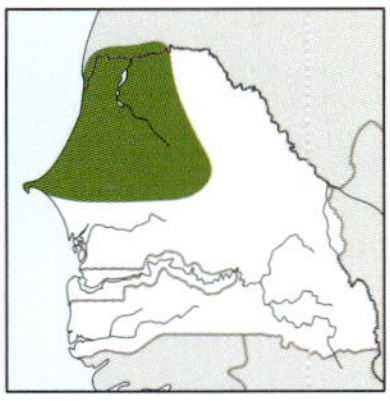

Desert Cisticola *Cisticola aridulus* 10–12 cm

C. a. aridulus. In all plumages very similar to Zitting Cisticola but noticeably paler, more sandy coloured. **HH** Fairly common in short dry grass and fallow farmland in semi-arid habitats, N Senegal. **Voice** Rapid repetition of clear, high-pitched note *teehu*, interspersed with dry clicking bill-snaps (only audible at short range) and occasional wing-snaps: *teehu (tk)-teehu (tk)-teehu (tk)-tktktktk-teehu (tk)-*... [CD10:74]

Black-backed Cisticola *Cisticola eximius* *c.*10 cm

C. e. occidens. **Male breeding** Top of head pale rufous-brown; head-sides pale greyish-buff; pale supraloral spot. Upperparts brown boldly streaked black; ***conspicuous orange-rufous rump***. Underparts whitish tinged buff on breast-sides and orange-rufous on lower flanks. **Male non-breeding/female** Crown streaked black; upperparts more rufous; flanks orangey. Brighter, overall more rufous than Zitting Cisticola, with pale supraloral spot (no supercilium). **HH** Very local; possibly more widespread (distribution inadequately known due to identification difficulties). In moist open grassland and grassy valleys. **Voice** Rather sharp, vibrant *tsree-tsree-tsree-*..., uttered from perch or in undulating display flight high above territory, sometimes accompanied by rhythmic noise, sounding like wing-snapping, and occasionally including a distinctive, sharp, dissonant *chereet-chereet*; latter also used in alarm. [CD10:72]

ad non-br
ad br
Winding Cisticola
ad br
ad br
ad non-br
Croaking Cisticola
ad non-br
Zitting Cisticola
ad br
ad non-br
Desert Cisticola
Black-backed Cisticola

***PRINIAS AND ALLIES** Rather slender, active warblers with long, graduated tails, occurring in pairs or small parties in grassy and bushy habitats.*

Tawny-flanked Prinia *Prinia subflava* 11–12 cm

Very common warbler with brownish upperparts, distinct ***buffish supercilium***, creamy underparts tinged pale tawny on breast-sides and flanks, and rather long, graduated tail. In savanna zone, plumage assumed in dry season is paler, rusty-brown above, with longer tail and horn-coloured bill. Plumage variation, with palest birds in driest zones, strongly clinal; two races occur: **nominate** (northern savanna) and ***melanorhyncha*** (southern savanna and forest zone). **Juvenile** Underparts tinged pale yellow; bill yellowish. **HH** Common resident. In pairs or small parties in wide variety of grassy and bushy habitats, incl. overgrown plantations, gardens, road edges and scrub at forest edges. Active and vocal, foraging low in rank herbage and shrubbery, with frequent jerky movements of cocked tail. Not shy. **Voice** Song, uttered from prominent perch, a monotonous, rhythmic series of a single note; speed of delivery and tone variable, e.g. sharp *tzreep tzreep tzreep tzreep…* and fast *plip-plip-plip-plip-…* Alarm a harsh *zbeee*. [CD11:2]

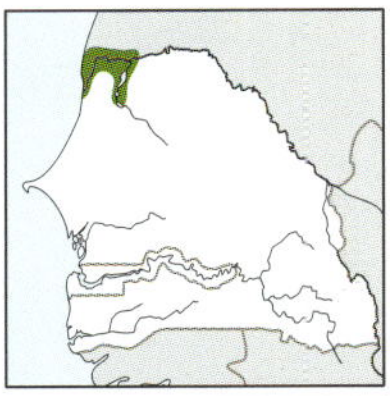

River Prinia *Prinia fluviatilis* *c.* 12 cm

As Tawny-flanked Prinia, but somewhat greyer above and whiter below. **HH** Locally common, NW Senegal and Gambia; distribution inadequately known. Inhabits waterside vegetation, such as reedbeds, rank herbage and bushes. Very active. Sings very low, near ground or in tops of bushes. **Voice** Song a rapid, rhythmic series of a single, high-pitched note. Not reminiscent of Tawny-flanked Prinia; notes more drawn-out, rhythm much less variable. [CD11:3]

Red-winged Warbler *Heliolais erythropterus* 12 cm

H. e. erythropterus. Prinia-like warbler with rufous wings and long, graduated tail. **Adult breeding** Diagnostic combination of ***grey head and upperparts, bright rufous wings***, and white throat merging into buff on rest of underparts. **Adult non-breeding** Pale vinous-rufous upperparts; horn-coloured bill. **Juvenile** Similar to adult non-breeding, but paler or duller. **HH** Uncommon to locally fairly common resident. In pairs or small family groups in wooded grasslands, restlessly foraging low in rank herbage. May perch conspicuously on bushes and fly to treetops. **Voice** Song a rapid, monotonous series of high *pseep* or *thu-weet* notes; in duet second bird utters dry chatter. Calls include thin *tseek* notes and little dry churring trill. [CD11:25]

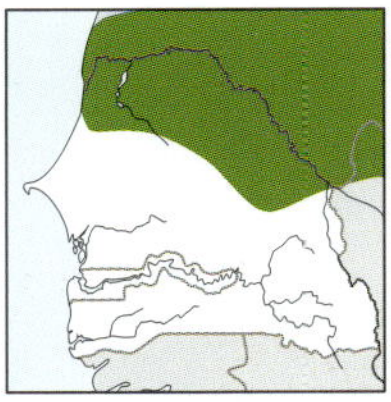

Cricket Warbler *Spiloptila clamans* *c.* 11.5 cm

Distinctive, delicately coloured, pale and long-tailed dry-country warbler with black-and-white pattern on wing-coverts. **Adult male** Hindcrown and nape pale grey. **Adult female** Hindcrown and nape cinnamon. **HH** Locally not uncommon resident in sparsely vegetated thorn scrub, N Senegal Usually in small, restless parties, constantly jerking tail in all directions. Not shy, often perching in the open or foraging on ground. When disturbed, dives for cover and prefers to escape by running rather than flying. **Voice** Song consists of dry, rhythmic, insect-like trills and series of single, high-pitched notes, both delivered in fast or slower tempo. Duets. [CD11:22]

***APALISES** Slender warblers with rather long and graduated tails. Active; often in mixed-species flocks of small insectivores. Song typically consists of rhythmic repetitions of a single note.*

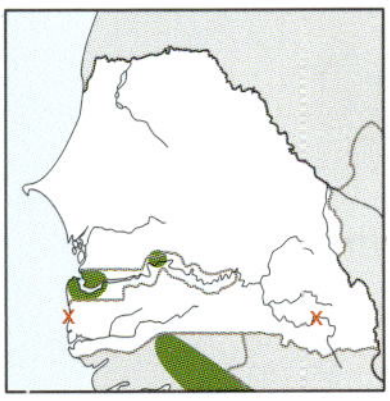

Yellow-breasted Apalis *Apalis flavida* 11.5 cm

A. f. caniceps. **Adult** Head grey, upperparts green, underparts white with yellow breast-band. **Juvenile** Head green, throat yellow, bill pale. **HH** Uncommon and local, Gambia; only recently discovered S Senegal (near Niokolo Koba, Dec 2004 & Feb 2005; Kafoutine, Oct 2006). Frequents gallery forest, thickets in wooded savanna, forest patches, bush, mangroves. In pairs or small family groups, from low down to treetops. Often in mixed-species flocks. **Voice** Male song a dry, rhythmic *churup-churup-churup-churup-…* very similar to one of the songs of Grey-backed Camaroptera. In duet, female utters sharp *kep-kep-kep-kep-…* [CD11:8]

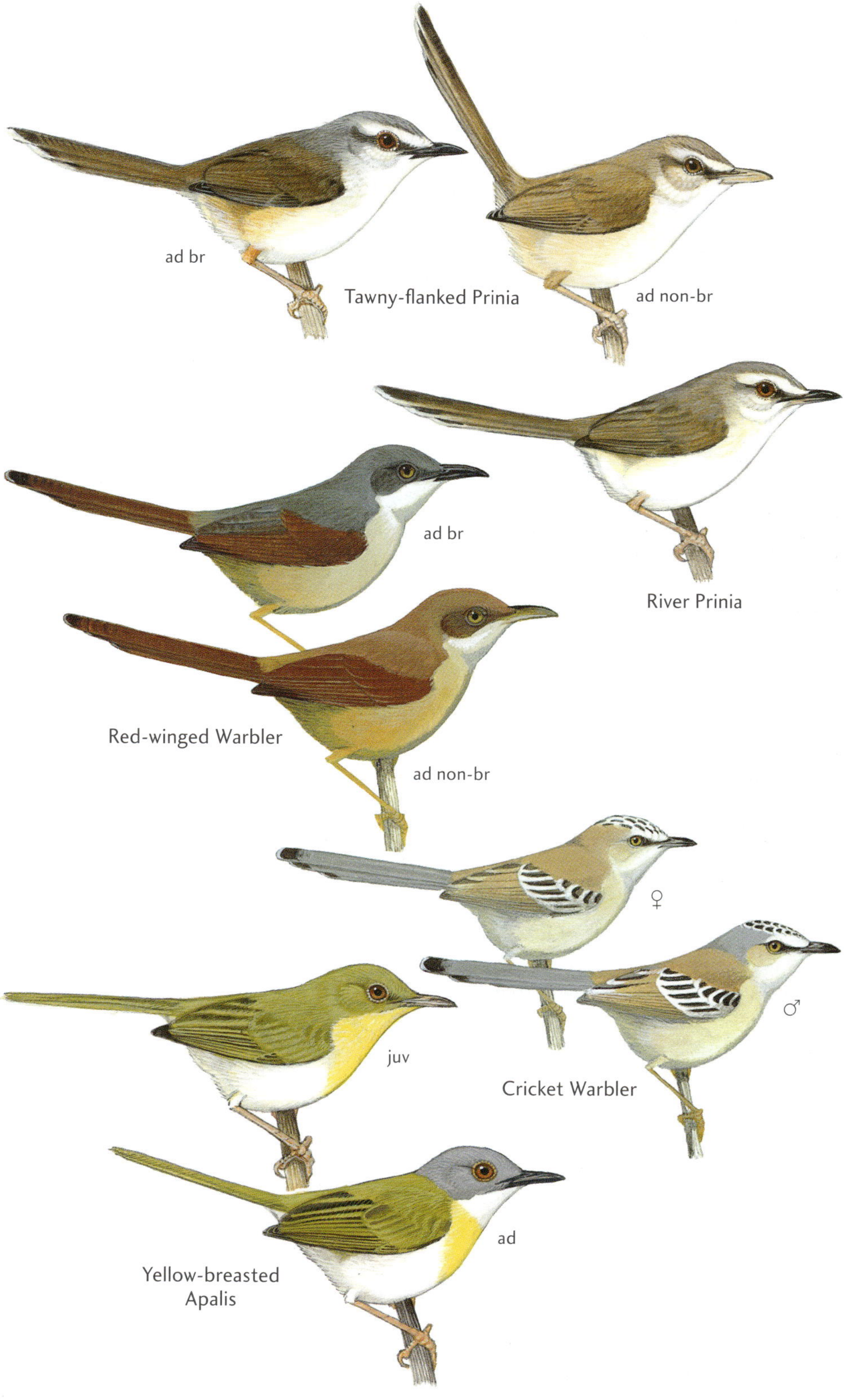
ad br
Tawny-flanked Prinia
ad non-br
River Prinia
ad br
Red-winged Warbler
ad non-br
♀
♂
Cricket Warbler
juv
ad
Yellow-breasted Apalis

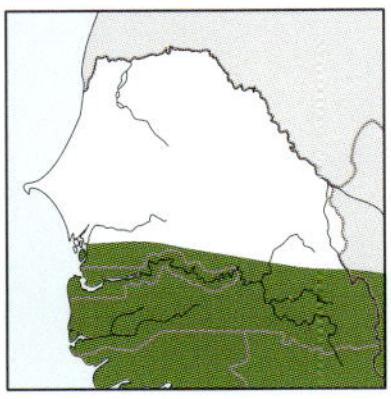

Yellow-bellied Hyliota *Hyliota flavigaster* 12.5 cm

H. f. flavigaster. Quite distinct, rather stocky, aberrant warbler with relatively short tail. **Adult male** Glossy blue-black above with large white wing patch, pale yellowish-buff below. **Adult female** Duller and paler, with dark grey-brown upperparts. **Juvenile** As adult female, but with some buffish mottling on upperparts and paler underparts. **HH** Uncommon to fairly common resident in savanna woodland south of 14°N. Perhaps declining in Gambia due to loss of woodland. Searches for insects on bark and among upper foliage, frequently upside-down. Singly, in pairs or small family parties; often in mixed-species flocks. **Voice** Usually silent. Call *twep twep*… and rather quiet *pli-ik* or *swit-it*. Song a repeated *tiu-wheep tiu-wheep*… or higher-pitched *tseep hweet tseep-tseep hweet*… Also dry trills. [CD11:58] **Note** Hyliotas recently proposed to be placed in their own family, Hyliotidae.

Green Hylia *Hylia prasina* 11.5 cm

H. p. prasina. **Adult** Dark olive-green above with ***creamy supercilium*** emphasised by blackish eye-stripe. Throat greenish-white becoming olivaceous-grey on rest of underparts. Bill short and black; legs olivaceous. **Juvenile** Greener overall, bill pale, legs pale pinkish. **HH** Locally common resident, SW Senegal; rare, W Gambia. Singly or in pairs in lowland forest and relict forest patches. Forages at all levels, mostly in mid-stratum. Often in mixed-species flocks. **Voice** A loud, clear, double whistle *hee-hee* and short dry rattle. [CD11:61] **Note** Affinities of this species uncertain.

Oriole Warbler *Hypergerus atriceps* 20 cm

Large, slender warbler, with unique plumage pattern, long, graduated tail and distinctive song. **Adult** Head, throat and upper breast black, feathers of crown and head-sides fringed silvery white (giving scaly appearance); upperparts and tail yellowish-olive; rest of underparts yellow. Slender, black bill. Eye reddish. **Juvenile** Duller, with black of head and throat washed green; eye grey. **HH** Locally fairly common resident. Singly or in pairs in undergrowth of gallery forest with stands of oil palms and bamboo in wooded savanna; also on coast, in mangroves and dense shrubbery at forest edge and clearings. Often difficult to see, but emerges in the open to inspect palm fronds. **Voice** Song a short, rapid and frequently repeated series of loud, melodious whistles *hu-hee-tee-teehu* and longer *hu-hee-tee-tu-hwu-hwuu* or *hu-heehee-heehu*, sometimes ending with descending *whee-whuu-whuu-whuu*; with variations. Female utters a *rikitikitikitik thuwthuwthuw* in duet. [CD11:36] **Note** Affinities of this species uncertain; currently placed in Cisticolidae.

FLYCATCHERS (Plates 117–118) Insectivorous and arboreal birds, occurring in all wooded habitats. Sexes similar in most. Juveniles duller and typically spotted. Most have broad, flattened bills with broad gape and rictal bristles. Catch insects in short sallying flights from perch, by pounding on prey on ground or by picking it from foliage in warbler-like manner. Mostly singly, in pairs or small family groups; some join mixed-species flocks.

Northern Black Flycatcher *Melaenornis edolioides* c.20 cm

M. e. edolioides. **Adult** Large and all black with long, slender tail. Female duller, washed slate-grey below. **Juvenile** Duller and heavily spotted buffish-brown. May recall drongos, but these are conspicuous, vocal species with harsh voice and red eyes. Tail of similar-sized Square-tailed Drongo broader and slightly notched. **HH** Fairly common resident. Singly or in pairs in various woodland types, incl. cultivated areas with trees. Hunts insects from perch, catching them in flight or on ground. Occasionally joins mixed-species flocks. **Voice** Rather silent. High-pitched *tseeeu*; also *tsik* and long churring note (alarm). Song a series of varied, mostly melodious phrases with some wheezy and scolding notes, delivered with frequent pauses. Also given by pair in duet. [CD11:68]

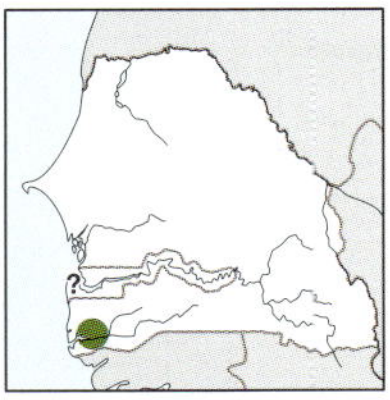

White-browed Forest Flycatcher *Fraseria cinerascens* c.17 cm

F. c. cinerascens. **Adult** Dark slate-grey above with ***white supraloral streak*** (usually conspicuous). Underparts white with indistinct dark grey crescentic bars on throat and breast. **Juvenile** Browner above densely spotted rufous-brown; breast and flanks washed rufous-brown. **HH** Uncommon resident, SW Senegal (Casamance); claimed, Gambia (Abuko). Singly or in pairs in forest along rivers, lagoons and creeks. Usually low; often perches silently on bough above margin of forest stream or on stilt roots in swampy areas. Unobtrusive. **Voice** High-pitched *tseeee*. Song a series of very high-pitched *tsreeee* notes, hard to detect. [CD11:70]

juv
♂
♀
Yellow-bellied Hyliota
Green Hylia
Oriole Warbler
Northern Black
Flycatcher
White-browed Forest
Flycatcher

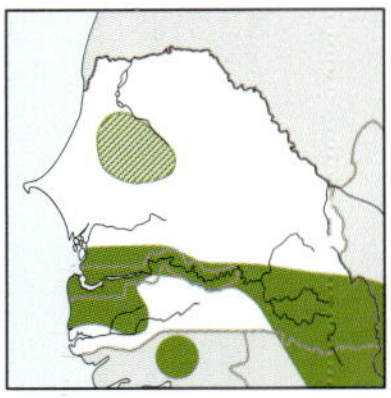

Pale Flycatcher *Bradornis pallidus* 15–17 cm

***B. p. pallidus*. Adult** Medium-sized, rather nondescript savanna flycatcher. Earth-brown above with pale buff eye-ring, very pale brown below becoming white on throat and centre of belly. **Juvenile** Spotted buffish above, mottled dark brown below. **HH** Rare to uncommon resident. Singly or in pairs in various types of woodland, cultivation with trees and orchard bush. Perches on low branches and seizes prey on ground. Occasionally joins mixed-species flocks. Not shy. **Voice** Mostly silent. Soft *churr* and *see-see*. Song a jumble of harsh churring notes. [CD11:63]

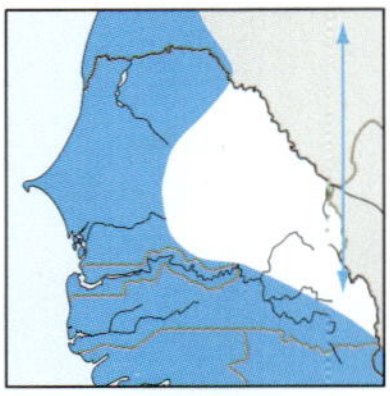

Spotted Flycatcher *Muscicapa striata* 13.5–14.5 cm

M. s. striata*.** Medium-sized, slender, greyish-brown flycatcher with ***faintly streaked crown and breast, and ***relatively long wings***. Latter evident in flight; when perched, ***wingtips extend beyond uppertail-coverts***. **HH** Uncommon Palearctic passage migrant (mainly Sep–Nov & Mar–May); a few overwinter. Singly, in all habitats with trees, incl. gardens. Characteristically perches upright on (often exposed) perch, making short sallies to catch flying insects. Frequently flicks wings and tail. **Voice** Mostly silent. Thin, scratchy *zreeht*; alarm *tik* or *tsee-tik*. [CD4:40]

Swamp Flycatcher *Muscicapa aquatica* 13 cm

***M. a. aquatica*. *Drab flycatcher found near water*. Adult** Brown above, white below, washed brownish on breast and flanks. **Juvenile** Spotted buff above; breast with some diffuse dark brown streaking. **HH** Fairly common resident, Gambia and S Senegal; also N Senegal. Singly or in pairs in trees, bushes or reeds bordering swamps, pools, lakes and rivers. Local. **Voice** Mostly silent. Sharp *tzitt*. Song a mixture of harsh and soft high-pitched notes, followed by a little trill. [CD11:74]

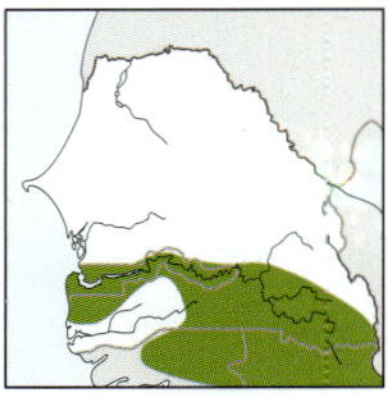

Lead-coloured Flycatcher *Myioparus plumbeus* 14 cm

M. p. plumbeus*.** Slender, warbler-like flycatcher with longish black tail and ***conspicuous white outer tail feathers. **Adult** Ashy-grey above; pale grey below becoming white on centre of belly. **Juvenile** Spotted rusty-buff above, mottled grey-buff and brown below. **HH** Uncommon and local resident. Singly or in pairs in wooded savanna, thickets, gallery forest and forest edges and clearings. Often cocks and slightly fans tail, displaying white outer feathers. Searches branches and gleans foliage for insects. **Voice** Song a soft, quavering *trruuu terree*, rising in pitch; unobtrusive, but quite distinctive once learnt. Variant: *truu tehee-hu-hee*... Call a high-pitched *heet*. [CD11:85]

Red-breasted Flycatcher *Ficedula parva* 11.5 cm

Small and active; ***brown above with bold white sides to base of blackish tail*** (diagnostic). **Adult male breeding** Orange-red throat, grey-brown head, narrow pale eye-ring. **Adult female/first-winter** Creamy-buff below, brown head. **HH** Palearctic vagrant, NW Senegal (Nov 1991 & Nov 1992). **Voice** Dry rattle *tzrrr*, single *tzk*, soft, plaintive *tulee*, weak *shrrr*. [CD4:39]

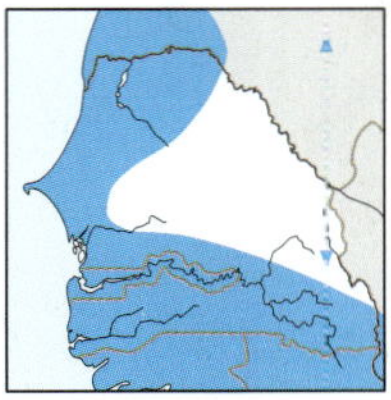

Pied Flycatcher *Ficedula hypoleuca* 13 cm

Rather compact, medium-sized flycatcher, mostly seen in non-breeding, female-like plumage. **Male breeding** Following moult (Jan–Feb) brownish to black above with ***white, elongated wing patch, small white patch on forehead*** (usually divided) and usually narrow white edges to outer tail feathers. Underparts white. ***Very small white patch at base of inner primaries appears as downward-pointing line***. *F. h. iberiae* differs from **nominate** in having usually larger white forehead patch, more white in wing, and less white in outer tail feathers. **Male non-breeding/female** Brownish above; white wingbar narrower. **HH** Common Palearctic passage migrant (Sep–Nov & Apr–May); some winter visitors in Gambia and S Senegal. Singly in wooded savanna and other rather open habitats with trees. Pumps tail up and down and flicks wings. **Voice** Sharp *whit*. [CD4:37]

Atlas Flycatcher *Ficedula (hypoleuca) speculigera* 13 cm

Male breeding As Pied Flycatcher but has larger forehead patch and more white in wing; some possess half-collar; tail generally all black. May not be safely separable from *F. h. iberiae* in the field. **Male non-breeding/female** Not safely separable from Pied (field characters inadequately known) but tends to have more white in wing, paler rump, less white in tail. **HH** Status unclear. Rare N African migrant/vagrant; recorded N Senegal. Habits and voice as Pied Flycatcher.

Pale Flycatcher
Spotted Flycatcher
Swamp Flycatcher
Lead-coloured Flycatcher
♀
♂ br
Red-breasted Flycatcher
♂ br
Pied Flycatcher
♀
1st-year ♂
♂ grey-brown variant
♂ br
♀
Atlas Flycatcher

MONARCHS Diverse family of insectivorous, arboreal species with crests or incipient crests, and relatively long, graduated tails. Very active, gleaning most prey from branches in middle and lower levels, often brushing foliage with fanned tail and partially open wings to dislodge insects. Often in mixed-species flocks.

PARADISE FLYCATCHERS Distinctive. Males have elongated median tail feathers when breeding. Females as males but lack tail streamers. Juveniles as females, but duller. Vocal and conspicuous. Call a rasping zwhee-zwhèh.

African Paradise Flycatcher *Terpsiphone viridis* 18 cm

***T. v. viridis*.** **Adult male** Head glossy blue-black with distinct crest; tail rufous with very long, ribbon-like streamers (projecting up to 10–18 cm beyond other tail feathers). Upperparts deep rufous with wingbar formed by white wing-coverts and white-edged secondaries. Throat and breast blue-black, undertail-coverts rufous. Bill, orbital ring and legs bright blue. **Adult female** Similar, with no white in wing and shorter tail. **Juvenile** Resembles adult female but more brownish, esp. on head and on rather pale underparts. **Immature** As adult female but duller and with shorter crest. **HH** Fairly common to locally common or scarce resident south of 15°N. Northward movements in rainy season recorded. In pairs or small family parties in savanna woodland, gallery forest, second growth and farmland. Mostly at mid-level. Active, vocal and conspicuous. Often joins mixed-species groups. **Voice** Song variable, typically loud and cheerful, e.g. *twee-twee-twee-twee twee-twee-twee-twee*. [CD12:28]

Red-bellied Paradise Flycatcher *Terpsiphone rufiventer* 18 cm

Adult male *rufiventer* distinguished from African Paradise Flycatcher by ***bright rufous underparts*** and shorter tail. ***T. r. nigriceps*** is slightly paler, more orange-rufous, with no crest and no white in wing, and much shorter tail (central feathers extending 2.5–7.0 cm beyond other rectrices). **Adult female** As adult male but rather duller, rufous extending up nape and lacking tail streamers. Nominate lacks white on wing. **Juvenile** As adult female, but duller, more brownish-rufous. Apparent hybrids with African Paradise Flycatcher known from Gambia; these usually have varying amounts of blue-black and rufous on underparts. **HH** Nominate is a locally fairly common resident, Gambia and S Senegal; a few claims of *T. r. nigriceps*. In mid-storey of forest, second growth, cocoa plantations, thickets in savanna. **Voice** Calls and songs typical of genus, including harsh, scolding *zwhee-zwhèh* and short, cheerful *tweedweedwee tweedweedwee*. [CD12:32]

African Blue Flycatcher *Elminia longicauda* 18 cm

E. l. longicauda*.** Unmistakable, pale, ***bright blue flycatcher with short crest and ***long, graduated tail***. **Juvenile** duller and faintly spotted buffish-grey above. **HH** Uncommon to locally frequent resident, Gambia and S Senegal. Graceful, tame and active, foraging with fanned tail and half-open wings among foliage in various types of woodland, gallery forest, edges, farmbush with trees and gardens, and mangroves. Occasionally descends to ground and hunts insects in the air. Singly, in pairs or small family groups. Joins mixed-species flocks. **Voice** Various little call notes, including *dzp* and *tsluip*. Song a jumble of sibilant notes. [CD12:21]

African Paradise Flycatcher
♀
♂
Red-bellied Paradise Flycatcher
♂
♀
hybrid Red-bellied x African Paradise Flycatcher
African Blue Flycatcher

WATTLE-EYES, BATISES AND ALLIES A family of well-marked, mostly rather short-tailed birds. Sexually dimorphic. Catch insects in flight or by gleaning from leaves and twigs. Occur singly, in pairs or small family groups. Several species readily join mixed-species flocks. Endemic to Africa and Madagascar.

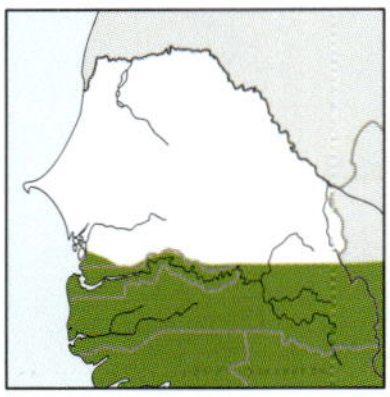

Common Wattle-eye *Platysteira cyanea* 13 cm

P. c. cyanea. **Adult male** Blue-black above with scarlet wattle above eye and long, narrow white wingbar. White below with black breast-band. **Adult female** Throat and upper breast dark chestnut. **Immature** Olive-grey above, rusty-buff wingbar, buffish wash on breast and flanks. **HH** Common resident in various wooded habitats, incl. gardens, gallery forest, coastal thickets and mangrove; absent from forest proper. **Voice** Song a diagnostic, far-carrying series of clear, melodious whistles *hee-hu-huu-ho hee-hu-ho*; also variations. Various harsh low calls, often included in song. [CD12:12]

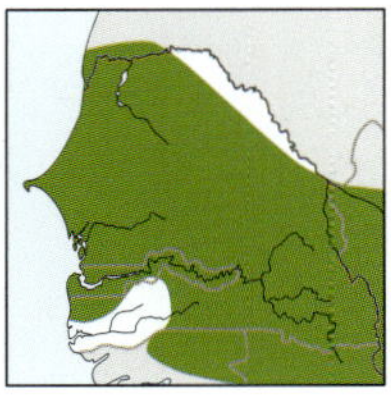

Senegal Batis *Batis senegalensis* 10.5 cm

Small, with relatively large head and short tail. **Adult male** Black mask, yellow eyes, long white supercilium, long white wingbar, and black breast-band. **Adult female** Supercilium and wingbar rusty-buff; pale chestnut breast-band. **Juvenile** As female, but spotted rusty above; breast mottled. **HH** Fairly common resident. Singly, in pairs or family groups in lower and middle strata of open woodland. Often in mixed-species flocks. Flight swift and bouncing, often on whirring wings. Snap bill loudly when catching insects. **Voice** Song a harsh buzzing note preceded or occasionally followed by 1–2 short notes *whut-tzeet* and *tzit-tzit-zheet* or *zheet-tit*, with variations. Calls include *tek-tek-tek-…*, *tlup-tlup-tlup-…* and *peew*, and various buzzy and croaking notes. [CD12:6]

Shrike Flycatcher *Megabyas flammulatus* 15–16 cm

M. f. flammulatus. Rather dumpy, with shrike-like bill and male plumage. **Adult male** Glossy black above with ***white rump***, white below. **Adult female** Earth-brown above with ***rufous rump, coarsely streaked brown below***. **HH** No certain records. Claimed as Afrotropical vagrant, Gambia (Abuko, Dec 1988). Frequents lowland forest; also farmland with scattered tall trees in forest zone and riparian woodland. Moves tail sideways. Often in mixed-species flocks in canopy. **Voice** Song: characteristic *chewee-cheweeet* or *chuah-chuwheesiu*, short, rather melodious trill and fast *whee-whee-whee-chu*, alternating with bill snapping. [CD12:1]

Black-and-white Flycatcher *Bias musicus* 16 cm

B. m. musicus. Distinctive, highly vocal species with prominent spiky crest, yellow eye, ***broad rounded wings*** and ***short tail***. **Adult male** Glossy black and white. White wing patch prominent in flight. **Adult female** Mainly rufous-chestnut above. **Juvenile** As female but duller. **HH** Afrotropical vagrant, Gambia (Seleti, May 1976). Upper levels of edges and clearings in evergreen and gallery forest. **Voice** Song, from perch or in flight, a loud, melodious *wheet-tee-tee-tiuw-tiuw* with variations. Call a harsh *tchèèèp* and *tk tk tk*. [CD12:2]

imm
♀
♂
Common Wattle-eye
♀
♂
Senegal Batis
♀
♂
Shrike Flycatcher
♂
♀
♂
Black-and-white Flycatcher

***BABBLERS** Sturdy and thrush-like birds, occurring in small, noisy groups, often on or close to ground.*

Brown Babbler *Turdoides plebejus* — 24 cm

T. p. platycirca. Greyish-brown above ***with contrasting pale buffish-brown rump*** (conspicuous in flight). Head-sides and underparts paler grey-brown with faint whitish squamations on breast. Eye bright orange or yellow. **Juvenile** has brownish eye. **HH** Common resident. In small, noisy groups of 4–12 in various bushy habitats, incl. gallery forest, and shrubby edges of farmland and gardens. Forages mainly on or near ground, usually in thick cover. Flies over short distances (next bush) with direct flight consisting of flapping wingbeats followed by a glide. Attracts attention by its frequent harsh chuckling and chattering calls. **Voice** Harsh and scolding calls, incl. a fast *chuk-chuk-chuk*, a single, often repeated *kuk* and a clashing *kaaa*. Fast series of chuckling notes uttered simultaneously by several birds result in excited, raucous chattering. [CD12:43]

Blackcap Babbler *Turdoides reinwardtii* — 25 cm

T. r. reinwardtii. Earth-brown above (no paler rump), with ***black head*** and contrasting creamy eye. Pale greyish-brown below becoming buffish-white on throat. **Juvenile** has dark eye. **HH** Fairly common resident. In various bushy habitats, often associated with damper areas, watercourses, etc. Habits as Brown Babbler. **Voice** Harsh and scolding calls, similar to Brown Babbler, but loud chorus of excited chattering interspersed with less harsh, rather nasal *ko-kwee ko-kwee*… [CD12:46]

Fulvous Babbler *Turdoides fulva* — 25 cm

T. f. buchanani. Unmistakable. Rather slender, entirely sandy-buff with long, graduated tail. Eye colour variable, from white to brown and red. **Juvenile** has more tawny upperparts and dark eye. **HH** Locally common to rare, N Senegal, Recent southward extension of range due to drought. Usually in small groups, in arid scrub. Unobtrusive and much less vocal than other babblers, running or flying from bush to bush. **Voice** Varied. Rather fast, slightly descending series of 6–9 whistled notes introduced by a more drawn-out one *peeeew peew-peew-peew-peew-peew-peew*. Also metallic trill *rirrrrrrrrrrr*, short *pip* and clear *pee*. Alarm a sharp *pwit*. [CD4:41]

Capuchin Babbler *Phyllanthus atripennis* — 24 cm

P. a. atripennis. Distinctive, stout, mainly ***chestnut*** species, with grey head and upper breast, conspicuous ***pale greenish-yellow bill*** and black lores. **Juvenile** as adult. **HH** Scarce resident, SW Senegal (Casamance); rare and local, Gambia. Usually in small groups within dense scrub and undergrowth of secondary and gallery forest, forest edges and forest patches in savanna. Occasionally in mixed-species flocks. Rather shy. **Voice** Short chuckling notes and loud, excited, raucous chattering similar to Brown Babbler. Also drawn-out whistle *hu-wheew*. [CD12:50]

Brown Illadopsis *Illadopsis fulvescens* — *c.* 16 cm

I. f. gularis. Dark fulvous-brown above becoming dark greyish on head-sides; rump brighter. ***Throat dirty white; rest of underparts pale fulvous-brown***. **Juvenile** as adult. **HH** Uncommon resident, SW Senegal (Casamance). In small groups within forest interior, thickets by roads in secondary forest and dense bush in clearings. Mostly perches low in vegetation (rarely on ground), but also higher (to lower canopy), esp. in dry season. Mainly active at dawn and dusk. When foraging, runs and climbs on branches, fallen logs and lianas, poking its bill into clusters of dead leaves. Readily joins mixed-species flocks. **Voice** Song a low, rather plaintive even whistle *wHEEEE* followed by pause and another, similar whistle on lower pitch, *wHUUUU*, the whole interspersed with a few short *huit* and *tew* notes (only audible at close range) by other group members. Drawn-out whistles easily imitated by human. Various harsh and rattling calls. [CD12:35]

Puvel's Illadopsis *Illadopsis puveli* — 18 cm

I. p. puveli. Large and robust, with strong pale legs and feet. Russet-brown above, esp. on wings and tail; white below washed olivaceous-buff on breast, flanks and undertail-coverts. **Juvenile** as adult. **HH** Scarce, SW Senegal (Casamance); rare, Gambia. Frequents forest edges, gallery forest and dense bush in forest–savanna mosaic. Mainly in pairs on or just above ground. **Voice** Song a rhythmic and fast series of a few short tuneless notes followed by 2–3 descending whistles; last two whistles sometimes repeated, e.g. *chk-chk-whit HEE-HU-HUU* or *chk-whit-HE-HU-HU-HE-HU*. [CD12:40]

Brown Babbler
Blackcap Babbler
Fulvous Babbler
Capuchin Babbler
Brown Illadopsis
Puvel's Illadopsis

TITS Rather small, arboreal birds. Active and agile, often hanging upside-down when searching for insects, nuts and seeds. Join mixed-species flocks. Vocal, attracting attention by rasping calls.

White-shouldered Black Tit *Parus (leucomelas) guineensis* 14 cm

Adult ***Wholly glossy blue-black with white, elongated wing patch*** and ***yellow eye***. **Juvenile** Duller; white on wings washed yellow; eye grey or brown. **HH** Fairly common to locally common resident, Senegal; uncommon to locally fairly common, Gambia. In pairs or small parties in open savanna woodland. Often joins mixed-species flocks. **Voice** Varied. Calls harsh and buzzing, incl. *zeet-zeet* and *churr*. Songs include short, often repeated, fast series of clear whistles *huwee-tEE-huwee-tee-huwèh* and *wheetu-tuleep wheetu-tuleep...* or *ptiu-weet ptui-weet ptiu-weet...*, and melodious series of various rolling notes (e.g. *tsrrrua tsrrrwèht*). [CD12:59]

PENDULINE TITS Tiny, mainly insectivorous, arboreal birds with short tails and short, sharp bills. Active but unobtrusive. Often in mixed groups with other small insectivores. Juvenile as adults but slightly duller.

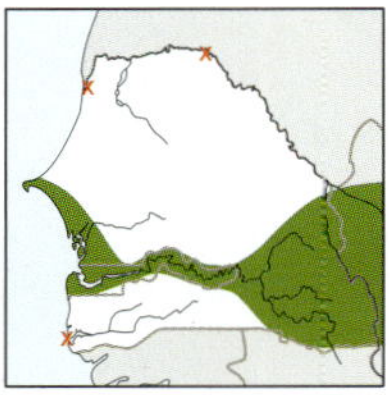

Yellow Penduline Tit *Anthoscopus parvulus* 7.5–8.0 cm

A. p. senegalensis. Olive-yellow above with yellow forehead and variably distinct white wingbar, ***bright yellow below***. Small black dots on forehead visible at close range. Yellow White-eye is larger, plainer and has obvious white eye-ring. **HH** Uncommon to locally fairly common resident. In pairs or small parties in wooded savanna and *Acacia* scrub. **Voice** Thin, quiet, tinkling *si sli-li-liii*, fast *slipilipilipliplipili* or *chipichipichipichipi...*, high-pitched, slightly hoarse *bzee-bzee-...*, and rhythmical buzzing. Also churs rather similar to Sennar Penduline Tit. [CD12:55]

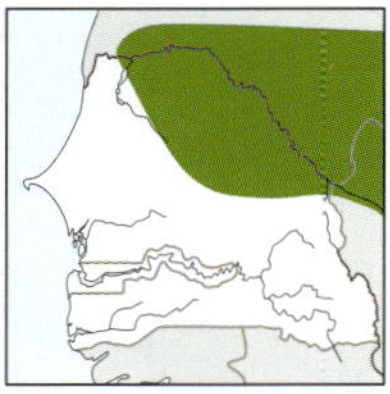

Sennar Penduline Tit *Anthoscopus punctifrons* 7.5–8.5 cm

Tiny, inconspicuous pale bird. Yellowish-olive above with pale yellowish forehead, buffish-white below. Small blackish dots on forehead only visible at close range. **HH** Uncommon to locally common resident, N Senegal. In pairs or small groups in dry *Acacia–Balanites* savanna. Rather easily overlooked, but vocalisations betray presence. **Voice** Very high-pitched *tsii* notes, churring *whut-chrrrrrrrrrrr* and harsh *tchuwhree-whree*, with variations. [CD12:57]

SPOTTED CREEPER Unmistakable, arboreal bird with long, decurved bill, short tail and short, strong legs.

Spotted Creeper *Salpornis spilonotus* 15 cm

S. s. emini. ***Cryptic brown plumage spotted and barred white***. **Juvenile** duller. **HH** Rare and inadequately known, Gambia (a few dry-season records) and SE Senegal (Niokolo Koba, Apr 1969). Singly or in pairs and usually in mixed-species flocks in wooded savanna. Clings to tree-trunks and branches, starting low down and working upwards in spirals. Feeds mainly on insects, also spiders. **Voice** Single, high-pitched *tseee*. Song a short series of clear, high-pitched notes *tsip-tsee-tsu-tuwee*, with slight variations. Also faster *tsitsutsitsutsitsitsu...* [CD12:66]

WHITE-EYES Small, arboreal, warbler-like birds with diagnostic white eye-ring. Bill short and pointed. Active.

Yellow White-eye *Zosterops senegalensis* 10–11 cm

Z. s. senegalensis. Small and yellow with ***dark eye surrounded by white ring*** (diagnostic). **Adult** Head and upperparts yellowish-olive with yellow forehead and black loral stripe. ***Entire underparts bright yellow***. **Juvenile** As adult but darker above. **HH** Fairly common resident. In pairs or small groups in various habitats, incl. thorn scrub, savanna woodland, forest, and gardens. Often joins other species. Feeds mainly on small insects, also nectar and fruit; forage principally by foliage gleaning, but also catch insects on the wing. **Voice** Quavering *ti-trrruutrrruu-ti-trrruuti*. Song, mainly given at sunrise, a series of burred notes, including calls. [CD13:26]

White-shouldered Black Tit

Yellow Penduline Tit

Sennar Penduline Tit

Spotted Creeper

Yellow White-eye

SUNBIRDS (Plates 123–124) *Distinctive passerines with long, slender, decurved bills (shortest and least decurved in* Anthreptes *and* Hedydipna*). Males often have brightly coloured and iridescent plumage; most females dull greenish or brownish. Juveniles usually as adult females. Some females and juveniles are difficult to distinguish and are best identified by the presence of the male. Feed on nectar, insects and spiders.*

Western Violet-backed Sunbird *Anthreptes longuemarei* — 13–14 cm

A. l. longuemarei. Distinctive sunbird with rather short and straight bill. **Adult male** Glossy violet above and on throat, pure white below. **Adult female** Rather warbler-like. Brown above with ***white supercilium***, white below with ***lemon-yellow belly, flanks and undertail-coverts***. **Juvenile** Resembles adult female, but more olive above, entirely pale yellow below. **HH** Uncommon resident in wooded savanna and gallery forest. Usually singly or in pairs; often joins mixed-species flocks. **Voice** Harsh *chep* and *tit*. Song a rapid, high-pitched jingle. [CD12:67]

Brown Sunbird *Anthreptes gabonicus* — 10 cm

Non-glossy, grey-brown above, whitish below, ***with short, almost straight bill, white lines above and below eye***, and ***rather warbler-like appearance***. Tail narrowly tipped white (visible from below). **Juvenile** more olive above, washed lemon below. **HH** Uncommon resident in mangroves and along forested river banks. Singly or in pairs, Unobtrusive. **Voice** Single, very high-pitched note, reminiscent of small kingfisher, and little sibilant notes, such as *sipsipsipsip*, uttered in flight. Song a series of sibilant and twittering notes. [CD12:68]

Green-headed Sunbird *Cyanomitra verticalis* — 13–14 cm

C. v. verticalis. **Adult male** ***Glossy blue-green head and throat***, olive-green upperparts and ***dusky-grey underparts***. Pectoral tufts lemon-yellow. **Adult female** As male but ***entire underparts pale grey***; no pectoral tufts. **Immature male** Forehead and throat blackish-grey. **HH** Uncommon to scarce resident in wooded savanna, gallery forest, thickets and gardens. Singly or in pairs; joins mixed-species flocks. **Voice** Vigorous *chuwee* and *chi-ep*, harsh *chee* and high-pitched *chip*. Songs a harsh, accelerating *chip-chip-chip-chichichichichichi* and high-pitched jingle. [CD12:79]

Olive Sunbird *Cyanomitra olivacea* — 13–15 cm

C. o. guineensis. ***Plain, olive-green***, non-glossy sunbird with ***long, decurved bill***. **Female** lacks yellow pectoral tufts; **juvenile** similar but strongly washed yellow below. **HH** Recorded north (vagrants only) and south of Gambia; locally common, SW Senegal (Casamance), occasionally in SE Senegal. Singly, in pairs or small family parties, frequenting mainly lower levels of forest interior; also at edges and in thickets. Regularly joins mixed-species flocks. **Voice** Vigorous *chip* or *cheep*, frequently and rapidly repeated. Also a clear, loud *tuweet-tuweet-*... often in long series. Song a series of clear, separate, piping notes, slightly accelerating and falling in pitch towards end. [CD12:84]

Collared Sunbird *Hedydipna collaris* — 10 cm

H. c. subcollaris. **Adult male** Bright glossy green upperparts and throat; bright yellow underparts. **Adult female** Entire underparts yellow. **Juvenile** Olive above with short yellowish stripes above and below eye, greyish and lemon-yellow below. **HH** Locally fairly common, Gambia; in Senegal, mainly in Casamance, where fairly common. In pairs or small family groups in variety of habitats, incl. forest edge, gallery forest, thickets, woodland, farmbush and gardens. Regularly in mixed flocks. **Voice** Vigorous *tsip* and *teew*, nasal *chee chee*. Song a series of high-pitched, separate notes, e.g. *swee swee swee swee swee see-see-see-see-see*... and *tsip-tsip-tsip-tsee-tsee tsee tsee tsee tsiup*, or *suweep-suweep-suweep-suweep -*... [CD12:77]

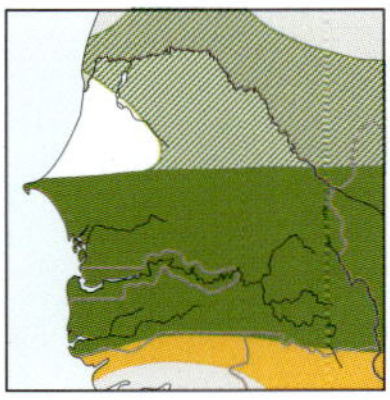

Pygmy Sunbird *Hedydipna platura* — 9–10 cm

Male breeding (dry season) Glossy coppery-green upperparts and throat; golden-yellow underparts; ***very long tail streamers*** (projecting up to 7 cm). **Male non-breeding** Grey-brown above with faint yellowish supercilium and some glossy green on wing-coverts, lemon-yellow below with some black on throat. No tail streamers. **Adult female** As non-breeding male, but lacks glossy feathers. **HH** Generally uncommon to locally and seasonally fairly common. In *Acacia* scrub, dry bush, wooded savanna and gardens. Singly or in pairs when breeding, gregarious outside breeding season. Intra-African migrant or locally dispersive, moving south to breed in dry season. **Voice** Call *cheep* or *twee*. Song a high-pitched jingle, including short trill. [CD12:78]

♀
♂
Western Violet-backed Sunbird
Brown Sunbird
♂ imm
♂
♀
Green-headed Sunbird
Olive Sunbird
♂ non-br
♀
♂
Collared Sunbird
juv
♀
♂ br
Pygmy Sunbird

Scarlet-chested Sunbird *Chalcomitra senegalensis* 14–15 cm

C. s. senegalensis. **Adult male** Bulky, blackish sunbird with ***brilliant red lower throat and breast*** and long, decurved bill. Forehead, forecrown and upper throat glossy green; breast with small glossy violet bars (difficult to see in field). Wings brownish-black and subject to bleaching, paling to milky brown. **Adult female** Mainly dark and dusky. Grey-brown above, dull yellow broadly streaked below with dusky throat. **Juvenile** As adult female but with sooty-grey throat; rest of underparts mottled and barred blackish. **HH** Common. In various habitats, incl. savanna woodland, riverine forest, gardens, etc. Joins mixed-species flocks. **Voice** Vigorous *chuw*, scolding *chep*, and *tiup*. Song a short, simple series of clear separate notes *tip tiu tip tiu* or *tip tiu tip tip* and *tip tiup hu* with variations. [CD12:89]

Olive-bellied Sunbird *Cinnyris chloropygius* 10.5 cm

C. c. kempi. **Adult male** Glossy green above and on throat, ***bright red breast***, rest of underparts dirty brownish-olive. Pectoral tufts yellow. (Compare Beautiful Sunbird that has not yet replaced dropped worn tail streamers.) **Adult female** Dark olive above with yellowish supercilium, plain, dirty yellow below. **Juvenile** Similar to adult female but with greyish throat. **HH** Uncommon, SW Senegal (Casamance); vagrant elsewhere. Usually in pairs, frequenting flowering trees and bushes in forest clearings, edges, gallery forest, thickets, second growth shrubbery, and gardens. Not shy. **Voice** Song: *tsup* or *tsip* notes stuttering into rapid, rising metallic jingle including many trills, descending at end. [CD13:1]

Beautiful Sunbird *Cinnyris pulchellus* 9–11 cm

Male breeding Easily identified by glossy green plumage, long tail streamers (projecting up to 6 cm) and red breast bordered on sides by yellow. **Male non-breeding** (dry season) as female but retains tail streamers and glossy green shoulders. **Adult female** Pale ashy-olive above, pale yellow below. **Juvenile** As adult female; young male has throat black. **HH** Common. Usually in pairs or small parties in wooded savanna and gardens. Largely resident but seasonal movements recorded (north with rains, south in dry season). **Voice** Vigorous *chip* or *tut*. Song a rapid series of similar short notes followed by short, high-pitched jingle, e.g. *chiupchupchipchitititit*. [CD13:8]

Variable Sunbird *Cinnyris venustus* 10 cm

C. v. venustus. **Adult male** Bright glossy green above with violet forecrown, throat and upper breast, and lemon-yellow belly. **Adult female** Grey-brown washed olive above, yellowish below. **Juvenile** Similar to adult female; immature male has blackish throat. **HH** Locally fairly common to uncommon. In various habitats, incl. wooded savanna, clearings, mangroves, coastal scrub, farmbush, gardens. **Voice** Song a series of *chuw* and *cheep* call notes accelerating into cheerful jingle. [CD13:13]

Splendid Sunbird *Cinnyris coccinigastrus* 14 cm

Fairly large and long-billed sunbird. **Adult male** Head and throat glossy purple, breast bright red mixed with glossy purple and blue, belly black. **Adult female** Brownish-olive above with yellowish supercilium, pale yellow below with dusky streaks on breast. **Juvenile** As adult female, but has black throat and upper breast; immature male has glossy purple throat. **HH** Fairly common. Singly or in pairs in savanna woodland, coastal thickets, shrubbery, large forest clearings and gardens. **Voice** Short *tsiup*, harsh *chew-chew-*..., and *hueet*. Song a series of 7–9 clear, separated notes *tip tiup tiup tiup*...; recalling song of Scarlet-chested Sunbird but louder. [CD13:15]

Copper Sunbird *Cinnyris cupreus* 12–13 cm

C. c. cupreus. **Male breeding** Distinctive glossy coppery and purple, although may appear all black at distance. **Male non-breeding** (dry season) Easily overlooked. As adult female, but retaining glossy wing- and uppertail-coverts, and some iridescent feathers on body; often has dark bib. **Adult female** Olivaceous-green above with yellowish supercilium, olive-yellow below. Very similar to female Beautiful Sunbird but less yellow. **Juvenile** Similar to adult female. Young male has dusky throat. **HH** Locally fairly common. Usually singly or in pairs, frequenting variety of habitats, incl. wooded grassland, coastal scrub, abandoned farmland, edges of gallery forest, and gardens. **Voice** Hoarse *chip-chip*. Song a rapid series of *chip* notes followed by short metallic jingle. Also a rattling alarm call. [CD13:19]

♀
♂
Scarlet-chested Sunbird
♀
♂
Olive-bellied Sunbird
♀
♂
Beautiful Sunbird
♂ imm
♀
♂
Variable Sunbird
♂ imm
♀
♂
Splendid Sunbird
♀
♂
Copper Sunbird

SHRIKES Small to medium-sized birds with strong, hooked bills and moderately long to very long tails. Bill and legs usually black(ish). Juveniles typically finely barred above and below; bill paler. Capture prey (insects, reptiles, young birds, small mammals) by pouncing from exposed perch. Some species impale prey on thorns. Flick, swing and fan tail when excited.

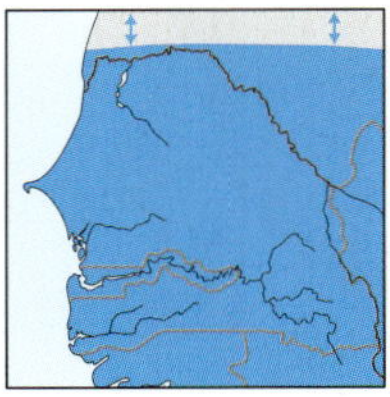

Woodchat Shrike *Lanius senator* 18–19 cm

Distinctive, medium-sized shrike with ***chestnut crown and nape*** and large white shoulder patch and rump. In fresh plumage has pale fringes to wing feathers. **Adult female** Slightly duller than male, with more white on forehead and lores. **Immature** has variable amount of buff wash and indistinct crescents on scapulars, breast and flanks. Iberian and N African race ***rutilans*** slightly smaller than **nominate**. W Mediterranean ***badius*** differs in having no white on primaries and less white on shoulders; male has narrower black forehead. **HH** Fairly common to locally common Palearctic passage migrant and winter visitor (Sep–Apr). Nominate throughout; *rutilans* and *badius* recorded from N Senegal. In various open habitats. **Voice** Hard *chak chak chak*. Mainly silent in winter quarters. [CD4:56]

Isabelline Shrike *Lanius isabellinus* 17–18 cm

Adult male *phoenicuroides* (Red-tailed Shrike) Rather plain, brownish shrike with rufous-tinged crown and ***contrasting rufous rump and tail***. Black mask bordered by narrow white supercilium. **Nominate** paler; crown and upperparts greyer (crown lacking rufous tinge); underparts usually distinctly cream coloured. **Adult female** Duller; usually with indistinct scaling on breast and flanks; white patch at base of primaries reduced or absent. **Immature** Resembles adult female but mask generally poorly developed; underparts with indistinct scaling; coverts and tertials tipped buffish with subterminal dark markings. Adult female and immature *isabellinus* doubtfully separable but tend to be plainer and more creamy below. **HH** Scarce to locally uncommon Palearctic migrant, N Senegal; rare, Gambia (respective status of races unclear). Frequents various open habitats. **Voice** Nasal *kihet* and short *kzi-ek* or *tzea*. Subdued song, occasionally uttered in winter quarters, varied and sustained. [CD13:35]

Southern Grey Shrike *Lanius meridionalis* 24–25 cm

Large, grey, white and black shrike with long, graduated tail. **Adult *leucopygos*** (Sahelian race) Pale grey above with black mask; wings black and white; rump white or greyish-white; underparts white washed pale buff. ***L. m. elegans*** (Saharan race) distinctly larger, with greyer rump and uppertail-coverts and pure white underparts. **Immature** Duller and tinged buffish. **HH** *L. m leucopygos* is a rare visitor (Nov–Apr), Gambia; *L. m. elegans* a scarce to uncommon dry-season visitor (also resident?), N Senegal. Perches prominently atop low trees and bushes. **Voice** Dry *trr-trr* and harsh *zzeh-zzeh*; also a subdued, varied warble, including imitations. Both sexes sing. [CD4:54, 13:38]

Yellow-billed Shrike *Corvinella corvina* 30–33 cm

C. c. corvina. Distinctive, large brown shrike with ***very long, graduated, often ragged tail*** and ***yellow bill***. Rufous wing-panel conspicuous in flight. **Juvenile** is barred and mottled (not streaked). **HH** Common resident in open savanna woodland. Usually in small, vocal parties. Display involves much calling, bowing and tail rotating. Breeds cooperatively. Group members roost together. **Voice** Various rasping, chirping and chattering calls, usually in series uttered simultaneously by several members of a group. Alarm a rapid, harsh *chwee-zhee-zhee*. Song includes variety of scratchy, rather shrill, but not unpleasant notes, typically also rolling *drrreeeooo*. Imitations of other birds have been recorded. [CD13:43]

♂ *senator*
imm *senator*
♂ *badius*
Woodchat Shrike
♀ *senator*
♂ *senator*
imm
♂ *isabellinus*
Isabelline Shrike
♀ *phoenicuroides*
♂ *phoenicuroides*
♂ *phoenicuroides*
ad
Yellow-billed Shrike
imm
ad
Southern Grey Shrike
ad *elegans*
ad *leucopygos*

PLATE 126: BUSH-SHRIKES

BUSH-SHRIKES *Arboreal birds with stout, hooked bills. Forage within foliage, preying on small vertebrates and invertebrates. Usually singly or in pairs. Most are highly vocal, with loud, ringing calls.*

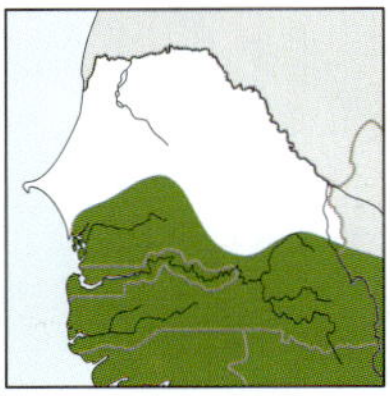

Grey-headed Bush-shrike *Malaconotus blanchoti* — 25 cm

***M. b. blanchoti*.** **Adult** Large, with massive bill, grey head, white lores, green upperparts and tail, and bright yellow underparts washed orange on breast. **Immature** Head mottled brown, underparts and bill paler. **HH** Fairly common resident. Singly or in pairs in upper and mid-levels of savanna woodland and gallery forest; occasionally lower. Inconspicuous, but call betrays presence. May join mixed-species flocks. **Voice** Far-carrying, drawn-out, single hollow whistle *whoooop* (easily imitated by human), occasionally rising at end (*whoooo-up*). Also various harsh and rasping notes, bill clicks and shorter whistles. [CD13:74]

Sulphur-breasted Bush-shrike *Chlorophoneus sulfureopectus* — 17–19 cm

***C. s. sulfureopectus*.** **Adult male** Forehead yellow, extending as narrow supercilium above black mask; forecrown moss-green; hindcrown to mantle grey; rest of upperparts and tail moss-green; wing and tail feathers tipped lemon-yellow. Underparts bright yellow with strong orange wash to breast. **Adult female** Mask less black, breast less orange. **Immature** Duller; adult plumage attained in c.1 year. **Juvenile** Heavily barred above and below. **HH** Fairly common resident. Singly or in pairs in savanna woodland. **Voice** Series of far-carrying, clear whistles, varying in speed and motif, e.g. *hu-hu-hu-hweet* or *hu-wheet hu-wheet hu-wheet* and *wheet-wheet-huuu*; also shorter *hwut-heee*. Call *tzzzzrr*, used with bill-snapping in alarm and occasionally also by female in duet; also *puwheet*. [CD13:65]

Tropical Boubou *Laniarius (aethiopicus) major* — 23 cm

Adult Black above with long white wingbar, white below. **Juvenile** Duller, with dusky barring on underparts; bill dark horn. **HH** No certain records. Claimed as Afrotropical vagrant, S Senegal (Toubakouta and Casamance 1987); presumably misidentified Northern Puffback. Normally singly or in pairs at edges of gallery forest and in savanna woodland, thickets, bush and gardens. **Voice** Duet consists of ringing *HOOO* in short series of variable speed (male), answered by similar notes or grating *gkrzzz* (female) , e.g. *hoo-hoo-hoo* or *hoo-hoo-HUloo*, fast *HUloo-HUloo-HUloo*, etc.; also *rrroh-gkzzz*. Call an explosive *KEK!*, often in long series. [CD13:53]

Yellow-crowned Gonolek *Laniarius barbarus* — 23 cm

***L. b. barbarus*.** Beautiful bush-shrike with distinctive coloration. **Adult** Black above with golden-yellow cap, crimson below. **Juvenile** Feathers of upperparts tipped buff; below, yellowish-buff barred black. **HH** Fairly common to common resident. Singly or in pairs, skulking in undergrowth of dense thickets, often near watercourses, in inland and coastal savanna, thorn scrub, mangroves and gardens. Occasionally perches in the open. **Voice** Male song a ringing double whistle, *WHEE-oo!*, likened to whip-lash, with variations e.g. *whuoo-wHEE* and *oo-WHEE* or *whow-WHEE*; also quivering *whiiiir*. Female usually responds with *kik-kik*. Various other, harsh calls. [CD13:56]

Northern Puffback *Dryoscopus gambensis* — 18–19 cm

***D. g. gambensis*.** **Adult male** Glossy black above with greyish-white scapulars, lower back and rump. Conspicuous orange-red eye. Soft elongated feathers on lower back and rump are puffed out to produce spectacular white ball in display. **Adult female** Very different from male; thus potentially confusing. Head grey, upperparts earth-brown, underparts tawny-buff. **Juvenile** As adult female but paler, more ashy-grey overall. Eye dark; bill horn. **HH** Fairly common to common resident. In pairs in canopy of savanna woodland and gallery forest edges; also forest clearings and mangroves. **Voice** Varied. Frequently repeated, loud, rasping *CHERP-CHERP-...* and harsh chattering, nasal and clicking sounds. [CD13:46]

Black-crowned Tchagra *Tchagra senegalus* — 20–22 cm

T. s. senegalus*.** ***Black crown bordered by broad buff-white supercilium and black eye-stripe. In flight, ***rufous wings*** and ***bold white tips to black, graduated tail*** conspicuous. **Juvenile** duller. **HH** Fairly common to common resident. Usually singly, low down or on ground in various wooded habitats (not forest). **Voice** Male song a series of vigorous, far-carrying, sweet but melancholy whistles, in display flight or from perch. Female may respond with *trrrrrrrrrrr*. Also various explosive, growling sounds, e.g. *tok-tok-tok-tok*, followed by rolling bubble and ending with *whu-heeuw*. [CD4:57]

ad
Grey-headed Bush-shrike
imm
♂
Sulphur-breasted Bush-shrike
ad
Tropical Boubou
juv
ad
Yellow-crowned Gonolek
♂ displaying
♀
♂
Northern Puffback
ad
Black-crowned Tchagra

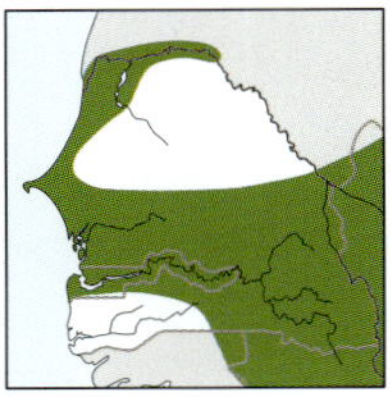

Brubru *Nilaus afer* 13–15 cm

N. a. afer. Small, active, pied bush-shrike. **Adult male** Black crown bordered by white forehead and supercilium; upperparts boldly mottled black, white and buff; chestnut on breast-sides and flanks. **Adult female** More brownish-grey above, flanks paler. **Juvenile** Speckled and mottled above, barred below. **HH** Fairly common. Singly or in pairs in savanna woodland. Forages at upper- and mid-levels, gleaning foliage and branches for insects, also hanging upside-down. Unobtrusive. Joins mixed-species flocks. Probably largely resident, but north–south movements recorded, north in rains, south in dry season. **Voice** Male gives a far-carrying, drawn-out trill *brrrruuuu* (reminiscent of referee's whistle). Female may respond with *wheeeu*. Also *tu* and *peep* (contact) and *chK-chK-...* (alarm). [CD13:44]

HELMETSHRIKES Gregarious birds with rather stout, hooked bills, boldly patterned plumage and brightly coloured eye-wattles. Brush-like feathers on forehead give 'helmeted' appearance. Conspicuous and vocal, constantly chattering and often snapping bill.

White Helmetshrike *Prionops plumatus* 19–23 cm

P. p. plumatus. Distinctive pied savanna species with ***long white crest***. In flight, characteristic pattern with ***broad, rounded wings having white band above and below***. **Adult** Yellow eye-wattle around yellow eye. **Juvenile** Duller, without crest nor eye-wattle and with dark eye. **HH** Fairly common resident. In parties in wooded savanna, mainly foraging on branches and trunks at mid- and lower levels; occasionally descending to ground. Feeds mainly on insects, also some fruit. Sometimes joins mixed-species flocks. Restless and bold, moving from tree to tree in low, undulating flight. Readily mobs predators such as Pearl-spotted Owlet. **Voice** Various frequently uttered, hoarse growling sounds, e.g. *krrreew* and *kreepkrw*, often accompanied by bill-snapping. Alarm a high-pitched *tzzee-tzzee*. [CD14:2a]

ORIOLES Robust, arboreal birds. In the two species occurring in Senegambia males are bright yellow and black and have a strong, reddish bill. Feed on insects and fruit. Mostly in canopy, where often difficult to observe. Voice loud, fluty and melodious. Flight strong and undulating.

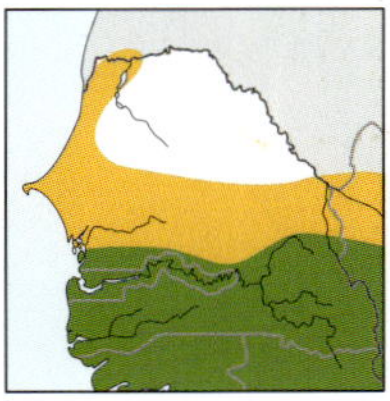

African Golden Oriole *Oriolus auratus* *c.*24 cm

O. a. auratus. **Adult male** ***Deep golden-yellow with black mask***; black wings and tail broadly edged yellow. Bill pink. **Adult female** Duller. More yellowish-olive above, lightly streaked dusky below. **Immature** Yellowish-olive above without mask and with yellow edges to wing-coverts and tertials; throat to belly yellowish-white heavily streaked blackish, flanks and undertail-coverts golden-yellow. Bill dusky. Eurasian Golden Oriole (both sexes and immature) lacks mask and yellow edges to wing feathers. **HH** Fairly common, in wooded savanna with large trees. Moves north to breed during rains; southward dispersal in dry season. **Voice** Various loud, melodious whistled phrases *wheetoliuw* and *tooleeoo* with longer variations; also harsh mewing *whrèèèh* or *mwaaarr*. [CD13:28]

Eurasian Golden Oriole *Oriolus oriolus* *c.*24 cm

O. o. oriolus. **Adult male** ***Golden-yellow with black lores, wings and tail***. In flight, black wings with bright yellow patch on wing-edge. **Adult female** Duller, more yellowish-olive above, narrowly streaked dusky below (streaks gradually disappear with age; females of 4 years and older may resemble males). **Immature** Greener above than immature African Golden Oriole and lacking yellow edges to wing feathers, whiter below. Full adult plumage apparently attained in *c.*2 years. **HH** Scarce to rare Palearctic migrant. Mainly on passage (esp. in spring), N Senegal (Oct & Apr–May); few records, Gambia (Dec, Mar–Apr). In wooded savanna. **Voice** Melodious *weehla-weeoo*; also a harsh note. Mostly silent in W Africa. [CD4:51]

juv
♀
♂
Brubru
juv
ad
White Helmetshrike
ad
♂
imm
♀
♂
African Golden Oriole
♂
imm
♀
♂
Eurasian Golden Oriole

***DRONGOS** Arboreal species with black, usually glossy, plumage and stout, slightly hooked bills. Eye red or orange in adults. Juveniles less glossy, with shorter tails. Hunt from perch, capturing insects on the wing like flycatchers. Conspicuous, bold, pugnacious and vocal. Voice varied and consisting of harsh, scolding notes interspersed by musical whistles.*

Square-tailed Drongo *Dicrurus ludwigii* — 19 cm

D. l. sharpei**.* Smallest drongo. All black slightly glossed purplish-blue with broad, ***slightly notched, often almost square tail. Inner webs of flight feathers blackish. Eye orange-red. Compare Northern Black Flycatcher (Plate 117). **HH** Fairly common resident, SW Senegal (Casamance); rare, Gambia. Singly, in pairs or small family parties in second growth, forest edges, gallery forest and thickets in savanna. **Voice** Call *whit whit*. Song: series of short, explosive notes; also characteristic, fast, melodious *whidididid* followed by nasal, weaver-like *jeeeeezz*; hard *rrrwee rrwee*, rapid *chichichi* and *chi-rrrwee*, and rasping notes. [CD14:4b]

Fork-tailed Drongo *Dicrurus adsimilis* — 22.5–25.0 cm

D. a. divaricatus**. **Wholly glossy blue-black with broad, distinctly forked tail. ***Inner webs of flight feathers ashy, giving pale, silvery appearance to wings in flight***. Eye red or brownish-red. In moult has double fork or 'ratchet' in tail. Female has less deeply forked tail. **Juvenile** Dull black with feathers of upper- and underparts variably tipped greyish, forming crescentic bars. Eye brownish. **HH** Common resident. Singly or in pairs in various types of woodland; not forest. Perches conspicuously and upright. Readily joins mixed-species flocks. Fearlessly mobs much larger birds, particularly raptors, owls and hornbills. **Voice** Vigorous, discordant, grating and twanging notes. Song a loud medley of harsh, creaking, metallic notes and clear whistles. Sometimes includes imitations. Also a rather subdued *tiueetiu-tiuee*, with some variations, uttered all day long. Dawn song a simple repetition of a few notes. [CD14:5b]

***CROWS** Medium-sized to large with stout bills and strong legs and feet. Omnivorous. Forage mainly on ground. Calls mostly loud and harsh.*

Brown-necked Raven *Corvus ruficollis* — 52–56 cm

Adult Black with brown tinge to head and breast (often difficult to see). Tail wedge shaped with central feathers often slightly protruding. With wear head and body become browner, contrasting with blacker wings and tail. **Juvenile** Duller. **HH** A species of desert and semi-desert, reaching N Senegal as scarce non-breeding visitor; occasionally in Gambia (a few coastal records, Jan–May). **Voice** Harsh *karr* or *aarg*. [CD4:65]

Pied Crow *Corvus albus* — 46–50 cm

Unmistakable, large, robust, black-and-white crow. White of lower breast to upper belly extends as broad collar on base of hindneck. In flight, white on underparts visible from great distance. **Juvenile** Duller; often with some dusky mottling within white. **HH** Common resident in all habitats except closed forest; typically near habitation and cultivation. Usually in pairs or small groups, sometimes in large flocks. Largely scavenges. Roosts communally. Parasitised by Great Spotted Cuckoo. **Voice** Harsh *kwaar*, deep, guttural croak *kaarrh* or *kroh*, etc. [CD14:7a]

Piapiac *Ptilostomus afer* — 35 cm

Slender and black ***with very long, stiff, steeply graduated tail*** and drab brown primaries. In flight, greyish webs to flight feathers give ashy appearance to underwings. Bill black in adult, pinkish with black tip in immature. **HH** Fairly common to common resident in wooded savanna, where partial to palms. Not shy; often common in or near villages. Frequently associates with domestic stock, occasionally perching on their backs and capturing insects disturbed by them. Usually in small flocks. **Voice** Shrill, squeaking and scolding calls (supposedly onomatopoeic). [CD14:6b]

ad
Square-tailed Drongo
ad
Fork-tailed Drongo
Brown-necked Raven
ad
ad
Pied Crow
Piapiac
ad
imm

STARLINGS (Plates 129–131) *Mainly arboreal species with strong and pointed bills and sturdy legs. Juveniles a dull version of adult or different. Omnivorous, most feeding on fruit and insects. Most forage in flocks, roost communally and nest in holes. Calls mostly harsh and grating but also including pleasing sounds.*

Purple Glossy Starling *Lamprotornis purpureus* — 24 cm

L. p. purpureus. Large, with noticeably ***longer bill and shorter tail*** than other glossy starlings, producing distinctive jizz. **Adult** ***Head, underparts and tail glossy purple*** becoming metallic blue on nape and rump, ***contrasting with iridescent blue-green upperparts***. ***Conspicuously large, golden-yellow eye***. In flight, broad wings and short tail impart rather compact appearance. Other savanna starlings have purple on head and breast replaced by glossy metallic green. **Juvenile** Duller; eye initially pale greenish, quickly becoming yellow. **HH** Common. In vocal flocks (of 10–100) in wooded savanna and thorn scrub. In pairs when breeding. Feeds on various fruits, particularly figs. Frequently walks on ground. **Voice** Various harsh, chattering and whistling notes. [CD14:17]

Bronze-tailed Glossy Starling *Lamprotornis chalcurus* — 21.5 cm

L. c. chalcurus. **Adult** Resembles Greater Blue-eared Starling but has ***noticeably shorter, mainly purple*** (not blue-green) ***tail***, contrasting with iridescent blue-green upperparts. Well defined ear-coverts and rump purple-blue. Tail may appear bronzy at some angles. ***Eye orange-yellow to reddish-orange***. Lesser Blue-eared Starling has blue-green tail and rump. **Juvenile** Duller; eye dark. Juvenile Lesser Blue-eared is paler and browner below. **HH** Locally common to fairly common, esp. in east; less frequent in coastal zone. In wooded savanna. Habits as Lesser Blue-eared Starling and both often occur together. In flight, wings make loud swishing sound. **Voice** Whining, drawn-out *weeaah*, abrupt *plip!* and various other nasal, chattering and whistling notes. [CD14:19]

Greater Blue-eared Starling *Lamprotornis chalybaeus* — 21–24 cm

L. c. chalybaeus. **Adult** Glossy metallic green with purple belly and more or less contrasting mask formed by black lores and poorly defined bluish ear-coverts; rump bluish; eye yellow (occasionally orange). Slightly smaller Lesser Blue-eared Starling is very similar and best distinguished by clear (not nasal) call. Bronze-tailed Glossy Starling has similar size but distinctly shorter, largely purple tail and, typically, redder eye. **Juvenile** As adult but duller; dark sooty-brown below with some iridescent feathers; eye dark. Darker, more blackish than juvenile Lesser Blue-eared. **HH** Common. Usually in flocks in thorn scrub with trees. Forages on ground and in fruiting trees. Moves south to northern savanna in dry season. **Voice** Nasal, whining *wèèh-aa-ah* or *skwee-aar*. Song a varied series of whistling, mewing, chirping and clicking notes, including call. [CD14:20]

Lesser Blue-eared Starling *Lamprotornis chloropterus* — 19–20 cm

L. c. chloropterus. **Adult** Strongly resembles Greater Blue-eared Starling but sharply defined ***ear-coverts blue-black, forming more contrasting and narrower mask***, tail relatively short, rump uniform with rest of upperparts, size smaller (noticeable when seen together), ***voice different***. Eye yellow to yellow-orange. Bronze-tailed Glossy Starling is slightly larger and has shorter, purplish (not metallic green) tail and usually more orangey or reddish eye. **Juvenile** Head and underparts ***greyish fawn-brown*** (almost 'café au lait') with some iridescent feathers; ear-coverts distinctly darker; upperparts and tail as adult but duller; eye initially dark. Paler below than juvenile Greater Blue-eared. **HH** Fairly common to common; less frequent near coast. In flocks in savanna woodland; typically more gregarious than Greater Blue-eared. Forages in fruiting trees and on ground. Occasionally associates with Bronze-tailed Glossy Starling. Seasonal movements reported but inadequately understood. **Voice** In flight a clear, distinctive *wirree-wirree*; from perch various rather similar calls, incl. *wreet* and *cherwee*. Song a rather simple, rhythmic series of 6–12 separated, chirruping and whistling syllables, e.g. *chirp chirp peelu chirp chrew whip*. [CD14:21]

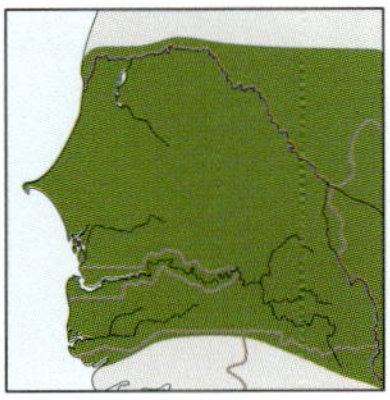

Long-tailed Glossy Starling *Lamprotornis caudatus* — c.51 cm (tail up to 33 cm)

Readily identified by ***very long, strongly graduated, supple tail***. **Juvenile** as adult but duller. **HH** Common. In small, lively and vocal flocks in open wooded savanna, farmland and thorn scrub. Often forages on ground. Flight slow and laborious on short rounded wings. **Voice** Various shrill and harsh notes. [CD14:25]

ad
Purple Glossy Starling
ad
Bronze-tailed Glossy Starling
juv
ad
Greater Blue-eared Starling
ad
juv
Lesser Blue-eared Starling
ad
Long-tailed Glossy Starling

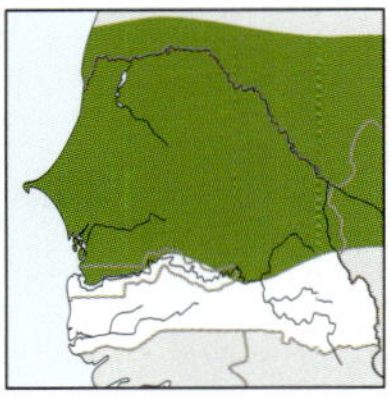

Chestnut-bellied Starling *Lamprotornis pulcher* 19 cm

Rather small, squat, dingy-looking starling with distinctive ***rufous-chestnut belly and undertail-coverts***. **Adult** Head greyish-brown, eye creamy-white, upperparts, throat and breast glossy oily green. In flight, creamy inner webs of primaries produce ***conspicuous pale wing patch*** (reminiscent of Northern Anteater Chat, Plate 105). **Juvenile** Duller, with ashy-brown throat and breast, pale bill and dark eye. **HH** Common resident in dry bushy country and farmland, often near villages. In small flocks. Feeds mainly on ground. Co-operative breeder. **Voice** Vigorous *whirrr*. Song a series of soft liquid notes. [CD14:27]

Violet-backed Starling *Cinnyricinclus leucogaster* 16–18 cm

C. l. leucogaster. Small, distinctive starling. Sexes markedly different. **Adult male** Unmistakable. ***Brilliant iridescent violet*** (may appear crimson or blue in bright light, or black when backlit) ***with contrasting pure white breast, belly and undertail-coverts***. **Adult female** Head, upperparts and tail brown with pale rufous-brown edges producing streaky and scale-like appearance to head and upperparts; ***underparts white heavily streaked dark brown***. **Juvenile** Similar to adult female. **HH** Seasonally fairly common to common (moving north with rains); irregular in some places, numbers varying from year to year. In Gambia, present in all months, but with marked increase in wet season. Frequents various wooded habitats, from savanna woodland to forest clearings. Arboreal, mainly foraging in trees and often perching in treetops. Usually in flocks, except when breeding; flocks often consist solely of females and young, or males. **Voice** Clear, high-pitched, drawn-out whistle *vfeeeee*. Also soft nasal call, usually on take-off. Song a short rapid series of whistling, twanging and clicking notes. [CD14:29]

Wattled Starling *Creatophora cinerea* 21.5 cm

Pale, drab-grey starling with black wings and tail and ***contrasting whitish rump*** (conspicuous in flight). **Male non-breeding** Bare black malar stripe; greater coverts whitish; primary-coverts white. Bill pale pinkish. **Male breeding** (this plumage unlikely to be seen in the region) Head loses feathers, exposing black skin and black wattles on forehead and throat, which contrast with bare bright yellow area on ear-coverts and hindcrown. **Adult female** Similar to male non-breeding, but greater and primary-coverts blackish. Worn wings all brownish. **Juvenile** As adult female but browner; bill dusky. **HH** Afrotropical vagrant, Gambia (up to three birds near Banjul, Dec–Feb 1975–76). Highly nomadic and gregarious; attracted to locusts. Usually in dry grassy savanna; occasionally in more wooded habitat. **Voice** Usually silent when not breeding. [CD14:30]

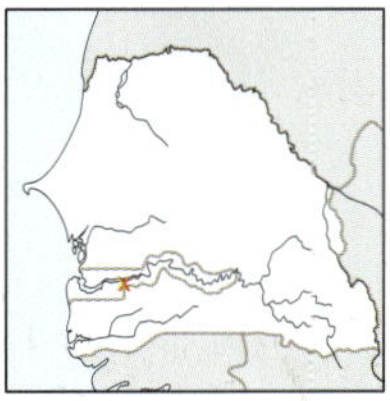

Common Starling *Sturnus vulgaris* 19–22 cm

S. v. vulgaris. **Adult non-breeding** Mainly glossy black, spotted buff above and white below. Bill long and pointed, brownish. In flight, characteristic silhouette with triangular wings and short, square tail. **HH** Palearctic vagrant (ship-assisted?), Gambia (Tendaba airfield, Jan 2007). Gait upright and confident. Flight fast and direct; also glides. **Voice** A grating *cheerr*; also various other harsh whistling sounds. [CD4:67]

juv
ad
Chestnut-bellied Starling
♀
♂
Violet-backed Starling
♂ br
♂ non-br
Wattled Starling
♂ non-br
Common Starling
ad

Splendid Glossy Starling *Lamprotornis splendidus* 27–30 cm

L. s. chrysonotis. **Adult** ***Large, highly glossed, noisy forest starling***. Has ***broad blackish band on wings and tail*** and ***creamy-white eye***. **Juvenile** Much duller, with dark eye. **HH** Locally fairly common, SW Senegal (Casamance); more uncommon and irregular further north along coast. Uncommon, Gambia, with numbers varying from year to year (records in all months). In flocks within canopy of forest, gallery forest and forest patches in savanna. Large numbers may congregate at fruiting trees and roosts. In flight, wings produce loud swishing noise. **Voice** Vocal. A variety of striking, harsh, piping and rasping calls with nasal and metallic quality, e.g. *niyar-èh; spiyok!; kwank!*; explosive *KYAH!* Makes considerable noise at roost. [CD14:24]

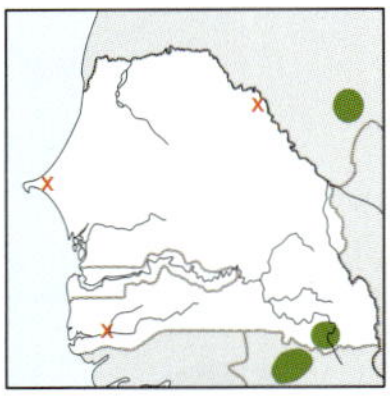

Neumann's Starling *Onychognathus neumanni* 28–33 cm

O. n. modicus. Large, with ***long graduated tail*** and black-tipped ***chestnut primaries forming conspicuous wing patch in flight***. Associated with rocky habitats. **Adult male** Glossy purplish-black with more greenish gloss on wings and tail. **Adult female** Head and throat ash-grey streaked blue-black. **Juvenile** Similar to adult male but duller. **HH** Scarce and local resident, SE Senegal; also recorded in atypical habitats in north-east and near Dakar. In pairs or flocks on rocky outcrops, inselbergs, crags and cliffs. Vocal, restless and conspicuous. **Voice** Various clear loud whistles, e.g. *twee-lee-uw, whutcheerrleeo* and *peeeo* etc., melodious and oriole-like; from perch and in flight. [CD14:12]

OXPECKERS *Medium-sized, mainly brown birds with moderately long, stiff tails, short legs and stout bills. Highly specialised, using their strong feet with sharp claws for clinging to herbivores, where they feed on parasites and wound tissue. Sexes similar. Juveniles duller with dusky bill. Usually forage in small flocks; roost communally. Nest in holes; breed cooperatively. Calls mainly harsh.*

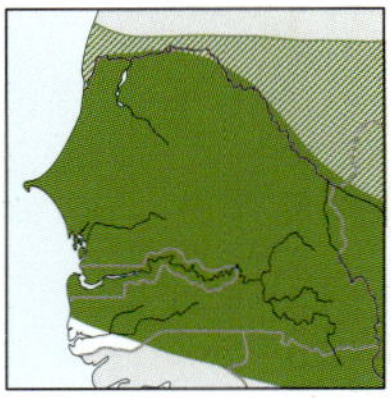

Yellow-billed Oxpecker *Buphagus africanus* 21–23 cm

B. a. africanus. Slender, dark brown bird associated with herbivores. **Adult** Head to throat, upperparts and tail earth-brown becoming paler brownish-buff on rump and rest of underparts. ***Bill bright yellow broadly tipped red***. Eye orange. In flight, ***contrasting pale rump and uppertail-coverts***; characteristic silhouette with rather long and pointed wings and tail. **Juvenile** Similar, but with dusky bill and dark eye. **HH** Widespread and generally fairly common, esp. in north; less common, S Senegal. Mainly in wooded savanna and bush. Clambers on wild or domestic herbivores, esp. buffalo and cattle (also donkeys, goats, sheep, ...), searching for ticks and other parasites, using stiff tail as prop. Perches in upright position on branches, with bill pointing skyward. Usually in small groups. In certain areas numbers are declining through improved veterinary care. **Voice** Hard rasping and hissing, metallic *krrizs* or *pszrr* in flight or when perched; also rattling *kzsririririri...* [CD14:31]

Splendid Glossy Starling
ad
♀
♂
Neumann's Starling
♂
juv
ad
Yellow-billed Oxpecker

PLATE 132: SPARROWS

SPARROWS Mostly brown and grey with short, conical bills. Bill of male typically turns from horn coloured to black in breeding season. Forage on or near the ground, feeding largely on seeds. Vocal, uttering a variety of rather harsh chirping calls.

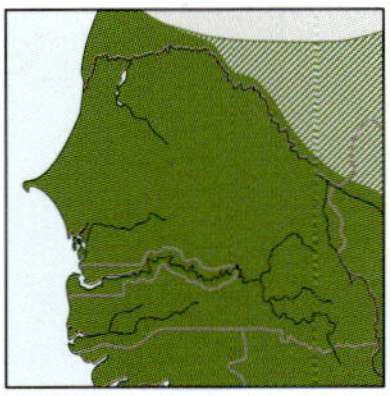

Northern Grey-headed Sparrow *Passer griseus* — 14 cm

P. g. griseus. Grey head and underparts; rich chestnut upperparts. **Juvenile** As adult but browner, with some dusky streaks on mantle, no white on wing-coverts, and horn-coloured bill. **HH** Common resident. In small, vocal groups or pairs, in variety of habitats, though mainly in towns and villages. Feeds on ground or in vegetation. **Voice** Monotonous *chirp* or *cheerp*. Also rattling *churrr-r-r-t-t* when alarmed or excited. Song a monotonous series of chirps, including call. [CD14:35]

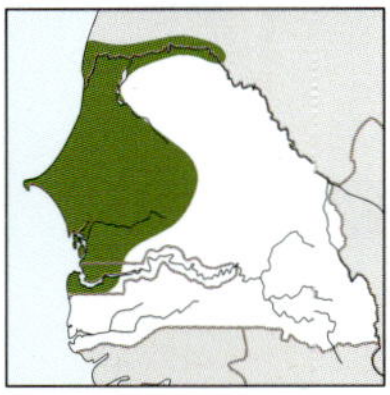

House Sparrow *Passer domesticus* — 14–15 cm

P. d. indicus. **Adult male** Grey crown and rump; black bib; ***whitish head-sides***. Mantle chestnut-brown streaked black; short wingbar formed by white-tipped median coverts; ***rump grey***. **Adult female** Dull brown with buff supercilium; horn-coloured bill. **Juvenile** Similar to adult female but duller; bill paler. **HH** Introduced in Senegal in 1970s, where now common in towns; spread to Gambia. Closely associated with man, occurring mainly in towns. Gregarious. Bold but wary. Flight fast and direct, with whirring wings. **Voice** Various chirps; rattling *churrr-r-r-t-t*. Song a monotonous series of chirping notes. [CD4:69, 14:33]

Desert Sparrow *Passer simplex* — 13.5 cm

Very pale. **Adult male** Unmistakable. Pale greyish above; buff-white on rump and underparts; black mask and bib; distinctive wing pattern. **Adult female** Pale sandy-buff without face markings. Wing pattern much reduced. Bill pale horn. **Juvenile** Similar to adult female, but wing pattern duller; bill pale pinkish. **HH** Vagrant, Gambia (Illisa, North Bank Division, Dec 2004; race undetermined but presumably *P. s. saharae*). Occurs in pairs or small groups at edge of desert in sandy areas with scattered trees or scrub, oases and human settlements. Feeds on ground. Partially nomadic depending on effects of rains on availability of grass. **Voice** Chirps similar to House Sparrow's. Also a soft melodious twitter, reminiscent of European Greenfinch. [CD4:73]

Sudan Golden Sparrow *Passer luteus* — 12–13 cm

Male breeding Unmistakable. ***Canary yellow with chestnut mantle, back and scapulars***, contrasting black bill and dark eye. **Male non-breeding** Plumage colours and pattern muted, thus somewhat resembling female. Bill horn. **Adult female** Head and upperparts buffish-brown faintly streaked dusky on mantle; pale supercilium; ***yellowish wash on head-sides and underparts***. **Juvenile** As adult female but paler, more buffish overall. **HH** Common N Senegal, wandering south to Gambia. Highly gregarious and nomadic, occurring in arid scrub, thornbush and dry savanna. Feeds on ground and in vegetation. Often in mixed flocks with other small seed-eaters, such as Red-billed Quelea. Sometimes in very large numbers. **Voice** Chirping *chilp* or *chirrup*. In flight also fast *che-che-che-...* Song a monotonous series of chirps, including call. [CD14:36]

Bush Petronia *Petronia dentata* — 13 cm

P. d. dentata. **Adult male** Dull brown above with ash-grey crown and ***chestnut stripe from eye around rear of ear-coverts***. Throat white with indistinct lemon-yellow patch (difficult to see in field; sometimes absent). **Adult female** Slightly paler with brown head and prominent ***buff supercilium***. White throat well demarcated; mantle streaked dark brown. Bill horn. **Juvenile** As adult female but browner. **HH** Common resident in wooded grassland. Most often in trees; rarely on ground. Singly, in pairs or small groups. Dry-season breeder, wandering widely when not breeding. **Voice** Sparrow-like chirps. [CD14:39]

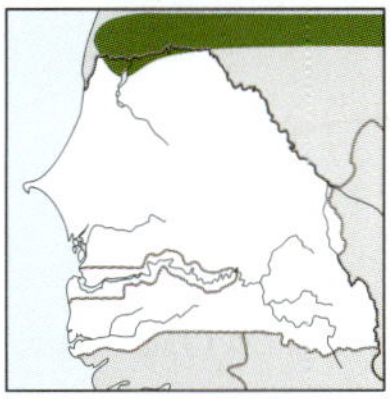

Yellow-spotted Petronia *Petronia pyrgita* — 15 cm

P. p. pallida. **Adult** Buffish-brown above, creamy-white below, small lemon patch on throat (difficult to see). **Juvenile** No throat patch. **HH** Scarce and local resident, N Senegal. Local movements outside breeding season reported. In pairs or small groups, in thorn scrub, dry open woodland and edges of cultivation. Forages mainly in trees; also on ground. Associates with other small seed-eaters. **Voice** Sparrow-like notes, *chirp*, *chillip* and *wurli*. Usually silent. [CD14:37]

Northern Grey-headed Sparrow
♀
♂ br
House Sparrow
♀
♂
Desert Sparrow
♀
♂ br
Sudan Golden Sparrow
♂ non-br
♂
Bush Petronia
♀
Yellow-spotted Petronia

WEAVERS (Plates 133–136) Large family of small to medium-sized birds with generally strong, conical bills. Many species gregarious, feeding and roosting in flocks and nesting in colonies. The name refers to their elaborate, woven nests, typically consisting of a domed structure with downward-pointing entrance, suspended from tree or bush.

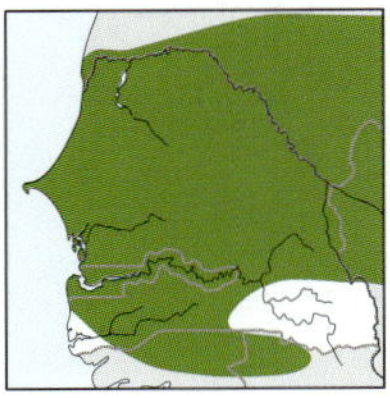

White-billed Buffalo Weaver *Bubalornis albirostris* 22–24 cm

Readily identified by large size, black plumage, massive bill and gregarious habits. Feathers reveal white base when ruffled, esp. on back and flanks. **Male breeding** Massive whitish bill. **Male non-breeding** Bill black. Female similar. **Juvenile** Blackish-brown; underparts mottled white. **HH** Common resident in dry savanna with *Acacia* and thorn scrub. In small groups. Forages on ground. Huge communal stick nest with many chambers very conspicuous and centre of activities year-round. Noisy at nest. **Voice** Various harsh, croaking and high-pitched, squeaky sounds, resulting in cackling chatter. [CD14:40]

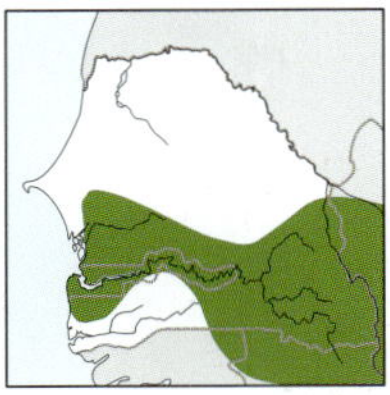

Chestnut-crowned Sparrow Weaver *Plocepasser superciliosus* 17 cm

Rather large, sparrow-like savanna weaver, with distinctive, well-marked head pattern. Rufous-chestnut crown, pale rufous-brown cheeks, conspicuous white supercilium, white line below eye and black malar stripe. **Juvenile** Crown duller chestnut. **HH** Uncommon or scarce to locally fairly common. In thorn scrub, wooded savanna and degraded savanna scrub. Arboreal. Usually in pairs; joins mixed-species flocks. Not shy but inconspicuous and easily overlooked. Local movements reported. **Voice** Distinctive, fast, metallic ticking trill *trrrrit* or *trrirr, hrt* somewhat recalling *Euplectes*. Song a short, continuous phrase *witsweeweetsweeuseesweeuseeswee*. [CD14:42]

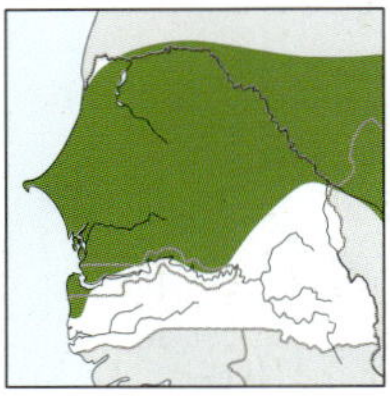

Speckle-fronted Weaver *Sporopipes frontalis* 12 cm

S. f. frontalis. Small; sparrow-like with conspicuous rufous nape. Forehead and moustachial streak black speckled white. **Juvenile** Rufous on crown and nape duller. **HH** Common in north, becoming less common further south. In small flocks in dry bush country; occasionally around villages. Forages on ground with characteristic hopping gait. **Voice** Short, silvery trill *sriiii* or *tsrrrrk* and series of mainly high-pitched, short notes *tsisit chep tsisisit che…*; rapid *tsip-tsip-tsip-tsip* when taking flight. [CD14:41]

Blue-billed Malimbe *Malimbus nitens* 17 cm

Adult All black with bright red patch from throat to upper breast and blue bill. Compare Western Bluebill (Plate 137). **Juvenile** Red area duller, more diffuse, and extending to throat, head-sides and forehead. Bill horn-coloured. **HH** Uncommon and local resident, SW Senegal (Casamance); possible vagrant, Gambia (recent claims lack adequate description, confusion with Western Bluebill cannot be eliminated). Singly, in pairs or small family parties in dense humid undergrowth of forest interior, clearings and edges with watercourses or pools, and swamp forest; also high mangrove and even dense savanna woodland and thickets, provided trees overhanging water available. Frequently in mixed-species flocks. Nests almost always above water; usually several (of which many are old) together. **Voice** Harsh *zheep*. Song a series of *zheet* followed by dissonant sounds and ending in harsh, weaver-like sizzling. [CD14:73]

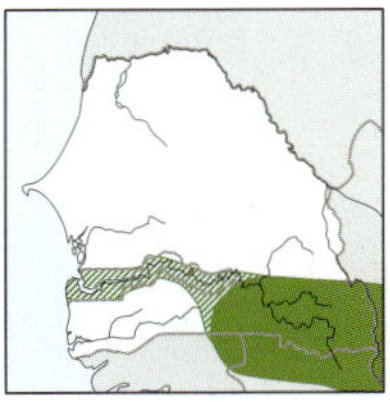

Red-headed Weaver *Anaplectes rubriceps* 14–15 cm

A. r. leuconotus. Distinctly coloured ***with rather slender, red bill.*** **Male breeding** Head to upper breast bright red surrounding black mask. Upperparts and tail grey-brown; mantle mottled red. Red edges to flight feathers form bright panel; tail also edged red. Rest of underparts white. Bill bright red or orange-red. **Adult female/male non-breeding** Differ in having greyish-brown head, white underparts washed greyish-buff on throat and breast, and pinkish bill. **Juvenile** Similar to adult female but head to breast washed yellowish-buff; edges of flight and tail feathers orangey. Bill dusky. **HH** Uncommon resident, S Senegal; scarce, Gambia. Singly, in pairs or small family parties in open woodland, scrub and *Acacia* savanna. Forages among foliage of trees and bushes, often hanging upside-down. Not gregarious. Joins mixed-species flocks. **Voice** Usually silent. At nest a rapid sizzling, typical of weavers, but distinctly high pitched and squeaky. Call a sharp *tzik*. [CD14:80]

White-billed Buffalo Weaver
♂ br
juv
♂ non-br
Chestnut-crowned
Sparrow Weaver
ad
Speckle-fronted
Weaver
ad
juv
ad
Blue-billed Malimbe
♀
♂
Red-headed
Weaver

TYPICAL WEAVERS (Genus **Ploceus***)* *Males in breeding dress mostly bright yellow; females dull and streaky (sparrow-like). In non-breeding dress (if any) males similar to females. Juveniles usually as females. Vocalisations typically consist of characteristic, drawn-out, wheezy, buzzy, chirping and chattering notes.*

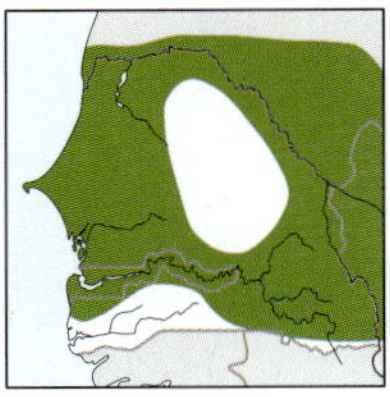

Little Weaver *Ploceus luteolus* — 11.5 cm

P. l. luteolus. Smallest savanna-dwelling typical weaver. **Male breeding** Large black mask, from forehead to throat. **Male non-breeding** No mask; crown yellow-green; head-sides to breast buff; mantle buff streaked dusky. **Female breeding** No mask; head-sides to breast pale yellow; mantle yellow-green streaked dusky. Non-breeding female duller; bill horn. **HH** Uncommon to locally fairly common resident. In pairs or small family groups in dry wooded savanna; also farmland with large trees and gardens. Joins mixed-species flocks. **Voice** Rather silent. Typical call a harsh *chep*. Also soft *tsssp*. Song a jumble of chattering and chirping notes, typical of weavers. [CD14:47]

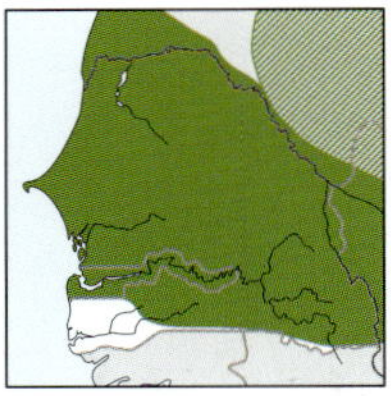

Vitelline Masked Weaver *Ploceus vitellinus* — 14 cm

P. v. vitellinus. **Male breeding** Distinguished by ***red eyes***, black mask ***with narrow band on forehead*** and variable ***chestnut wash to crown and lower throat***. **Male non-breeding/adult female** No mask; upperparts brownish-olive broadly streaked dusky on mantle; throat pale lemon-yellow becoming buff on breast and white on belly; eyes reddish. **HH** Generally common resident, becoming less common towards coast. In pairs near water in dry woodland. Joins mixed-species flocks. **Voice** Call a sharp *pzit!* or *tzik!* Song of chattering and wheezy notes. [CD14:58]

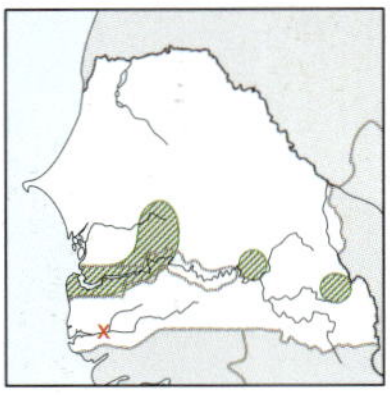

Heuglin's Masked Weaver *Ploceus heuglini* — 14 cm

Male breeding Differs from Vitelline Masked Weaver in having black mask extending in point to upper breast (but *not* onto forehead), ***no chestnut wash*** on head and throat, and ***yellow eye***. **Adult female** As female Vitelline Masked but with less distinct streaking on mantle and yellowish eyes. Bill brown to black (breeding) or horn (non-breeding). Non-breeding male similar but yellower below. **Juvenile** Similar to non-breeding female. **HH** Uncommon (overlooked?). In pairs or small colonies in wooded savanna and farmland. Often breeds in small colonies close to raptors', wasps' and ants' nests. **Voice** Jumble of chattering and wheezy notes; more varied and including more melodious notes than most other *Ploceus*. [CD14:60]

Black-necked Weaver *Ploceus nigricollis* — 17 cm

P. n. brachypterus (Swainson's Weaver). **Adult male** Head golden-chestnut becoming bright yellowish-olive on hindcrown; black stripe through and around pale eye, forming small mask. Upperparts and tail ***bright yellowish-olive***. Throat black bordered golden-chestnut; rest of underparts golden-yellow. **Adult female** No black on throat; top of head yellowish-olive. **HH** Fairly common to locally common resident. In pairs or small (family?) groups within forest clearings and edges, gallery forest, wooded savanna, gardens, etc. May join mixed-species flocks. Forages for insects in leafy trees. **Voice** Harsh *chet-chet* and a twittering wheeze. Usually rather silent. [CD14:49] **Note** English name based on extralimital races, but inappropriate for *brachypterus*.

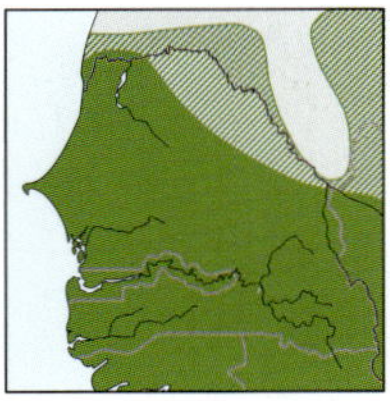

Village Weaver *Ploceus cucullatus* — 15.0–17.5 cm

P. c. cucullatus. The ubiquitous, well-known, gregarious weaver, common in towns and villages. **Male breeding** Black head; chestnut collar; black V bordering mantle. Eye red. **Male non-breeding** Lemon-yellow throat; underparts variable (yellow, or white with buffish breast and flanks). **Female breeding** Yellow-olive head contrasting with brownish-olive upperparts; lemon-yellow below; belly white. Non-breeding female duller; bill paler. **HH** Common resident. In dense, noisy colonies, in various habitats; mainly near habitation. **Voice** Harsh chattering and twittering notes at colony. Song a jumble of rasping, chattering notes ending in drawn-out rattling wheeze, *chit-chit chit-t-t-t-t shirrrzzzzwrrerr*. Alarm a sharp *zip*. In flight a short *chuk-chuk* or *chit-chit*. [CD14:63]

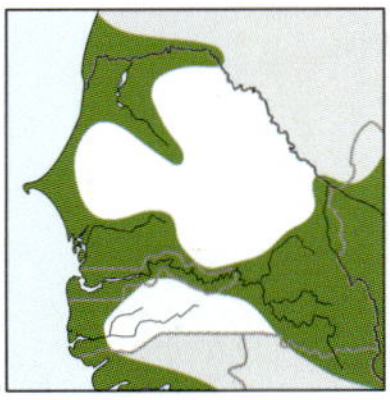

Black-headed Weaver *Ploceus melanocephalus* — 14–15 cm

P. m. capitalis. **Male breeding** Black head; golden nuchal collar; yellowish-olive upperparts. Eye dark brown. **Male non-breeding** Brownish-olive head; rest of plumage mainly brownish, buff and white. Female similar but smaller. **HH** Common resident. In pairs (males sometimes polygynous) near water within semi-arid savanna. Frequents high grass and bushes along river banks, margins of lakes, marshes and swamps. **Voice** Discordant chattering and twittering; variety of calls, incl. hoarse, nasal *chè-èp, tshk* and *tseew*. [CD14:64]

♂ non-br
♀ br
Little Weaver
♂ br
♂ non-br /♀
♂ br
Vitelline Masked Weaver
♀
Heuglin's Masked Weaver
♂ br
♀
♂
Black-necked Weaver
♀ br
♂ non-br
♂ br
Village Weaver
♂ non-br
♂ br
Black-headed Weaver

PLATE 135: WEAVERS III

Vieillot's Black Weaver *Ploceus nigerrimus* 17 cm

P. c. castaneofuscus (Chestnut-and-black Weaver). **Adult male** Easily identified, black and chestnut weaver with yellow eye. **Adult female** Brownish-olive above streaked dusky on mantle, buffish-yellow below with some rufous wash. **Juvenile** Similar to adult female. **HH** Afrotropical vagrant, Gambia; occurrence SW Senegal (Casamance) unproven. Frequents forest clearings, villages, marshy habitats, and edges of gallery forest. Normally in noisy colonies, often in association with Village Weaver. **Voice** Shrill jumble of chattering notes, similar to that of other social weavers, esp. at breeding colony. [CD14:61-62]

Compact Weaver *Pachyphantes superciliosus* 15 cm

Distinctive stocky weaver with short, stout bill. **Male breeding** Black mask, chestnut forehead becoming yellow on crown, dark upperparts, and dark eye. **Female breeding** As male, but forehead and centre of crown black. **Adult non-breeding** Dark earth-brown above with blackish top of head and eye-stripe; head-sides and underparts pale cinnamon. **Juvenile** As non-breeding adult but head-sides and underparts washed pale yellow. Bill horn. **HH** Scarce, SW Senegal (Casamance). Locally in grasslands within forest and savanna zones, and in rank grass in moist areas and near rivers. In small flocks or, when breeding, in pairs. Subject to movements, apparently without seasonal pattern. **Voice** Rapid, abrupt, crackling calls in flight. Also harsh *cheee* and repeated *pink*. Song a series of hard, abrupt notes followed by typical weaver sizzling *trrrip, trrp trrp trip trip, tru-tru-huu-tzzzzz*. [CD14:67]

Grosbeak Weaver *Amblyospiza albifrons* 18 cm

A. a. capitalba. Robust with massive bill. **Adult male** Head to mantle and upper breast chestnut with ***white forehead and primary patch*** (latter conspicuous in flight). Rest of upperparts and tail brownish-black. Lower breast and belly sooty-grey. **Adult female** Dark brown above, ***whitish below heavily streaked brown***; bill dark horn. **Juvenile** As adult female, but bill yellowish-horn. **HH** Afrotropical vagrant, SW Senegal (three birds, Casamance, Jul 1977). Usually in small scattered colonies in rank grass, reedbeds and tall vegetation in swamps, marshy areas and along rivers; occasionally in tall herbage within dry areas. Forages in forest canopy. Subject to local movements. **Voice** Chirping and twittering calls in flight and at breeding colony. Song a vigorous, fast jumble of hard, rasping, chattering and churring notes. [CD14:98]

Red-headed Quelea *Quelea erythrops* 11–12 cm

Male breeding Easily identified by red head and throat contrasting with buff-brown, blackish-streaked upperparts; bill black. **Adult female** Lacks any red in plumage, face washed with yellow; bill brownish-horn. **Male non-breeding** As adult female; often with red wash on head-sides and forehead. **Juvenile** As adult female but with pale sandy edges to feathers of mantle, wing-coverts and tertials. Female Yellow-crowned Bishop differs from female Red-headed Quelea by longer, more conspicuous, yellow supercilium. **HH** Uncommon wet-season breeder. Movements unclear as hard to identify in non-breeding plumage. In small or large flocks, sometimes with other small seed-eaters, in moist grasslands, ricefields and coastal thickets. **Voice** Flocks utter a squeaky chattering, esp. at breeding colony. Song a slightly rising series of churring notes ending in drawn-out wheeze. [CD14:82]

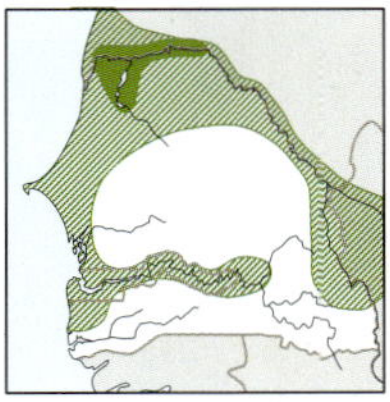

Red-billed Quelea *Quelea quelea* 11–13 cm

Q. q. quelea. **Male breeding** Distinguished by strong, conical ***red bill***. Head pattern variable, most having black or dark brown mask of variable shape (usually from forehead over ear-coverts to throat). Mask sometimes lacking, replaced by white or buff-brown. Crown and nape pale to dark buff, upperparts streaked black and buff-brown, underparts buffish-white variably washed tawny or pinkish on breast and belly. **Male non-breeding/adult female** Streaky above with long off-white supercilium and greyish cheeks. Resemble Red-headed Quelea but have ***red bill***. **HH** Common along Senegal River, uncommon elsewhere in Senegal; uncommon dry-season visitor, Gambia. In small or large flocks in grasslands, ricefields, open savanna and dry *Acacia* country. Itinerant breeder. Nomadic outside breeding season. **Voice** Flocks utter rasping and squeaky chattering, esp. at breeding colony. [CD14:83]

♀
Vieillot's Black Weaver
♂
♀ br
ad non-br
Compact Weaver
♂ br
♂
Grosbeak Weaver
♀
♂
Red-headed Quelea
♂ non-br
♀
♂ br
juv
♂ non-br/ ♀
♂ br
♂ br
♂ br
♂ br
Red-billed Quelea

PLATE 136: WIDOWBIRDS AND BISHOPS

WIDOWBIRDS AND BISHOPS Occur in open grassland and edges of marshy areas. Breeding males conspicuous in plumage and behaviour, and easily identified. In non-breeding season males resemble the nondescript, sparrow-like females, but are often noticeably larger and, in some species, may retain coloured shoulder patches. Juveniles similar to females. Females and many males in non-breeding plumage hard to identify; some not safely separable. Most breed during rains, forming mixed-species flocks in off-season.

Yellow-mantled Widowbird *Euplectes macroura* ♂19–21 cm, ♀13–14 cm

E. m. macroura. **Male breeding** Unmistakable. ***Black with contrasting yellow mantle and shoulders, and long, broad, graduated tail***. **Male non-breeding** As female, but larger, ***with yellow shoulder patches***. **Adult female** Streaked buff-brown and blackish above, lesser coverts tipped yellow. Below pale buff, faintly streaked dusky on breast and flanks. Bill horn. **Juvenile** As adult female but tinged yellow above. **HH** Uncommon resident. In moist grassland, rank herbage on fallow farmland, marshy areas and large grassy clearings. Male displays in low cruising flight above territory with ruffled neck and rump plumes, uttering song. **Voice** Rhythmic series of thin, buzzing *zeet* and *tsweep* notes, in display flight and from perch. [CD14:92]

Red-collared Widowbird *Euplectes ardens* 12–14 cm

E. a. ardens. **Male breeding** Unmistakable. ***All black with very long, supple, graduated tail*** (of up to 20 cm). **Male non-breeding** As female, but larger and has ***black wing feathers***. **Adult female** Head-sides and throat washed yellow; rest of underparts ***unstreaked*** fulvous washed yellow on breast and flanks and becoming white on central belly. **Juvenile** As adult female but usually has brighter yellow wash to supercilium and throat. Female and juvenile Black-winged Bishop are more yellowish and faintly streaked on throat and breast. **HH** Status uncertain; no recent records. Frequents grassy plains and hill sides. Display of male consists of slow, bouncing flight above territory with rapid wingbeats and depressed tail, uttering song; also short, low flight, with intermittent drops into grass. **Voice** Rapid series of buzzes, chirps and ticks, in display flight and from perch. [CD14:94] **Note** English name based on red in plumage of extralimital populations, but inappropriate in W Africa.

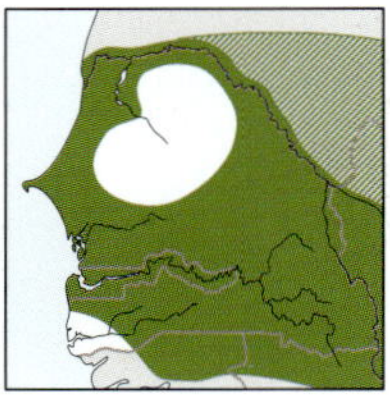

Northern Red Bishop *Euplectes franciscanus* 11–12 cm

E. f. franciscanus. **Male breeding** ***Bright scarlet*** with ***black crown and head-sides***, and black lower breast and belly. Wings and tail brown. Scarlet upper- and undertail-coverts very long, covering short tail. **Male non-breeding/adult female** Smaller than Black-winged Bishop, with smaller bill. Female Yellow-crowned Bishop very similar but more distinctly streaked on breast and flanks. **HH** Common resident. In moist grassland, seasonal rank vegetation and reedbeds. Male displays from perch or in slow, bouncing cruising flight with rapidly beating, rustling wings and puffed-out plumage, uttering song. **Voice** Fast jumble of twittering, churring and wheezy notes, uttered in display flight and from perch. Call a sharp *chiz*. [CD14:88]

Yellow-crowned Bishop *Euplectes afer* 11 cm

E. a. afer. Small bishop. **Male breeding** Unmistakable. ***Bright yellow with large black mask***, mottled black collar on mantle, and black lower breast and belly. **Female breeding** Broad yellowish-white supercilium bordered by narrow dusky eye-stripe; sparse streaking on breast and flanks. Non-breeding male and female similar but more distinctly streaked on breast and flanks. Northern Red Bishop slightly larger with less streaky breast. **HH** Locally fairly common resident. Prefers moister habitats than red bishops: swampy areas, floodplains, rank vegetation near rivers, lakes and ponds, ricefields. Display of male consists of low, bouncing flight over territory with puffed-out feathers, uttering song. **Voice** Series of high-pitched chirping and buzzing notes, e.g. *szit-szit-szit-*..., esp. in display flight. Call a sharp *tsip*. [CD14:85]

Black-winged Bishop *Euplectes hordeaceus* 13–14 cm

Male breeding Bright scarlet with contrasting ***black mask***, breast, belly, wings and tail. Undertail-coverts buff. **Male non-breeding** As other *Euplectes*, but with ***black flight feathers*** (conspicuous in flight). **Adult female** Streaky above, with yellowish supercilium; breast-sides faintly streaked. **HH** Fairly common resident. In rank herbage within savanna, edges of cultivation, fallow farmland and large forest clearings. Male displays in slow, bouncing cruising flight with rapidly beating, rustling wings and puffed-out plumage, uttering wheezy song. **Voice** Fast series of twittering, churring and buzzing notes, interspersed by occasional whistles. [CD14:86]

♂ non-br
Yellow-mantled Widowbird
♀
♂ br
♂ non-br
♀
♂ br
Red-collared Widowbird
moulting ♂
♂ non-br/♀
♂ br
Northern Red Bishop
♀
♂ br
Yellow-crowned Bishop
♂ non-br
♀
♂ br
Black-winged Bishop

ESTRILDID FINCHES (Plates 137–140) Small passerines with short, generally conical bills. Predominantly granivorous, but Nesocharis *and forest-dwelling* Nigrita *(this plate) largely or entirely insectivorous. Open country estrildids may gather in flocks, especially outside breeding season; forest species usually in pairs or small parties. Principally sedentary, but local movements may occur in drier parts of range.*

Chestnut-breasted Negrofinch *Nigrita bicolor* 11–12 cm

N. b. bicolor. **Adult** Slate-grey above, dark chestnut below and on forehead and head-sides. **Juvenile** Pale grey-brown above, rich buff-brown below. **HH** Uncommon resident, SW Senegal (Casamance); rare, Gambia (Abuko; no records since 1986). Singly, in pairs or small family groups, frequenting all levels of lowland forest and forest outliers in savanna. Sometimes joins mixed-species flocks. **Voice** Series of cheerful whistles, e.g. *weet-weet-teeu weet-weet-teeu weet-weet-tiutiutiuhuu tiutiuhuu*, or rising *tuwee-tuwee-twee-twee* and *tititi-tiu-tiu-teeu tiu-tiu-teeu*. [CD15:4]

Grey-headed Oliveback *Nesocharis capistrata* 12–13 cm

Adult Thickset with characteristic head pattern: white face, grey crown and nape, and black throat extending as a line below ear-coverts. Upperparts and tail olive-green; underparts grey with conspicuous yellow flanks. Bill short and stubby, dark grey. **Juvenile** Duller, overall more dusky, lacking white face; bill whitish. **HH** Status uncertain. One old specimen from Gambia (details unknown) and possible sighting (Abuko, May 1968). No records Senegal. Occurs in moist grassland and bush, forest edges and riparian woodland. Singly or in pairs. Skulking and usually keeping low, though may search foliage at all heights in rather warbler-like manner. **Voice** Usually silent. Thin, high-pitched *tsip*. Song a fast, descending series of high-pitched *tsi* or *tseee* notes. [CD15:8]

Western Bluebill *Spermophaga haematina* 15 cm

S. h. haematina. Stout black-and-red estrildid with heavy, mainly blue, bill and black tail. **Adult male** Black with red throat, breast and flanks; bill blue with red tip. **Adult female** Face washed red; red uppertail-coverts; lower underparts densely spotted white. **Juvenile** Dark brownish-slate; dull reddish uppertail-coverts. **HH** Uncommon resident, SW Senegal (Casamance) and SW Gambia (Abuko). Singly or in pairs in forest patches, gallery forest, thickets, edges, clearings and overgrown cultivation. Low in undergrowth, occasionally mid-level. Sometimes in mixed-species flocks. Skulking and unobtrusive. **Voice** Metallic *tswink-tswink-...*, short *tsik*, high-pitched *tsee*, sharp *tak*. Song a variable series of clear whistles, trills and/or rattles. [CD15:17]

Crimson Seedcracker *Pyrenestes sanguineus* 13–14 cm

Stout, with distinctive heavy, triangular bill and large, rounded head. **Adult male** Entire head to breast and upper flanks, rump and tail bright red. Upperparts and rest of underparts warm earth-brown. White, broken, orbital ring. Bill steel-blue. **Adult female** Red areas more restricted; rest of plumage warm brown. **Juvenile** All brown with dull red rump and tail. **HH** Scarce to rare resident, Gambia and S Senegal. Singly, in pairs or small parties, frequenting marshes, ricefields, farmbush, rank vegetation at edges of forest and gallery forest, and suburban gardens. Shy and unobtrusive, usually foraging low near dense cover. May fly high into trees when flushed. **Voice** A sharp *tsu-it* or *tsut*. Song a fast, short, pleasant warble, one version ending with drawn-out, sharp *tsee-ik*. [CD15:16]

Red-cheeked Cordon-bleu *Uraeginthus bengalus* 13 cm

U. b. bengalus. Immediately identified by pale blue underparts and tail. **Adult male** Top of head and upperparts pale brown; rump, tail, head-sides to breast and flanks pale cerulean-blue; ***bright crimson patch on ear-coverts***; rest of underparts pale buffish-brown. Bill pinkish. **Adult female** Lacks crimson cheek-patch and is slightly duller. **Juvenile** Similar to female, but duller and paler; blue on underparts more confined to throat. Bill dark slate. **HH** Common resident in savanna grassland, edges of cultivation, gardens and villages. In pairs or small flocks; often with firefinches and waxbills. Tame. **Voice** Thin, high-pitched *sweep-sweep*; alarm *chchchrrrt*. Song a thin *tsee-tsee-tsu-tsuu*, with variations. [CD15:42]

Chestnut-breasted Negrofinch

Grey-headed Oliveback

Western Bluebill

Crimson Seedcracker

Red-cheeked Cordon-bleu

WAXBILLS *Small, active and agile estrildids, which feed mainly on grass seeds taken from growing plants. In pairs or small flocks. Parasitised by Pin-tailed Whydah.*

Lavender Waxbill *Estrilda caerulescens* — 10 cm

Unmistakable. **Adult** ***Wholly soft grey*** with contrasting ***red rump and tail***. **Juvenile** Paler and duller; lower flanks lack speckles. **HH** Locally fairly common resident. In various habitats in savanna zone, incl. grassland with scattered trees and bushes, edges of thickets and cultivation, roadsides and gardens. From ground to high up in trees. Joins mixed-species flocks in savanna. **Voice** Clear *tsee* and *tseeu*; occasionally dry *tik-tik…* and *skweeep* (alarm). Song a series of thin notes *see-see-swee-see*. [CD15:31]

Orange-cheeked Waxbill *Estrilda melpoda* — 10 cm

E. m. melpoda. In all plumages, ***orange face*** diagnostic. **Adult** Head pale grey with bright orange face and red bill; rump red. **Juvenile** Duller and buffier, with pale orange face, black bill and paler rump. **HH** Fairly common to locally common resident in rank grass in savanna, large forest clearings and swampy areas, farmbush, and gardens. **Voice** High-pitched *tsee-tsee*, frequently repeated; also short trill *tsiririririt*. [CD15:36]

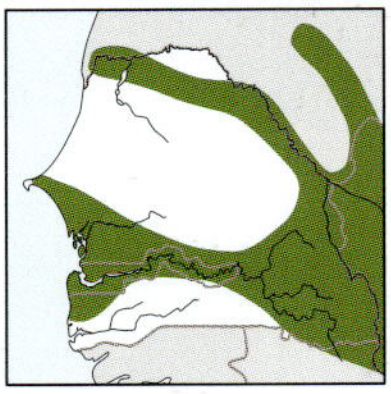

Black-rumped Waxbill *Estrilda troglodytes* — 10 cm

Adult Distinguished by bright red mask and bill, ***black rump*** and ***white undertail-coverts***. Underparts pale buffish washed pink. Sexes similar, but female less pink below. **Juvenile** Duller, lacking red mask and with black bill. **HH** Uncommon to locally fairly common resident in dry grassy savanna, scrub, edges of thickets and farmbush. Flicks tail from side to side. **Voice** Vigorous *cheeeu* and *chup*, frequently repeated; in flight, flocks utter a constant *chiup-chiup-chiup*. Song of male a disyllabic *che-cheer* or *chu-weee*; that of female a shorter *pwitch pwitch* or *pwitch cheee…*[CD15:37]

Dybowski's Twinspot *Euschistospiza dybowskii* — 12 cm

Distinctive, dark finch. **Adult male** Crimson from mantle to uppertail-coverts; slate-grey from head to breast; white spots on black flanks. **Adult female** Generally paler grey on head and breast; rest of underparts greyer. **Juvenile** Dull slate-grey underparts without spots; dull rusty upperparts. **HH** Rare and local resident, SE Senegal. In pairs or small parties in tall grass and bushes in wooded savanna and at forest edges. **Voice** Call *chip-chip-…*; also *see, tsit-tsit-* and *tswink-tswink-…* Song a varied medley of trills, melodious whistles and bubbling notes. [CD15:21]

PYTILIAS *Medium-sized estrildids with red rumps and tails, and rather slender bills. Frequently in trees and bushes, but feed on ground. Parasitised by paradise whydahs.*

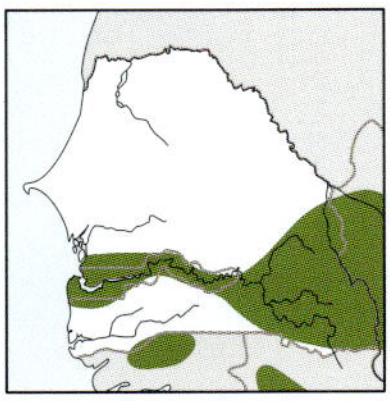

Red-winged Pytilia *Pytilia phoenicoptera* — 12.5–13.0 cm

P. p. phoenicoptera. **Adult male** Grey with red wings, rump and tail. Lower breast to undertail-coverts delicately barred grey and white. Bill greyish-black. **Adult female** As male but browner grey. **Juvenile** Similar to female but browner and duller with fainter barring on underparts. **HH** Uncommon; numbers augmented in wet season. Singly, in pairs or small groups in various types of open wooded savanna, thickets and grassland; also edges of farmland with tall grass and bushes. Occasionally with mixed-species flocks or other small granivores. Parasitised by Exclamatory Paradise Whydah. **Voice** Single *twink*. Song a short, variable phrase, consisting of some hard whistles followed by a rolling, croaking sound. [CD15:12]

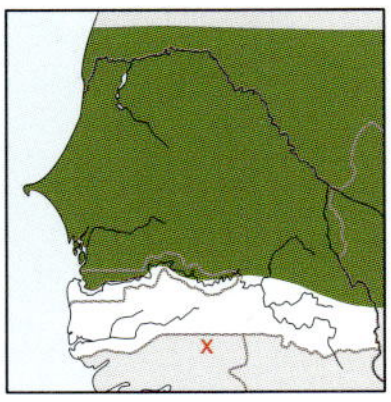

Green-winged Pytilia *Pytilia melba* — 12–13 cm

P. m. citerior. **Adult male** Red face and bill, olive-green upperparts, yellow upper breast, olive barring on lower breast and flanks. **Adult female** Red on head replaced by grey; barred grey and white below. Bill dull reddish with brownish culmen. **Juvenile** Dull olive above, unbarred buffish-brown below; bill dark. **HH** Fairly common in north, becoming less frequent towards the south. Usually in pairs or small family parties in *Acacia* woodland and thorn scrub, also in savanna woodland and edges of cultivation. Parasitised by Sahel Paradise Whydah. **Voice** Usually silent. Thin, penetrating whistle *see-eh* or *heee*, often monotonously repeated and difficult to locate; also short *wik*. Song a variable mix of buzzing and croaking sounds interspersed with call notes. [CD15:9]

ad
Lavender Waxbill
ad
juv
Orange-cheeked Waxbill
♂
juv
Black-rumped Waxbill
♀
juv
Dybowski's Twinspot
♂
♂
♀
Red-winged Pytilia
juv
Green-winged Pytilia
♂
♀

FIREFINCHES *Small, ground-feeding estrildids with much crimson or wine-red in male plumage and usually some small white spots on breast-sides. All males have red rump and tail-base. Usually in pairs or small groups. Associate with other firefinches and waxbills. Parasitised by indigobirds.*

Red-billed Firefinch *Lagonosticta senegala* — 10 cm

Very common firefinch, associated with habitation. Both sexes have ***red bill***. **Adult male** Readily identified by ***entirely red head*** and ***yellowish orbital ring***. Upperparts brown washed reddish. ***L. s. rhodopsis*** slightly paler than **nominate**. **Adult female** Mainly buff-brown above, paler below; red loral spot and rump. **Juvenile** As female but lacks red on lores and has black bill. **HH** Common resident throughout in towns and villages, farmland, scrub and open savanna. *L. s. rhodopsis* in N Senegal. Parasitised by Village Indigobird. **Voice** Soft, high-pitched *dwee* or *pfweet*; alarm *chep*. Song a short *tsi-tu-tuuuuit* or a rising series of fluting notes, e.g. *tu-tweet-tweet-...* [CD15:24]

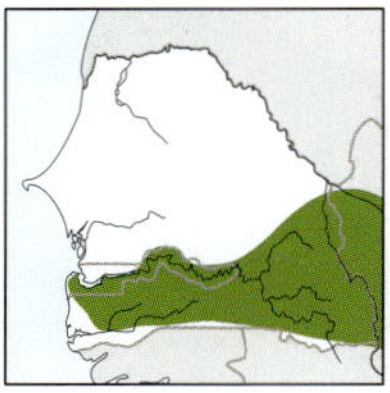

Bar-breasted Firefinch *Lagonosticta rufopicta* — 11 cm

L. r. rufopicta. Sexes similar, with ***red bill***. **Adult** Face pinkish-red; above earth-brown; rump and base of tail crimson; below pinkish-red becoming greyish on lower belly and dirty cream on undertail-coverts. Breast-sides and upper flanks have small white crescents. **Juvenile** Dull grey-brown washed pinkish-red on face and breast; rump dull crimson. Bill dusky. **HH** Uncommon to scarce resident in various open habitats, incl. grassy forest clearings, rank vegetation near water, open wooded savanna, farmland, and gardens. Parasitised by Wilson's Indigobird. **Voice** High-pitched *chip* or *pik*. Song a fast jumble of short, mostly clear notes. [CD15:22]

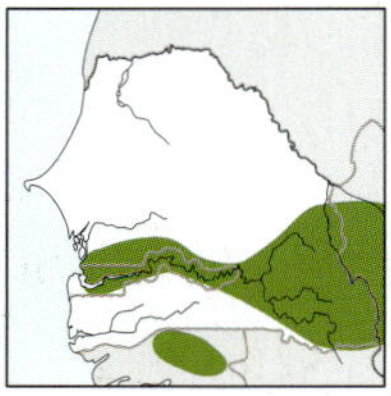

Black-faced Firefinch *Lagonosticta larvata* — 11.5 cm

L. l. vinacea. **Adult male** Distinctive firefinch with ***black face***. Pinkish-mauve above, bright pink below; rump and central tail deep red. **Adult female** Head pale grey-brown, throat buff tinged greyish, rest of underparts paler than male. **Juvenile** Dull brown with dull red rump; no white specks on sides. **HH** Uncommon resident. In wooded grassland, rank herbage near water and moist thickets. Parasitised by Barka Indigobird. **Voice** Piercing seesee and hard *twit-it-it* (alarm). Song a variable series of short notes ending in a clear, rising double note, e.g. *pwit twet twet-te-twet twiweet*, with variations; also a fast *tweetweetweetwee*. [CD15:30]

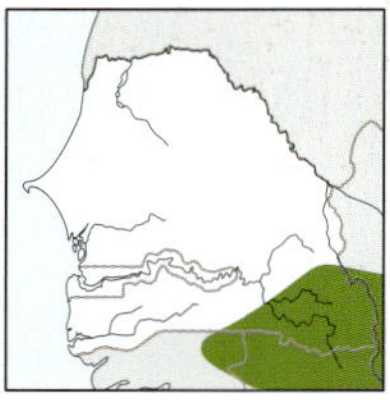

Black-bellied Firefinch *Lagonosticta rara* — 10 cm

L. r. forbesi. **Adult male** Mainly ***deep wine-red*** with ***black of undertail-coverts extending to central lower breast;*** bill two-toned. **Adult female** ***Head greyish*** with red patch in front of eye; upperparts dark brown washed wine-red; breast and upper flanks dull pinkish-carmine; rest of underparts and bill as male. **Juvenile** Dull buff-brown above, paler below; bill with pinkish base below. **HH** Uncommon resident, SE Senegal. In grassy savanna, edges of gallery forest and abandoned farmland. Parasitised by Cameroon Indigobird. **Voice** Short *pwit, pwit*, hard *chep*, nasal *cheeay* and *chew*, and short, pleasant trill. Song *tiu-tiu-tiu-tiu-tiu* and *tsiuw-tsiuw-tsiuw*. [CD15:25]

Kulikoro (Mali) Firefinch *Lagonosticta virata* — 10–11 cm

Distinguished by ***bluish bill*** and ***black undertail-coverts***. **Adult male** Top of head and upperparts grey-brown; rump and tail-base deep carmine-red. Face and underparts deep red. Bill deep bluish with black tip (often all black above), straight and narrow; narrow orbital ring pinkish. As Blue-billed Firefinch but slightly duller and with fewer white spots on breast-sides; best distinguished by voice. **Adult female** As male but usually slightly paler. **Juvenile** Pale grey-brown above; paler, more buffish below; rump dull red. **HH** Two records, SE Senegal (specimen, May 1966; pair, Sabodala, Nov 2009). Main range in Mali. In rocky areas with bushes, thickets and grass, and scrubby rocky hillsides; also dry forest on rocks. **Voice** Distinctive calls include a low-pitched, nasal, slightly harsh *chew* or *kyah* (contact); a drawn-out, plaintive whistle *feeee*, sometimes rising or falling in pitch *tseeeu* or *tsueee* (long-distance contact); a rattling trill *chrrrrrrr*, often uttered in series of three *chrrr, chrr, chrr* (alarm) . Also whistled trills similar to Blue-billed Firefinch's. [CD15:29]

Blue-billed Firefinch *Lagonosticta rubricata* — 10.0–11.5 cm

L. r. polionota. **Adult male** As Kulikoro Firefinch but slightly brighter. **Adult female** Paler than male; belly buff-brown. **Juvenile** Dull brownish; dark bill. **HH** No certain records. Claimed, SW Senegal (Casamance). In forest–savanna mosaic. **Voice** Twittering *trrrrrrrr-t*, loud *chew-chew-chew*, dry *prttt* and *prrititit* (alarm) and rapid *wink-wink-wink* or *whitwhitwhitwhit...* [CD15:26]

juv
♂
♀
Red-billed Firefinch
ad
juv
Bar-breasted Firefinch
♀
juv
♂
Black-faced Firefinch
♂
♀
juv
Black-bellied Firefinch
♂
Kulikoro (Mali) Firefinch
♀
juv
Blue-billed Firefinch
♂
♀

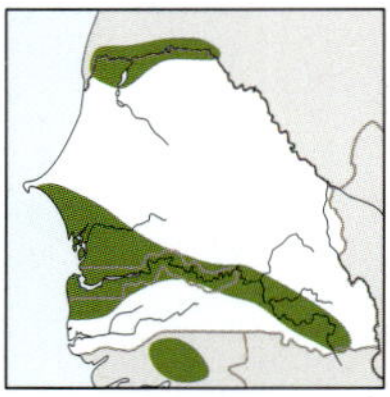

Black-faced Quailfinch *Ortygospiza atricollis* 9.5–10.0 cm

O. a. atricollis. Small, dumpy and unobtrusive; usually first noticed when flushed. **Adult male** Brown above with black face and throat. ***Tail short*** with ***white corners*** (conspicuous in flight). Bill red (breeding) or reddish with dusky culmen. **Adult female** Duller; head greyer without black mask. **Juvenile** Similar to female but even duller and paler. **HH** Locally fairly common. In small flocks in open grassy habitats, often near water. Entirely terrestrial; never perches. When flushed, flies with short jerky movements, usually dropping into nearby cover. **Voice** Metallic *krreep-krreep*, in flight; distinctive but easily passing unnoticed. Song a rapid series of *klik-klak-kloik-kluk*. [CD15:45]

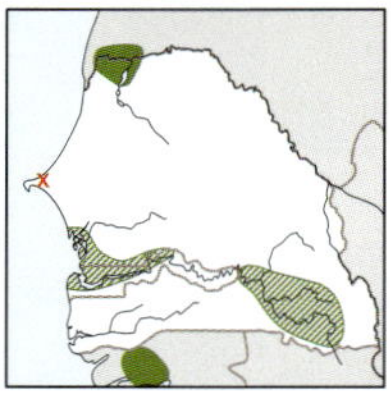

Zebra Waxbill *Sporaeginthus subflavus* 9–10 cm

S. s. subflavus. Very small and active with, in male, diagnostic ***bright orange breast***. **Adult male** Crown and upperparts olive-brown with bright red supercilium, rump and uppertail-coverts. Throat yellow; rest of underparts orange, barred olivaceous and yellow on flanks. Bill bright red. **Adult female** Duller; no red supercilium. **Juvenile** Dull brownish above with rump washed orange, buffish below; bill blackish. **HH** Locally uncommon, N Senegal; scarce elsewhere. Usually in small flocks (larger when not breeding) in various grassy habitats, often near water. **Voice** Squeaky *cheep*; in flight a fast *trip-trp-*… and soft metallic twittering. [CD15:43]

Bronze Mannikin *Spermestes cucullatus* 9 cm

S. c. cucullatus. Very small, dumpy, grey-brown and white. **Adult** Dark brown above, white below; barred rump and flanks. Metallic bronze-green shoulder patch (*not* conspicuous). Bill black above, blue-grey below. **Juvenile** Entirely buff-brown. **HH** Common resident. Usually in small flocks in farmbush, scrub, rank grass and gardens. Often with other mannikins and waxbills. Very tame. **Voice** Rather soft, rolling *trreep*, frequently uttered in flight and sounding like wheezy twittering *chrreep-rreep-chrrepchreepchrewprreep*… when coming from flock. [CD15:48]

Black-and-white Mannikin *Spermestes bicolor* 9.5–10.5 cm

S. b. bicolor. **Adult** Glossy black and pure white with ***heavy, pale grey-blue*** bill. **Juvenile** Dull brown above, paler below; bill dark. **HH** Claimed, Gambia (Brufut, Jul 2010) and SW Senegal (Casamance, Jan 1977). Occurs in rank grass within forest clearings, along edges, near water and in cultivation. Usually in small groups. **Voice** A clear, soft *kip* or *tsik* and drawn-out *tseew*. [CD15:49]

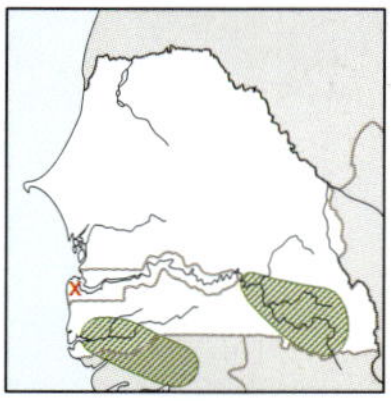

Magpie Mannikin *Spermestes fringilloides* 11.5–12.0 cm

Resembles Bronze Mannikin but ***noticeably larger*** and ***has heavier, weaver-like bill***. **Adult** Diagnostic ***black patch on breast-sides*** and black rump. **Juvenile** Dusky brown and buffish. **HH** Scarce, S Senegal; possibly extinct, Gambia. In rank grass and bushes within forest clearings, along edges, and in cultivation, esp. ricefields. Partial to indigenous bamboo seeds. Usually in small groups; often with other mannikins. **Voice** Fairly loud *teeoo*, frequently repeated; alarm a thin *cheep* or *tsek*. Song a simple series of similar notes *teeo-teeo-twee teew-tee twee-twee-twee*… [CD15:50]

African Silverbill *Euodice cantans* 10 cm

E. c. cantans. Small, stout-billed, slender, pale estrildid with black rump and black, pointed tail. **Adult** Pale sandy-brown above, paler below. Bill blue-grey. Some very fine barring above only visible at close range. **Juvenile** Lacks fine barring; tail edged brown. **HH** Common in north, becoming uncommon in south of range. Usually in small flocks in dry *Acacia* savanna. **Voice** Short *tsik*. Flocks give twittering *chip-chi-chi-chi-chip*… in flight. Song a soft, fast series of gentle rising and descending muffled sounds. [CD15:47]

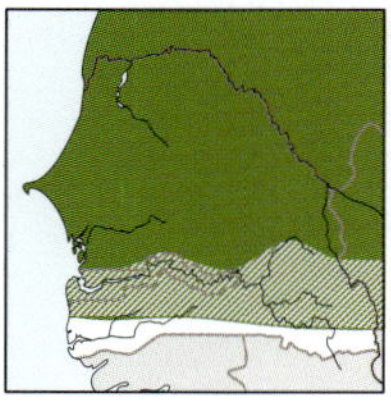

Cut-throat Finch *Amadina fasciata* 11–12 cm

A. f. fasciata. Sandy coloured and scaly looking. **Adult male** ***Red band across throat*** extending to ear-coverts; chestnut on belly. **Adult female** No red on throat; face barred; no chestnut on belly. **Juvenile** As female, but mantle usually unbarred and underparts poorly marked. **HH** Uncommon to locally fairly common; mainly dry-season visitor to south of range. In pairs or small flocks in dry woodland, thorn scrub and edges of cultivation. Often with other estrildids outside breeding season. **Voice** Sparrow-like *chilp*; also nasal *dzèèp*. In flight, thin *eee-eee-*… Song a fast medley of churring, ticking, rattling, chirping and clear whistling notes. [CD15:51]

Black-faced Quailfinch
♂
♀
♂
♀
juv
Zebra Waxbill
juv
ad
Bronze Mannikin
juv
ad
Black-and-white Mannikin
juv
ad
Magpie Mannikin
♀
African Silverbill
ad
♂
Cut-throat Finch

CUCKOO FINCH, INDIGOBIRDS AND WHYDAHS (Plates 141–142) Small, seed-eating finches, occurring mainly in savanna. Forage on the ground. Polygamous, mostly species-specific brood parasites, laying eggs in nests of Cisticolidae (Cuckoo Finch) or estrildids (indigobirds and whydahs).

Cuckoo Finch *Anomalospiza imberbis* 13 cm

Small with short tail and diagnostic, ***small, conical bill*** (unlike any *Ploceus* weaver). **Adult male** Head and underparts bright yellow; upperparts yellow-green streaked dark brown. Lower flanks streaked dusky. Bill black. Plumage becomes brighter and yellower with wear. **Adult female** Head and upperparts buff-brown boldly streaked black; buff supercilium. Pale ear-coverts and lack of eye-stripe impart plain-faced appearance. Underparts buffish becoming whitish on belly and variably streaked dark brown on breast and flanks. Head-sides to throat washed yellow in some. Bill dark horn above, paler below. **Juvenile** As adult female, but more tawny or sandy-buff. **HH** Afrotropical vagrant, Gambia (near Banjul, Sep 1969). Frequents open or lightly wooded grassland and cultivation. Usually singly or in pairs. Parasitises cisticolas and prinias. **Voice** Song a fast series of squeaky, sibilant notes *tslee-tslee-tslee-…*, from perch or in display flight, occasionally followed by a drawn-out weaver-like wheeze *shweeeeeoooo*. Flocks give soft chattering notes in flight. [CD14:97]

INDIGOBIRDS Often only distinguishable in the field if song of foster species is known. Songs include non-mimetic chattering and clear, whistled notes mimicking those of hosts. Status and distribution imperfectly known; presence of host an indication of species possibly involved. Given the presence of host species, indigobirds not yet identified in Senegambia may occur (e.g. Jambandu Indigobird V. raricola, *parasite of Zebra Waxbill, and Cameroon Indigobird V.* camerunensis, *thought to parasitise Black-bellied Firefinch and Dybowski's Twinspot).*

Village Indigobird *Vidua chalybeata* 11.5 cm

V. c. chalybeata*.** **Male breeding** The only indigobird with ***orange to pinkish-red legs. Plumage glossy green; black flight feathers narrowly edged whitish. Bill white or pinkish-white. **Male non-breeding/adult female** Top of head has broad buff median crown-stripe bordered by bold dark brown lateral stripes; buff supercilium; upperparts buff-brown streaked dark brown; underparts pale buffish-brown becoming whitish on throat and belly. Bill white; ***legs orange***. **Juvenile** Plain dusky-brown crown. **HH** Fairly common resident throughout. Generally tame, occurring near human habitation and in farmland. Male displays with hovering and bouncing flight before female. **Voice** Mimics and parasitises Red-billed Firefinch. Song consists of alarm call *chik* followed by a series of simple, clear whistled notes *pea-pea-pea*. Flight call a hoarse *cheh cheh cheh…* [CD15:52]

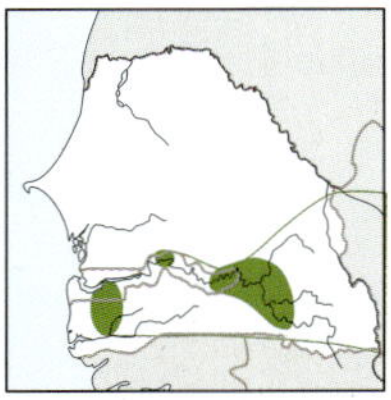

Wilson's Indigobird *Vidua wilsoni* 11.5 cm

Male breeding Glossy purplish with dark brown flight feathers narrowly edged whitish. Bill white; legs pale purplish. **Male non-breeding/adult female** As Village Indigobird but with pale purplish legs. **HH** Recorded S Senegal; some Gambian records probably referable to this species. **Voice** Mimics and parasitises Bar-breasted Firefinch. Mimicry song includes a sharp *pik* or *tzet* (alarm) and rapid series of harsh metallic and low nasal notes. [CD15:58]

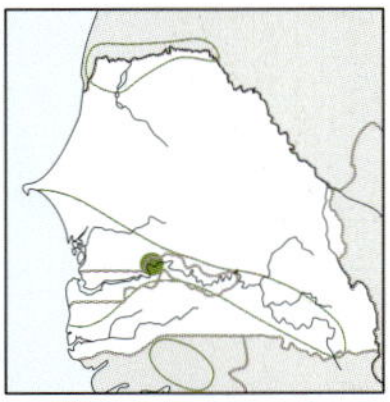

Quailfinch Indigobird *Vidua nigeriae* 11.5 cm

Male breeding As Wilson's Indigobird but plumage dull green (not distinctive in field). **Male non-breeding/adult female** As Wilson's Indigobird. **HH** A few old records, Gambia. **Voice** Mimics and parasitises Black-faced Quailfinch. Mimicry song includes metallic *tink* and *trillink* and rapid series of *klik-klak-kloik-kluk* notes. [CD15:59]

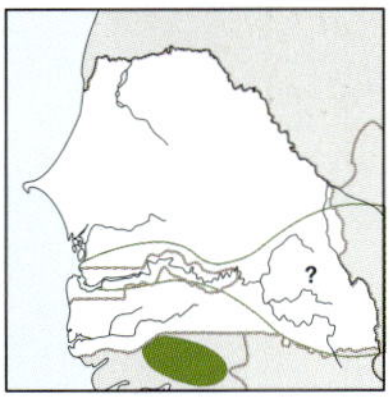

Barka Indigobird *Vidua larvaticola* 11.5 cm

Male breeding As Wilson's Indigobird but plumage glossy blue to green-blue. **Male non-breeding/adult female** As Wilson's Indigobird. **HH** Status unclear. May occur in Senegambia. **Voice** Mimics and parasitises Black-faced Firefinch. Mimicry song includes short series of slow, slurred whistles *whuu-ee whuu-ee* and *whee-uu whee-uu*, sharp alarm *dwit-it-it* and shrill contact call seesee. [CD15:55]

juv
♂ br
Cuckoo Finch
♀
juv
♂ non-br/♀
♂ br
Village Indigobird
♂ br
Wilson's Indigobird
♂ br
♂ br
Quailfinch Indigobird
Barka Indigobird

WHYDAHS Males in breeding plumage have four greatly elongated central tail feathers. The two paradise whydahs occurring in Senegambia are very similar but differ mainly in width and shape of tail (often difficult to assess due to moult or abrasion). Status and distribution inadequately known. Parasitise waxbills Estrilda *and pytilias* Pytilia.

Pin-tailed Whydah *Vidua macroura* 12.5 cm

Male breeding Unmistakable. Black and white with four long ribbon-like tail streamers of up to 20 cm and red bill. **Female breeding** Mainly buffish-brown with boldly striped black and tawny head, distinctly streaked breast-sides, mainly dark brown bill, and dark legs. **Male/female non-breeding** As breeding female but ***bill pinkish***. **Juvenile** Plain pale brown above, pale buffish below; dusky bill. Female and non-breeding male indigobirds distinguished by whitish bills and pale or, in one species, orange legs. Paradise whydahs in corresponding plumages have less extensive dark markings on head-sides and only little and faint streaking on breast-sides. **HH** Widespread but rather uncommon resident in various open habitats, incl. savanna, farmland, and gardens. Male in breeding season very conspicuous, singing in low hovering and bouncing display flight over territory. Parasitises mainly waxbills (incl. Orange-cheeked and Black-rumped); also Bronze Mannikin. **Voice** Call *tsip-tsip*. Song, uttered in display flight and at rest, a vigorous jerky series of high-pitched variations on the call note *tsip tsweep tswup tsweeu tswip...* [CD15:61]

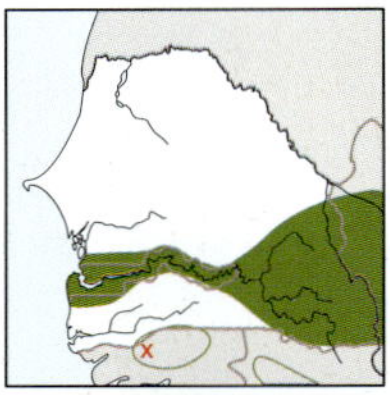

Exclamatory Paradise Whydah *Vidua interjecta* 12.5 cm (+tail of >27 cm)

Male breeding Striking black, chestnut and creamy-buff with very long tail. **Male non-breeding/adult female** Boldly striped head; pale orange bill and pinkish-grey (male) or pinkish (female) legs. **HH** Uncommon to fairly common resident. In small flocks, in open savanna woodland and adjacent cultivation. Displaying male reaches considerable height, circles with flapping wings, then dives to perch. Often perches conspicuously on high treetop. Parasitises Red-winged Pytilia. **Voice** Mimicry song includes *pik* calls and *to-wit-to-wit* song of host. [CD15:63]

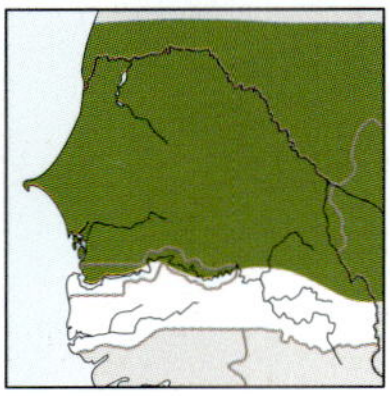

Sahel Paradise Whydah *Vidua orientalis* 12.5 cm (+tail of up to *c.*24 cm)

V. o. aucupum. **Male breeding** Differs from Exclamatory P.W. in having shorter tail streamers. **Male non-breeding/adult female** Similar to Exclamatory P.W. but with ***dark bill and legs***. **Juvenile** Plain earth-brown above; pale buffish-brown below. Bill dark horn. **HH** Probably rather uncommon resident. Parasitises Green-winged Pytilia. **Voice** Sharp *chip*. Mimicry song includes vocalisations of Green-winged Pytilia. [CD15:62]

TRUE FINCHES Small seed-eaters with stout, conical bills. Arboreal or terrestrial. Have remarkable, melodious songs.

White-rumped Seedeater *Serinus leucopygius* 10.0–11.5 cm

S. l. riggenbachi. Small and drab coloured with contrasting ***white rump*** (conspicuous in flight). **Juvenile** paler or warmer brown, and more heavily streaked. **HH** Uncommon to rare resident, from Gambia north, in *Acacia* scrub, farmland and grassland near water; often near or in villages. **Voice** Call *twee-eet*. Song a varied series of clear, sweet whistled and twittering phrases, resembling (but superior to) Yellow-fronted Canary's. [CD15:69]

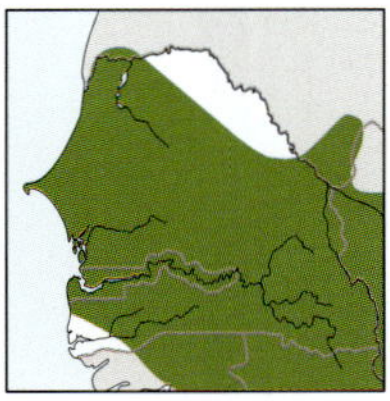

Yellow-fronted Canary *Serinus mozambicus* 11–13 cm

S. m. caniceps. Small with distinctive face pattern consisting of ***yellow forehead, supercilium and cheeks*** contrasting with grey crown, ***dark eye-stripe*** and ***dark moustachial stripe***. ***Yellow rump*** contrasts with greenish upperparts (conspicuous in flight). Underparts bright yellow. **Female** duller than male, with or without necklace of dark spots. **HH** Common resident throughout (except in extreme north) in open woodland, edges of cultivation and near villages. Forages mostly on ground. **Voice** Single or double *tseeu* or *tsssp* and *tuwu-tsilip* or *tuwu-tuwee*. Song a fairly simple but melodious, rapid jumble of twittering and whistled phrases, delivered from treetop. [CD15:71]

Common Linnet *Carduelis cannabina* 14 cm

Adult male Greyish head; chestnut upperparts. In flight, white patches in wings and tail-sides. **Adult female** More heavily streaked. **HH** Palearctic vagrant, N Senegal (a male trapped, Mar 1971). **Voice** Dry stuttering *knutnutnutnut*, usually in flight. [CD4:79]

juv
ad
non-br
♂ br
♀
♀ br
Pin-tailed Whydah
Exclamatory
Paradise Whydah
♂ br
juv
♂ non-br/♀
Sahel
Paradise Whydah
♂ br
ad
White-rumped
Seedeater
♀
♂
♂
Yellow-fronted
Canary
♀
Common Linnet

BUNTINGS *Small, mainly seed-eating species with stout, conical bills. Predominantly terrestrial. Flight typically undulating. Usually singly or in pairs, but some gregarious outside breeding season. Vocalisations usually rather simple.*

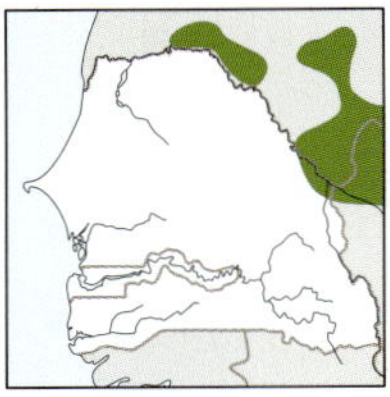

Golden-breasted Bunting *Emberiza flaviventris* 15–16 cm

E. f. flavigaster. **Adult male** Bold black-and-white head pattern and golden-yellow underparts. Upperparts chestnut with ***white wingbars*** and ***grey rump***. **Adult female** Head markings browner. **Juvenile** Duller and browner than adult; head pattern brown and buff. **HH** Afrotropical vagrant, E Senegal (male, near Kidira, Aug 1985); probably overlooked. Singly or in pairs in *Acacia* steppe and dry open savanna. Nearest breeding areas in Mauritania and Mali. **Voice** Subdued *chruteeu*; in flight a soft *chup*. Song a rapid succession of variable series of 4–10 similar syllables, e.g. *tsee-ep tsee-ep tsee-ep...*, melodious *tueet-tueet-tueet...*, faster *tsewtsewtsewtsew...*, drawn-out sibilant *hseeuw-hseeuw-hseeuw-...* etc.. [CD15:81]

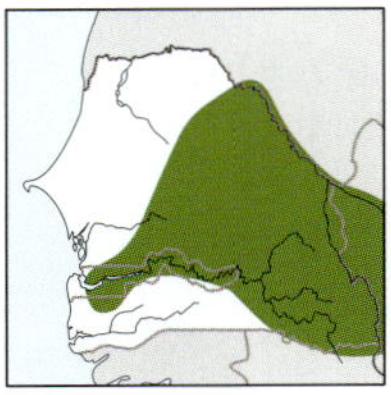

Brown-rumped Bunting *Emberiza affinis* 13–14 cm

E. a. nigeriae. **Adult male** ***Bold white and black head pattern*** and ***golden-yellow underparts***. Upperparts chestnut, indistinctly streaked on mantle, and separated from head by greyish collar; wings lack white. **Adult female** Head markings browner. **Juvenile** Duller; head washed rusty-buff, stripes brown. **HH** Frequent resident, S & E Senegal (rare/absent in NW), in open wooded savanna and farmland; uncommon to locally frequent, Gambia. Singly or in pairs; joins mixed-species flocks. **Voice** Liquid, melodious *pidrewdrlwi*; in flight a short *chip*. Song, uttered from perch, a rather short, slightly harsh warble *rwidji-drrrwidji-drr-djirree*, with variations. [CD15:83]

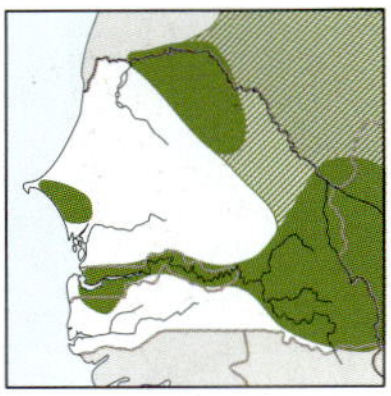

Cinnamon-breasted Rock Bunting *Emberiza tahapisi* 13–14 cm

E. t. goslingi. Rufous-brown with distinctly striped head. **Adult male** ***Head striped black and white***. Upperparts rufous-brown boldly streaked black; secondaries rufous (forming bright panel); tail blackish-brown, without white. Throat grey; rest of underparts plain cinnamon-brown. Bill dark above, yellowish below. **Adult female** Duller; head striped buffish-white and blackish-brown. **Juvenile** Similar to adult female but duller with less contrasting head pattern; throat paler with some dusky streaking. Bill grey. **HH** Resident and partial migrant. In dry season, widespread in open savanna with rocky outcrops E Senegal; in wet season, regular visitor in woodland in north. In Gambia uncommon to locally frequent. **Voice** Distinctive nasal *wee-eh*; alarm a thin *sweeee*. Song short and rapid, with emphasis on final note, e.g. *chrr-trr-erl-CHEEP*. [CD15:80]

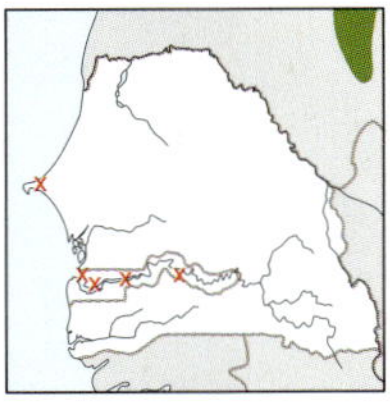

House Bunting *Emberiza (striolata) sahari* 13–14 cm

Adult male Grey head streaked black; almost uniformly rufous body. **Adult female** Paler and duller; head more grey-brown; no streaks on breast. **HH** Vagrant, Senegal (one, near Dakar, Apr 1983) and Gambia (pair, Albreda, Dec 1926; singles, Barra, Feb 1976 and Nov 1982; Kaif, Tendaba & Bansang, Nov 2004). Normally in pairs and small groups in rocky wadis, cultivation and villages in desert and Sahel. Nearest breeding areas in Mauritania and N Mali. **Voice** Nasal *chzwee* and thin *chik*. Song a rapid, simple and rather monotonous series of similar syllables. [CD4:92]

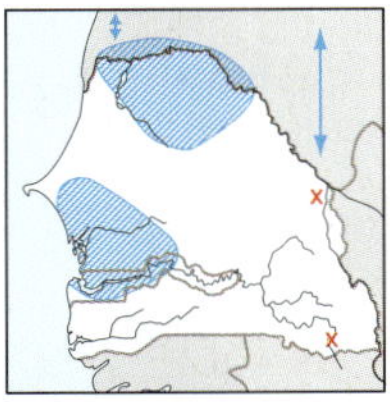

Ortolan Bunting *Emberiza hortulana* 15–16.5 cm

In all plumages distinguished by ***sulphur-yellow moustachial stripe, throat and eye-ring***, and ***pink bill***. **Adult male** Head to upper breast olive-grey, contrasting with orange-rufous underparts. Upperparts rufous-brown boldly streaked black. Outer tail feathers with large amount of white. **Adult female** Slightly duller; head more olive-brown and finely streaked on crown; breast distinctly streaked dark. **Immature** As adult female but duller and more distinctly and extensively streaked, with narrow dark streaks on flanks. **HH** Palearctic migrant, Senegal (uncommon; probably throughout) and Gambia (one, Banjul, Nov 1955; male, Tendaba, Apr 1981). Forages mainly on ground in open habitats, but also in trees. **Voice** Calls, mostly in flight, include short *twit* and clear metallic *seee*. [CD4:91]

Corn Bunting *Emberiza calandra* 16–19 cm

Bulky, brown, streaked bunting with heavy, horn-coloured bill and no white in tail. **HH** Palearctic vagrant, NW Senegal (three, Senegal delta, Feb 1974). **Voice** Dry, hard *kwit* and *kwitit* on take-off and in flight. [CD4:96]

♀
juv
♂
Golden-breasted Bunting
♂
juv
Brown-rumped
Bunting
♀
♂
Cinnamon-breasted Rock Bunting
♀
♂
House Bunting
♀
imm
♂
Ortolan Bunting
ad
Corn Bunting

CHECKLIST OF THE BIRDS OF SENEGAL AND THE GAMBIA

This checklist consists of all the species reliably recorded in Senegal and The Gambia and presents species in taxonomic sequence.

IUCN Threat Status Species of global conservation concern (BirdLife International 2000, 2008 & updates) :
CR = Critically Endangered, **EN** = Endangered, **VU** = Vulnerable, **NT** = Near Threatened

Status Key

R = Resident, **M** = Intra-African migrant, **P** = Palearctic migrant (including a few species of Nearctic origin), **V** = Vagrant, **§** = Unsubstantiated record(s)

Sen	Gam	Scientific name	English name	French name	IUCN	Status	
						Sen	Gam
Ostriches STRUTHIONIDAE							
		Struthio camelus	Ostrich	Autruche d'Afrique		R	
Petrels and shearwaters PROCELLARIIDAE							
		Fulmarus glacialis	Northern Fulmar	Fulmar boréal		V	
		Bulweria bulwerii	Bulwer's Petrel	Pétrel de Bulwer		M	
		Calonectris diomedea	Cory's Shearwater	Puffin cendré		P	
		Calonectris edwardsii	Cape Verde Shearwater	Puffin du Cap-Vert	NT	M	
		Puffinus gravis	Great Shearwater	Puffin majeur		M	
		Puffinus griseus	Sooty Shearwater	Puffin fuligineux	NT	M	
		Puffinus puffinus	Manx Shearwater	Puffin des Anglais		P	V
		Puffinus mauretanicus	Balearic Shearwater	Puffin des Baléares	CR	P	
		Puffinus baroli	Macaronesian Shearwater	Puffin de Macaronésie		P	V
Storm-petrels HYDROBATIDAE							
		Oceanites oceanicus	Wilson's Storm-petrel	Océanite de Wilson		M	M
		Pelagodroma marina	White-faced Storm-petrel	Océanite frégate		M	
		Hydrobates pelagicus	European Storm-petrel	Océanite tempête		P	P
		Oceanodroma castro	Madeiran Storm-petrel	Océanite de Castro		M	
		Oceanodroma leucorhoa	Leach's Storm-petrel	Océanite culblanc		P	V
Grebes PODICIPEDIDAE							
		Tachybaptus ruficollis	Little Grebe	Grèbe castagneux		R	M
		Podiceps cristatus	Great Crested Grebe	Grèbe huppé		V	
		Podiceps nigricollis	Black-necked Grebe	Grèbe à cou noir		P	
Tropicbirds PHAETHONTIDAE							
		Phaethon aethereus	Red-billed Tropicbird	Phaéton à bec rouge		R	V
Gannets and boobies SULIDAE							
		Sula bassana	Northern Gannet	Fou de Bassan		P	P
		Sula leucogaster	Brown Booby	Fou brun		M	V
Cormorants PHALACROCORACIDAE							
		Phalacrocorax carbo	Great Cormorant	Grand Cormoran		R/P	M
		Phalacrocorax africanus	Long-tailed Cormorant	Cormoran africain		R	R
Darter ANHINGIDAE							
		Anhinga rufa	African Darter	Anhinga d'Afrique		R	R

Sen	Gam	Scientific name	English name	French name	IUCN	Status	
						Sen	Gam
Pelicans PELECANIDAE							
		Pelecanus onocrotalus	Great White Pelican	Pélican blanc		**R/M**	**M**
		Pelecanus rufescens	Pink-backed Pelican	Pélican gris		**R**	**R/M**
Frigatebirds FREGATIDAE							
		Fregata magnificens	Magnificent Frigatebird	Frégate superbe			**V**
Bitterns, herons and egrets ARDEIDAE							
		Botaurus stellaris	Great Bittern	Butor étoilé		**V**	**V**
		Ixobrychus minutus	Little Bittern	Blongios nain		**P/R**	**M**
		Ixobrychus sturmii	Dwarf Bittern	Blongios de Sturm		**M**	**M**
		Tigriornis leucolopha	White-crested Tiger Heron	Onoré à huppe blanche		**R**	**R**
		Gorsachius leuconotus	White-backed Night Heron	Bihoreau à dos blanc		**R**	**R**
		Nycticorax nycticorax	Black-crowned Night Heron	Bihoreau gris		**R/P**	**R/P**
		Ardeola ralloides	Squacco Heron	Crabier chevelu		**R/P**	**R/P**
		Bubulcus ibis	Cattle Egret	Héron garde-bœufs		**R/M**	**R/M**
		Butorides striata	Green-backed Heron	Héron strié		**R**	**R**
		Egretta ardesiaca	Black Heron	Aigrette ardoisée		**R**	**R**
		Egretta gularis	Western Reef Egret	Aigrette des récifs		**R**	**R**
		Egretta garzetta	Little Egret	Aigrette garzette		**R/P**	**R/P**
		Egretta intermedia	Intermediate Egret	Aigrette intermédiaire		**R**	**R**
		Egretta alba	Great Egret	Grande Aigrette		**R**	**R**
		Ardea purpurea	Purple Heron	Héron pourpré		**R/P**	**R/P**
		Ardea cinerea	Grey Heron	Héron cendré		**R/P**	**R/P**
		Ardea melanocephala	Black-headed Heron	Héron mélanocéphale		**R**	**R/M**
		Ardea goliath	Goliath Heron	Héron goliath		**R**	**R**
Hamerkop SCOPIDAE							
		Scopus umbretta	Hamerkop	Ombrette africaine		**R**	**R**
Storks CICONIIDAE							
		Mycteria ibis	Yellow-billed Stork	Tantale ibis		**R/M**	**R/M**
		Anastomus lamelligerus	African Openbill Stork	Bec-ouvert africain		**M**	
		Ciconia nigra	Black Stork	Cigogne noire		**P**	**V**
		Ciconia abdimii	Abdim's Stork	Cigogne d'Abdim		**M**	**V**
		Ciconia episcopus	Woolly-necked Stork	Cigogne épiscopale		**M**	**M/R**
		Ciconia ciconia	White Stork	Cigogne blanche		**P**	**V**
		Ephippiorhynchus senegalensis	Saddle-billed Stork	Jabiru d'Afrique		**R**	**R**
		Leptoptilos crumenifer	Marabou Stork	Marabout d'Afrique		**R**	**R/M**
Ibises and spoonbills THRESKIORNITHIDAE							
		Plegadis falcinellus	Glossy Ibis	Ibis falcinelle		**P/R?**	**P**
		Bostrychia hagedash	Hadada Ibis	Ibis hagedash		**R**	**R**
		Threskiornis aethiopicus	Sacred Ibis	Ibis sacré		**R**	**R**
		Geronticus eremita	Northern Bald Ibis	Ibis chauve	**CR**	**V**	
		Platalea leucorodia	Eurasian Spoonbill	Spatule blanche		**P**	**P**
		Platalea alba	African Spoonbill	Spatule d'Afrique		**R**	**R**

Sen	Gam	Scientific name	English name	French name	IUCN	Status Sen	Status Gam
Flamingos PHOENICOPTERIDAE							
		Phoenicopterus roseus	Greater Flamingo	Flamant rose		**R/P**	**M/P**
		Phoeniconaias minor	Lesser Flamingo	Flamant nain	**NT**	**M**	**M**
Ducks and geese ANATIDAE							
		Dendrocygna bicolor	Fulvous Whistling Duck	Dendrocygne fauve		**R**	**M**
		Dendrocygna viduata	White-faced Whistling Duck	Dendrocygne veuf		**R**	**R**
		Dendrocygna autumnalis	Black-bellied Whistling Duck	Dendrocygne à bec rouge			**V**
		Thalassornis leuconotus	White-backed Duck	Dendrocygne à dos blanc		**R**	
		Branta bernicla	Brent Goose	Bernache cravant		**V**	
		Alopochen aegyptiaca	Egyptian Goose	Ouette d'Égypte		**R**	**M**
		Tadorna tadorna	Common Shelduck	Tadorne de Belon		**P**	
		Plectropterus gambensis	Spur-winged Goose	Oie-armée de Gambie		**R**	**R**
		Sarkidiornis melanotos	Knob-billed Duck	Canard à bosse		**M**	**M**
		Nettapus auritus	African Pygmy Goose	Anserelle naine		**R**	**R**
		Anas penelope	Eurasian Wigeon	Canard siffleur		**P**	**V**
		Anas americana	American Wigeon	Canard à front blanc		**V**	
		Anas strepera	Gadwall	Canard chipeau		**P**	**§**
		Anas crecca	Common Teal	Sarcelle d'hiver		**P**	**P**
		Anas platyrhynchos	Mallard	Canard colvert		**V**	
		Anas acuta	Northern Pintail	Canard pilet		**P**	**P**
		Anas querquedula	Garganey	Sarcelle d'été		**P**	**P**
		Anas discors	Blue-winged Teal	Sarcelle à ailes bleues		**V**	
		Anas clypeata	Northern Shoveler	Canard souchet		**P**	**P**
		Marmaronetta angustirostris	Marbled Duck	Sarcelle marbrée	**VU**	**P**	
		Aythya ferina	Common Pochard	Fuligule milouin		**P**	**P**
		Aythya nyroca	Ferruginous Duck	Fuligule nyroca	**NT**	**P**	**P**
		Aythya fuligula	Tufted Duck	Fuligule morillon		**P**	**P**
Osprey PANDIONIDAE							
		Pandion haliaetus	Osprey	Balbuzard pêcheur		**P**	**P**
Hawks, vultures and eagles ACCIPITRIDAE							
		Aviceda cuculoides	African Cuckoo Hawk	Baza coucou		**M**	**M**
		Pernis apivorus	European Honey Buzzard	Bondrée apivore		**P**	**P**
		Macheiramphus alcinus	Bat Hawk	Milan des chauves-souris		**R?**	**M**
		Elanus caeruleus	Black-shouldered Kite	Elanion blanc		**R**	**M/R**
		Chelictinia riocourii	African Swallow-tailed Kite	Elanion naucler		**M**	**M**
		Milvus migrans	Black Kite	Milan noir		**P**	**P**
		Milvus migrans parasitus	Yellow-billed Kite	Milan d'Afrique		**M/R**	**M/R**
		Milvus milvus	Red Kite	Milan royal	**NT**		**V**
		Haliaeetus vocifer	African Fish Eagle	Pygargue vocifère		**R**	**R**
		Gypohierax angolensis	Palm-nut Vulture	Palmiste africain		**R**	**R**
		Neophron percnopterus	Egyptian Vulture	Vautour percnoptère	**EN**	**R/M**	**M**
		Necrosyrtes monachus	Hooded Vulture	Vautour charognard	**EN**	**R**	**R**
		Gyps africanus	African White-backed Vulture	Vautour africain	**NT**	**R**	**R**
		Gyps rueppellii	Rüppell's Griffon Vulture	Vautour de Rüppell	**NT**	**R**	**R**

Sen	Gam	Scientific name	English name	French name	IUCN	Status	
						Sen	Gam
		Gyps fulvus	Eurasian Griffon Vulture	Vautour fauve		P	P
		Aegypius monachus	Eurasian Black Vulture	Vautour moine	NT	V	
		Torgos tracheliotus	Lappet-faced Vulture	Vautour oricou	VU	R	M
		Trigonoceps occipitalis	White-headed Vulture	Vautour à tête blanche	VU	R	R
		Circaetus gallicus	Short-toed Snake Eagle	Circaète Jean-le-Blanc		P	P
		Circaetus beaudouini	Beaudouin's Snake Eagle	Circaète de Beaudouin	VU	R	R
		Circaetus cinereus	Brown Snake Eagle	Circaète brun		R	R
		Circaetus cinerascens	Western Banded Snake Eagle	Circaète cendré		R	R
		Terathopius ecaudatus	Bateleur	Bateleur des savanes	NT	R	R
		Polyboroides typus	African Harrier Hawk	Gymnogène d'Afrique		R	R
		Circus macrourus	Pallid Harrier	Busard pâle	NT	P	P
		Circus pygargus	Montagu's Harrier	Busard cendré		P	P
		Circus aeruginosus	Western Marsh Harrier	Busard des roseaux		P	P
		Micronisus gabar	Gabar Goshawk	Autour gabar		R	R
		Melierax metabates	Dark Chanting Goshawk	Autour sombre		R	R
		Accipiter tachiro	African Goshawk	Autour tachiro		R	R
		Accipiter badius	Shikra	Épervier shikra		R/M	R
		Accipiter erythropus	Red-thighed Sparrowhawk	Épervier de Hartlaub		R	R
		Accipiter ovampensis	Ovambo Sparrowhawk	Épervier de l'Ovampo		V	M
		Accipiter nisus	Eurasian Sparrowhawk	Épervier d'Europe			V
		Accipiter melanoleucus	Black Sparrowhawk	Autour noir		R	M
		Butastur rufipennis	Grasshopper Buzzard	Busautour des sauterelles		M	M
		Kaupifalco monogrammicus	Lizard Buzzard	Autour unibande		R	R
		Buteo buteo	Common Buzzard	Buse variable		P	P
		Buteo rufinus	Long-legged Buzzard	Buse féroce		P	V
		Buteo auguralis	Red-necked Buzzard	Buse d'Afrique		M	M
		Aquila rapax	Tawny Eagle	Aigle ravisseur		R	R
		Aquila fasciata	Bonelli's Eagle	Aigle de Bonelli		P	
		Aquila spilogaster	African Hawk Eagle	Aigle fascié		R	R
		Hieraaetus wahlbergi	Wahlberg's Eagle	Aigle de Wahlberg		M	M
		Hieraaetus pennatus	Booted Eagle	Aigle botté		P	P
		Hieraaetus ayresii	Ayres's Hawk Eagle	Aigle d'Ayres		§	M
		Lophaetus occipitalis	Long-crested Eagle	Aigle huppard		R	R
		Stephanoaetus coronatus	Crowned Eagle	Aigle couronné		R	V
		Polemaetus bellicosus	Martial Eagle	Aigle martial	NT	R	R
Secretarybird SAGITTARIIDAE							
		Sagittarius serpentarius	Secretarybird	Messager sagittaire	VU	V	V
Falcons FALCONIDAE							
		Falco naumanni	Lesser Kestrel	Faucon crécerellette		P	P
		Falco tinnunculus	Common Kestrel	Faucon crécerelle		R/P	P
		Falco alopex	Fox Kestrel	Crécerelle renard		V	V
		Falco ardosiaceus	Grey Kestrel	Faucon ardoisé		R	R
		Falco chicquera	Red-necked Falcon	Faucon chicquera		R	R
		Falco vespertinus	Red-footed Falcon	Faucon kobez	NT	P	V

Sen	Gam	Scientific name	English name	French name	IUCN	Status Sen	Status Gam
		Falco eleonorae	Eleonora's Falcon	Faucon d'Éléonore		P	
		Falco columbarius	Merlin	Faucon émerillon		V	
		Falco subbuteo	Eurasian Hobby	Faucon hobereau		P	P
		Falco cuvierii	African Hobby	Faucon de Cuvier		R	R
		Falco biarmicus	Lanner Falcon	Faucon lanier		R	M/R?
		Falco cherrug	Saker Falcon	Faucon sacre	VU	V	
		Falco peregrinus	Peregrine Falcon	Faucon pèlerin		P/R?	P
		Falco pelegrinoides	Barbary Falcon	Faucon de Barbarie		V	V
Guineafowl NUMIDIDAE							
		Numida meleagris	Helmeted Guineafowl	Pintade de Numidie (P. commune)		R	R
Quails, partridges and francolins PHASIANIDAE							
		Coturnix coturnix	Common Quail	Caille des blés		P	P
		Coturnix delegorguei	Harlequin Quail	Caille arlequin		V/M?	
		Ptilopachus petrosus	Stone Partridge	Poule de roche		R	R
		Francolinus albogularis	White-throated Francolin	Francolin à gorge blanche		R	R
		Francolinus ahantensis	Ahanta Francolin	Francolin d'Ahanta		R	R
		Francolinus bicalcaratus	Double-spurred Francolin	Francolin à double éperon		R	R
Buttonquails TURNICIDAE							
		Ortyxelos meiffrenii	Quail-plover	Turnix à ailes blanches		R	V
		Turnix sylvaticus	Common Buttonquail	Turnix mugissant (T. d'Andalousie)		R/M	R/M
Rails, flufftails and gallinules RALLIDAE							
		Sarothrura pulchra	White-spotted Flufftail	Râle perlé		R	R
		Crex egregia	African Crake	Râle des prés		M	M
		Rallus caerulescens	African Water Rail	Râle bleuâtre			V
		Porzana parva	Little Crake	Marouette poussin		P	V
		Porzana pusilla	Baillon's Crake	Marouette de Baillon		V	V
		Porzana porzana	Spotted Crake	Marouette ponctuée		P	V
		Amaurornis flavirostra	Black Crake	Râle à bec jaune		R	R
		Porphyrio alleni	Allen's Gallinule	Talève d'Allen		M/R?	M
		Porphyrio porphyrio	Purple Swamphen	Talève sultane		R	M
		Gallinula chloropus	Common Moorhen	Gallinule poule-d'eau		P/R?	P/R?
		Gallinula angulata	Lesser Moorhen	Gallinule africaine		R	R
		Fulica atra	Eurasian Coot	Foulque macroule		P	V
Cranes GRUIDAE							
		Grus grus	Common Crane	Grue cendrée		V	
		Balearica pavonina	Black Crowned Crane	Grue couronnée	VU	R	M/R?
Finfoot HELIORNITHIDAE							
		Podica senegalensis	African Finfoot	Grébifoulque d'Afrique		R	R
Bustards OTIDIDAE							
		Neotis denhami	Denham's Bustard	Outarde de Denham	NT	M	M
		Ardeotis arabs	Arabian Bustard	Outarde arabe		R	V
		Lophotis savilei	Savile's Bustard	Outarde de Savile		R	R
		Eupodotis senegalensis	White-bellied Bustard	Outarde du Sénégal		R	R?

Sen	Gam	Scientific name	English name	French name	IUCN	Status Sen	Status Gam
		Lissotis melanogaster	Black-bellied Bustard	Outarde à ventre noir		**R**	**R**
Jacanas JACANIDAE							
		Actophilornis africanus	African Jacana	Jacana à poitrine dorée		**R**	**R**
		Microparra capensis	Lesser Jacana	Jacana nain		**V**	
Painted-snipe ROSTRATULIDAE							
		Rostratula benghalensis	Greater Painted-snipe	Rhynchée peinte		**R**	**R**
Oystercatchers HAEMATOPODIDAE							
		Haematopus ostralegus	Eurasian Oystercatcher	Huîtrier pie		**P**	**P**
		Haematopus moquini	African Black Oystercatcher	Huîtrier de Moquin	**NT**	**V**	
Stilts and avocets RECURVIROSTRIDAE							
		Himantopus himantopus	Black-winged Stilt	Échasse blanche		**R/P**	**P**
		Recurvirostra avosetta	Pied Avocet	Avocette élégante		**P**	**P**
Thick-knees BURHINIDAE							
		Burhinus oedicnemus	Stone-curlew	Œdicnème criard		**P**	
		Burhinus senegalensis	Senegal Thick-knee	Œdicnème du Sénégal		**R**	**R**
		Burhinus vermiculatus	Water Thick-knee	Œdicnème vermiculé		**M**	
		Burhinus capensis	Spotted Thick-knee	Œdicnème tachard		**R**	**R**
Coursers and pratincoles GLAREOLIDAE							
		Pluvianus aegyptius	Egyptian Plover	Pluvian fluviatile		**R/M**	**M**
		Cursorius cursor	Cream-coloured Courser	Courvite isabelle		**P/R**	**V**
		Cursorius temminckii	Temminck's Courser	Courvite de Temminck		**R**	**M**
		Rhinoptilus chalcopterus	Bronze-winged Courser	Courvite à ailes bronzées		**M/R?**	**R?**
		Glareola pratincola	Collared Pratincole	Glaréole à collier		**M/P**	**M/P**
Plovers and lapwings CHARADRIIDAE							
		Charadrius dubius	Little Ringed Plover	Petit Gravelot		**P**	**P**
		Charadrius hiaticula	Common Ringed Plover	Grand Gravelot		**P**	**P**
		Charadrius pecuarius	Kittlitz's Plover	Gravelot pâtre		**M**	**M**
		Charadrius forbesi	Forbes's Plover	Gravelot de Forbes		**M**	**V**
		Charadrius alexandrinus	Kentish Plover	Gravelot à collier interrompu		**P**	**P**
		Charadrius marginatus	White-fronted Plover	Gravelot à front blanc		**R**	**R**
		Charadrius leschenaultii	Greater Sand Plover	Gravelot de Leschenault		**V**	**§**
		Charadrius morinellus	Eurasian Dotterel	Pluvier guignard			**V**
		Pluvialis dominica	American Golden Plover	Pluvier bronzé		**V**	**V**
		Pluvialis apricaria	European Golden Plover	Pluvier doré		**V**	**V**
		Pluvialis squatarola	Grey Plover	Pluvier argenté		**P**	**P**
		Vanellus senegallus	African Wattled Lapwing	Vanneau du Sénégal		**R**	**R**
		Vanellus albiceps	White-headed Lapwing	Vanneau à tête blanche		**R/M**	**M**
		Vanellus tectus	Black-headed Lapwing	Vanneau à tête noire		**R**	**R**
		Vanellus spinosus	Spur-winged Lapwing	Vanneau éperonné		**R**	**R**
		Vanellus lugubris	Lesser Black-winged Lapwing	Vanneau terne		**M**	**V**
		Vanellus vanellus	Northern Lapwing	Vanneau huppé		**P**	**V**

Sen	Gam	Scientific name	English name	French name	IUCN	Status Sen	Status Gam
Sandpipers and allies SCOLOPACIDAE							
		Calidris canutus	Red Knot	Bécasseau maubèche		P	P
		Calidris alba	Sanderling	Bécasseau sanderling		P	P
		Calidris minuta	Little Stint	Bécasseau minute		P	P
		Calidris temminckii	Temminck's Stint	Bécasseau de Temminck		P	P
		Calidris bairdii	Baird's Sandpiper	Bécasseau de Baird		V	V
		Calidris melanotos	Pectoral Sandpiper	Bécasseau tacheté		V	
		Calidris ferruginea	Curlew Sandpiper	Bécasseau cocorli		P	P
		Calidris alpina	Dunlin	Bécasseau variable		P	P
		Tryngites subruficollis	Buff-breasted Sandpiper	Bécasseau rousset		V	V
		Philomachus pugnax	Ruff	Combattant varié		P	P
		Lymnocryptes minimus	Jack Snipe	Bécassine sourde		P	P
		Gallinago gallinago	Common Snipe	Bécassine des marais		P	P
		Gallinago media	Great Snipe	Bécassine double	NT	P	P
		Limnodromus griseus/scolopaceus	Short-/Long-billed Dowitcher	Bécassin à bec court/à long bec			V
		Limosa limosa	Black-tailed Godwit	Barge à queue noire	NT	P	P
		Limosa lapponica	Bar-tailed Godwit	Barge rousse		P	P
		Numenius phaeopus	Whimbrel	Courlis corlieu		P	P
		Numenius arquata	Eurasian Curlew	Courlis cendré	NT	P	P
		Tringa erythropus	Spotted Redshank	Chevalier arlequin		P	P
		Tringa totanus	Common Redshank	Chevalier gambette		P	P
		Tringa stagnatilis	Marsh Sandpiper	Chevalier stagnatile		P	P
		Tringa nebularia	Common Greenshank	Chevalier aboyeur		P	P
		Tringa flavipes	Lesser Yellowlegs	Chevalier à pattes jaunes		V	V
		Tringa solitaria	Solitary Sandpiper	Chevalier solitaire			V
		Tringa ochropus	Green Sandpiper	Chevalier culblanc		P	P
		Tringa glareola	Wood Sandpiper	Chevalier sylvain		P	P
		Xenus cinereus	Terek Sandpiper	Chevalier bargette			V
		Actitis hypoleucos	Common Sandpiper	Chevalier guignette		P	P
		Actitis macularius	Spotted Sandpiper	Chevalier grivelé		V	
		Arenaria interpres	Ruddy Turnstone	Tournepierre à collier		P	P
		Phalaropus lobatus	Red-necked Phalarope	Phalarope à bec étroit		V	V
		Phalaropus fulicarius	Red Phalarope	Phalarope à bec large		P	P
Skuas STERCORARIIDAE							
		Stercorarius pomarinus	Pomarine Skua	Labbe pomarin		P	P
		Stercorarius parasiticus	Arctic Skua	Labbe parasite		P	P
		Stercorarius longicaudus	Long-tailed Skua	Labbe à longue queue		V	
		Stercorarius skua	Great Skua	Grand Labbe		V	
		Stercorarius maccormicki	South Polar Skua	Labbe de McCormick		V	
Gulls LARIDAE							
		Larus melanocephalus	Mediterranean Gull	Mouette mélanocéphale		P	P
		Larus atricilla	Laughing Gull	Mouette atricille		V	V
		Larus pipixcan	Franklin's Gull	Mouette de Franklin		V	V
		Larus minutus	Little Gull	Mouette pygmée		V	V

Sen	Gam	Scientific name	English name	French name	IUCN	Status	
						Sen	Gam
		Larus cirrocephalus	Grey-headed Gull	Mouette à tête grise		R	R
		Larus ridibundus	Black-headed Gull	Mouette rieuse		P	P
		Larus philadelphia	Bonaparte's Gull	Mouette de Bonaparte		V	
		Larus genei	Slender-billed Gull	Goéland railleur		R/P	R/P
		Larus audouinii	Audouin's Gull	Goéland d'Audouin	NT	P	P
		Larus delawarensis	Ring-billed Gull	Goéland à bec cerclé		V	V
		Larus canus	Common Gull	Goéland cendré		V	V
		Larus dominicanus	Kelp Gull	Goéland dominicain		M	M
		Larus fuscus	Lesser Black-backed Gull	Goéland brun		P	P
		Larus (cachinnans) michahellis	Yellow-legged Gull	Goéland leucophée		P	P
		Xema sabini	Sabine's Gull	Mouette de Sabine		P	V
		Rissa tridactyla	Black-legged Kittiwake	Mouette tridactyle		V	V
Terns STERNIDAE							
		Gelochelidon nilotica	Gull-billed Tern	Sterne hansel		P/R	P
		Sterna caspia	Caspian Tern	Sterne caspienne		R/P	R/P
		Sterna maxima	Royal Tern	Sterne royale		R/M	R/M
		Sterna bengalensis	Lesser Crested Tern	Sterne voyageuse		P	P
		Sterna sandvicensis	Sandwich Tern	Sterne caugek		P	P
		Sterna dougallii	Roseate Tern	Sterne de Dougall		P	P
		Sterna hirundo	Common Tern	Sterne pierregarin		P	P
		Sterna paradisaea	Arctic Tern	Sterne arctique		P	P
		Sterna anaethetus	Bridled Tern	Sterne bridée		M	V
		Sterna fuscata	Sooty Tern	Sterne fuligineuse		M	V
		Sterna albifrons	Little Tern	Sterne naine		R/P	P/M
		Chlidonias hybrida	Whiskered Tern	Guifette moustac		P	P
		Chlidonias niger	Black Tern	Guifette noire		P	P
		Chlidonias leucopterus	White-winged Tern	Guifette leucoptère		P	P
		Anous minutus	Black Noddy	Noddi noir			V
		Anous stolidus	Brown Noddy	Noddi brun			V
Skimmers RYNCHOPIDAE							
		Rynchops flavirostris	African Skimmer	Bec-en-ciseaux d'Afrique	NT	M/R	M
Auks ALCIDAE							
		Cepphus grylle	Black Guillemot	Guillemot à miroir		V	
Sandgrouse PTEROCLIDIDAE							
		Pterocles exustus	Chestnut-bellied Sandgrouse	Ganga à ventre brun		R	M/R?
		Pterocles lichtensteinii	Lichtenstein's Sandgrouse	Ganga de Lichtenstein		V	
		Pterocles quadricinctus	Four-banded Sandgrouse	Ganga quadribande		M/R	R
Pigeons and doves COLUMBIDAE							
		Treron calvus	African Green Pigeon	Colombar à front nu		R	R
		Treron waalia	Bruce's Green Pigeon	Colombar waalia		R	R
		Turtur tympanistria	Tambourine Dove	Tourtelette tambourette		R	
		Turtur afer	Blue-spotted Wood Dove	Tourtelette améthystine		R	R
		Turtur abyssinicus	Black-billed Wood Dove	Tourtelette d'Abyssinie		R	R
		Oena capensis	Namaqua Dove	Tourterelle masquée		R/M	M

Sen	Gam	Scientific name	English name	French name	IUCN	Status	
						Sen	Gam
		Columba guinea	Speckled Pigeon	Pigeon roussard		R	R
		Columba livia	Rock Dove	Pigeon biset		R	
		Streptopelia semitorquata	Red-eyed Dove	Tourterelle à collier		R	R
		Streptopelia decipiens	African Mourning Dove	Tourterelle pleureuse		R	R
		Streptopelia vinacea	Vinaceous Dove	Tourterelle vineuse		R	R
		Streptopelia roseogrisea	African Collared Dove	Tourterelle rieuse		R	M
		Streptopelia turtur	European Turtle Dove	Tourterelle des bois		P	P
		Streptopelia hypopyrrha	Adamawa Turtle Dove	Tourterelle de l'Adamaoua		R	R?
		Streptopelia senegalensis	Laughing Dove	Tourterelle maillée		R	R
Parrots PSITTACIDAE							
		Poicephalus robustus	Brown-necked Parrot	Perroquet robuste		R	R
		Poicephalus senegalus	Senegal Parrot	Perroquet youyou		R	R
		Psittacula krameri	Rose-ringed Parakeet	Perruche à collier		R	R
Turacos MUSOPHAGIDAE							
		Tauraco persa	Green Turaco	Touraco vert		R	R
		Musophaga violacea	Violet Turaco	Touraco violet		R	R
		Crinifer piscator	Western Grey Plantain-eater	Touraco gris		R	R
Cuckoos and coucals CUCULIDAE							
		Oxylophus jacobinus	Jacobin Cuckoo	Coucou jacobin		M	M
		Oxylophus levaillantii	Levaillant's Cuckoo	Coucou de Levaillant		M	M
		Clamator glandarius	Great Spotted Cuckoo	Coucou geai		R/P	R/P
		Cuculus solitarius	Red-chested Cuckoo	Coucou solitaire		M	M
		Cuculus clamosus	Black Cuckoo	Coucou criard		V	V
		Cuculus canorus	Common Cuckoo	Coucou gris		P	P
		Cuculus gularis	African Cuckoo	Coucou africain		M	M
		Chrysococcyx cupreus	African Emerald Cuckoo	Coucou foliotocol		M	M
		Chrysococcyx klaas	Klaas's Cuckoo	Coucou de Klaas		R/M	R/M
		Chrysococcyx caprius	Didric Cuckoo	Coucou didric		M	M
		Ceuthmochares aereus	Yellowbill	Malcoha à bec jaune		R	R
		Centropus leucogaster	Black-throated Coucal	Coucal à ventre blanc		R	
		Centropus grillii	Black Coucal	Coucal de Grill		M	M
		Centropus senegalensis	Senegal Coucal	Coucal du Sénégal		R	R
Barn owls TYTONIDAE							
		Tyto alba	Barn Owl	Effraie des clochers		R	R
Typical owls STRIGIDAE							
		Otus scops	European Scops Owl	Petit-duc scops		P	P
		Otus senegalensis	African Scops Owl	Petit-duc africain		R	R
		Ptilopsis leucotis	Northern White-faced Owl	Petit-duc à face blanche		R	R
		Bubo (bubo) ascalaphus	Desert Eagle Owl	Grand-duc ascalaphe		V	
		Bubo (africanus) cinerascens	Greyish Eagle Owl	Grand-duc du Sahel		R	R
		Bubo lacteus	Verreaux's Eagle Owl	Grand-duc de Verreaux		R	R
		Scotopelia peli	Pel's Fishing Owl	Chouette-pêcheuse de Pel		R	R
		Glaucidium perlatum	Pearl-spotted Owlet	Chevêchette perlée		R	R
		Strix woodfordii	African Wood Owl	Chouette africaine		R	R

Sen	Gam	Scientific name	English name	French name	IUCN	Status	
						Sen	Gam
		Asio flammeus	Short-eared Owl	Hibou des marais		P	V
		Asio capensis	Marsh Owl	Hibou du Cap		M	M
Nightjars CAPRIMULGIDAE							
		Caprimulgus natalensis	Swamp Nightjar	Engoulevent du Natal			V
		Caprimulgus climacurus	Long-tailed Nightjar	Engoulevent à longue queue		R/M	R/M
		Caprimulgus nigriscapularis	Black-shouldered Nightjar	Engoulevent à épaulettes noires		R	V
		Caprimulgus inornatus	Plain Nightjar	Engoulevent terne		M	V
		Caprimulgus eximius	Golden Nightjar	Engoulevent doré		R	
		Caprimulgus aegyptius	Egyptian Nightjar	Engoulevent du désert		P	
		Caprimulgus ruficollis	Red-necked Nightjar	Engoulevent à collier roux		V	P
		Caprimulgus europaeus	European Nightjar	Engoulevent d'Europe		V	P
		Macrodipteryx longipennis	Standard-winged Nightjar	Engoulevent à balanciers		R/M	R/M
Swifts APODIDAE							
		Telacanthura ussheri	Mottled Spinetail	Martinet d'Ussher		R	R
		Cypsiurus parvus	African Palm Swift	Martinet des palmes		R	R
		Apus pallidus	Pallid Swift	Martinet pâle		P	P
		Apus apus	Common Swift	Martinet noir		P	P
		Apus caffer	White-rumped Swift	Martinet cafre		M/R?	M
		Apus affinis	Little Swift	Martinet des maisons		R	R
		Tachymarptis aequatorialis	Mottled Swift	Martinet marbré		V	
		Tachymarptis melba	Alpine Swift	Martinet à ventre blanc		P	V
Mousebirds COLIIDAE							
		Urocolius macrourus	Blue-naped Mousebird	Coliou huppé		R	V
Trogons TROGONIDAE							
		Apaloderma narina	Narina's Trogon	Trogon narina		R	
Kingfishers ALCEDINIDAE							
		Halcyon leucocephala	Grey-headed Kingfisher	Martin-chasseur à tête grise		R/M	R/M
		Halcyon malimbica	Blue-breasted Kingfisher	Martin-chasseur à poitrine bleue		R	R
		Halcyon senegalensis	Woodland Kingfisher	Martin-chasseur du Sénégal		M/R	M/R
		Halcyon chelicuti	Striped Kingfisher	Martin-chasseur strié		R	R
		Ceyx pictus	African Pygmy Kingfisher	Martin-pêcheur pygmée		R/M	R/M
		Alcedo cristata	Malachite Kingfisher	Martin-pêcheur huppé		R	R
		Alcedo quadribrachys	Shining-blue Kingfisher	Martin-pêcheur azuré		R	M
		Megaceryle maxima	Giant Kingfisher	Martin-pêcheur géant		R	R
		Ceryle rudis	Pied Kingfisher	Martin-pêcheur pie		R	R
Bee-eaters MEROPIDAE							
		Merops pusillus	Little Bee-eater	Guêpier nain		R	R
		Merops hirundineus	Swallow-tailed Bee-eater	Guêpier à queue d'aronde		R/M	R/M
		Merops bulocki	Red-throated Bee-eater	Guêpier à gorge rouge		R	R
		Merops albicollis	White-throated Bee-eater	Guêpier à gorge blanche		M	M
		Merops orientalis	Little Green Bee-eater	Guêpier d'Orient		R	R
		Merops persicus	Blue-cheeked Bee-eater	Guêpier de Perse		R/P	M/P
		Merops apiaster	European Bee-eater	Guêpier d'Europe		P	P
		Merops nubicus	Northern Carmine Bee-eater	Guêpier écarlate		M	M

Sen	Gam	Scientific name	English name	French name	IUCN	Status	
						Sen	Gam
Rollers CORACIIDAE							
		Coracias naevius	Rufous-crowned Roller	Rollier varié		**R/M**	**R/M**
		Coracias cyanogaster	Blue-bellied Roller	Rollier à ventre bleu		**R**	**R**
		Coracias abyssinicus	Abyssinian Roller	Rollier d'Abyssinie		**M**	**M**
		Coracias garrulus	European Roller	Rollier d'Europe	**NT**	**P**	**P**
		Eurystomus glaucurus	Broad-billed Roller	Rolle violet		**R/M**	**R/M**
Wood-hoopoes PHOENICULIDAE							
		Phoeniculus purpureus	Green Wood-hoopoe	Irrisor moqueur		**R**	**R**
		Rhinopomastus aterrimus	Black Scimitarbill	Irrisor noir		**R**	**R**
Hoopoes UPUPIDAE							
		Upupa epops	Hoopoe	Huppe fasciée		**P/R?**	**P/M**
Hornbills BUCEROTIDAE							
		Bucorvus abyssinicus	Abyssinian Ground Hornbill	Bucorve d'Abyssinie		**R**	**R**
		Tockus (erythrorhynchus) kempi	(Western) Red-billed Hornbill	Calao à bec rouge		**R**	**R**
		Tockus fasciatus	African Pied Hornbill	Calao longibande		**R**	**R**
		Tockus nasutus	African Grey Hornbill	Calao à bec noir		**R/M**	**R**
		Bycanistes fistulator	Piping Hornbill	Calao siffleur		**R**	
		Bycanistes subcylindricus	Black-and-white-casqued Hornbill	Calao à joues grises			**V?**
		Ceratogymna elata	Yellow-casqued Hornbill	Calao à casque jaune	**NT**	**R**	
Barbets and tinkerbirds CAPITONIDAE							
		Pogoniulus atroflavus	Red-rumped Tinkerbird	Barbion à croupion rouge		**R**	
		Pogoniulus bilineatus	Yellow-rumped Tinkerbird	Barbion à croupion jaune		**R**	**R**
		Pogoniulus chrysoconus	Yellow-fronted Tinkerbird	Barbion à front jaune		**R**	**R**
		Tricholaema hirsuta	Hairy-breasted Barbet	Barbican hérissé		**R**	
		Lybius vieilloti	Vieillot's Barbet	Barbican de Vieillot		**R**	**R**
		Lybius dubius	Bearded Barbet	Barbican à poitrine rouge		**R**	**R**
Honeyguides INDICATORIDAE							
		Prodotiscus regulus	Wahlberg's Honeybird	Indicateur de Wahlberg			**V**
		Indicator maculatus	Spotted Honeyguide	Indicateur tacheté		**R**	**R**
		Indicator indicator	Greater Honeyguide	Grand Indicateur		**R**	**R**
		Indicator minor	Lesser Honeyguide	Petit Indicateur		**R**	**R**
Woodpeckers and wrynecks PICIDAE							
		Jynx torquilla	Eurasian Wryneck	Torcol fourmilier		**P**	**P**
		Campethera punctuligera	Fine-spotted Woodpecker	Pic à taches noires		**R**	**R**
		Campethera abingoni	Golden-tailed Woodpecker	Pic à queue dorée		**R**	**R**
		Campethera maculosa	Little Green Woodpecker	Pic barré		**R**	**V**
		Campethera nivosa	Buff-spotted Woodpecker	Pic tacheté		**R**	**R**
		Dendropicos elachus	Little Grey Woodpecker	Pic gris		**R**	**V**
		Dendropicos fuscescens	Cardinal Woodpecker	Pic cardinal		**R**	**R**
		Dendropicos goertae	Grey Woodpecker	Pic goertan		**R**	**R**
		Picoides obsoletus	Brown-backed Woodpecker	Pic à dos brun		**R**	**R**

Sen	Gam	Scientific name	English name	French name	IUCN	Status	
						Sen	Gam
Larks ALAUDIDAE							
		Mirafra cantillans	Singing Bush Lark	Alouette chanteuse		**R**	**M**
		Mirafra cordofanica	Kordofan Lark	Alouette du Kordofan		**R**	
		Mirafra rufocinnamomea	Flappet Lark	Alouette bourdonnante		**R**	**R**
		Alaemon alaudipes	Greater Hoopoe Lark	Sirli du désert		**M**	
		Calandrella brachydactyla	Greater Short-toed Lark	Alouette calandrelle		**P**	**V**
		Galerida modesta	Sun Lark	Cochevis modeste		**R**	**R**
		Galerida cristata	Crested Lark	Cochevis huppé		**R**	**R**
		Eremopterix leucotis	Chestnut-backed Sparrow Lark	Moinelette à oreillons blancs		**R/M**	**M**
		Eremopterix nigriceps	Black-crowned Sparrow Lark	Moinelette à front blanc		**R/M**	**V**
		Eremophila bilopha	Temminck's Horned Lark	Alouette bilophe			**V**
Swallows and martins HIRUNDINIDAE							
		Psalidoprocne obscura	Fanti Saw-wing	Hirondelle fanti		**R/M**	**R/M**
		Riparia paludicola	Plain Martin	Hirondelle paludicole		**M**	**M**
		Riparia riparia	Common Sand Martin	Hirondelle de rivage		**P**	**P**
		Riparia cincta	Banded Martin	Hirondelle à collier			**V**
		Pseudhirundo griseopyga	Grey-rumped Swallow	Hirondelle à croupion gris		**V**	**M**
		Cecropis semirufa	Rufous-chested Swallow	Hirondelle à ventre roux		**M**	**M**
		Cecropis senegalensis	Mosque Swallow	Hirondelle des mosquées		**R/M**	**R/M**
		Cecropis abyssinica	Lesser Striped Swallow	Hirondelle striée		**M**	**M**
		Cecropis daurica	Red-rumped Swallow	Hirondelle rousseline		**P**	
		Cecropis daurica domicella	West African Swallow	Hirondelle ouest-africaine		**R**	**R**
		Ptyonoprogne fuligula	Rock Martin	Hirondelle isabelline		**R**	
		Ptyonoprogne rupestris	Crag Martin	Hirondelle de rochers		**P**	**V**
		Hirundo smithii	Wire-tailed Swallow	Hirondelle à longs brins		**R**	**R**
		Hirundo leucosoma	Pied-winged Swallow	Hirondelle à ailes tachetées		**R**	**R**
		Hirundo aethiopica	Ethiopian Swallow	Hirondelle d'Éthiopie		**V**	
		Hirundo lucida	Red-chested Swallow	Hirondelle de Guinée		**R**	**R**
		Hirundo rustica	Barn Swallow	Hirondelle rustique		**P**	**P**
		Delichon urbicum	Common House Martin	Hirondelle de fenêtre		**P**	**P**
Wagtails and pipits MOTACILLIDAE							
		Motacilla flava	Yellow Wagtail	Bergeronnette printanière		**P**	**P**
		Motacilla citreola	Citrine Wagtail	Bergeronnette citrine		**V**	
		Motacilla cinerea	Grey Wagtail	Bergeronnette des ruisseaux		**V**	**V**
		Motacilla alba	White Wagtail	Bergeronnette grise		**P**	**P**
		Motacilla aguimp	African Pied Wagtail	Bergeronnette pie		**R**	**M**
		Anthus campestris	Tawny Pipit	Pipit rousseline		**P**	**P**
		Anthus similis	Long-billed Pipit	Pipit à long bec		**M**	
		Anthus leucophrys	Plain-backed Pipit	Pipit à dos uni		**R**	**R**
		Anthus trivialis	Tree Pipit	Pipit des arbres		**P**	**P**
		Anthus cervinus	Red-throated Pipit	Pipit à gorge rousse		**P**	**P**
		Macronyx croceus	Yellow-throated Longclaw	Sentinelle à gorge jaune		**R**	**R**

Sen	Gam	Scientific name	English name	French name	IUCN	Status	
						Sen	Gam
Cuckooshrikes CAMPEPHAGIDAE							
		Campephaga phoenicea	Red-shouldered Cuckooshrike	Échenilleur à épaulettes rouges		**R/M**	**R/M**
		Coracina pectoralis	White-breasted Cuckooshrike	Échenilleur à ventre blanc		**R**	**R**
Bulbuls PYCNONOTIDAE							
		Andropadus virens	Little Greenbul	Bulbul verdâtre		**R**	**R**
		Andropadus gracilirostris	Slender-billed Greenbul	Bulbul à bec grêle		**R**	
		Andropadus latirostris	Yellow-whiskered Greenbul	Bulbul à moustaches jaunes		**R**	
		Chlorocichla simplex	Simple Leaflove	Bulbul modeste		**V**	
		Chlorocichla flavicollis	Yellow-throated Leaflove	Bulbul à gorge claire		**R**	**R**
		Thescelocichla leucopleura	Swamp Palm Bulbul	Bulbul des raphias		**R**	**V**
		Pyrrhurus scandens	Leaflove	Bulbul à queue rousse		**R**	**R**
		Phyllastrephus albigularis	White-throated Greenbul	Bulbul à gorge blanche		**R**	
		Bleda canicapillus	Grey-headed Bristlebill	Bulbul fourmilier		**R**	**R**
		Criniger calurus	Red-tailed Greenbul	Bulbul à barbe blanche		**R**	
		Pycnonotus barbatus	Common Bulbul	Bulbul des jardins		**R**	**R**
		Nicator chloris	Western Nicator	Bulbul nicator		**R**	**V**
Thrushes, chats and allies TURDIDAE							
		Erithacus rubecula	European Robin	Rougegorge familier			**V?**
		Luscinia megarhynchos	Common Nightingale	Rossignol philomèle		**P**	**P**
		Luscinia svecica	Bluethroat	Gorgebleue à miroir		**P**	**P**
		Cossypha niveicapilla	Snowy-crowned Robin Chat	Cossyphe à calotte neigeuse		**R**	**R**
		Cossypha albicapilla	White-crowned Robin Chat	Cossyphe à calotte blanche		**R**	**R**
		Alethe diademata	White-tailed Alethe	Alèthe à huppe rousse		**R**	
		Cercotrichas galactotes	Rufous Scrub Robin	Agrobate roux		**R/P**	**R**
		Cercotrichas podobe	Black Scrub Robin	Agrobate podobé		**R**	
		Phoenicurus ochruros	Black Redstart	Rougequeue noir		**P**	
		Phoenicurus phoenicurus	Common Redstart	Rougequeue à front blanc		**P**	**P**
		Saxicola torquatus	African Stonechat	Tarier pâtre		**R**	
		Saxicola rubetra	Whinchat	Tarier des prés		**P**	**P**
		Oenanthe oenanthe	Northern Wheatear	Traquet motteux		**P**	**P**
		Oenanthe hispanica	Black-eared Wheatear	Traquet oreillard		**P**	**V**
		Oenanthe deserti	Desert Wheatear	Traquet du désert		**P**	**§**
		Oenanthe isabellina	Isabelline Wheatear	Traquet isabelle		**P**	**V**
		Cercomela familiaris	Familiar Chat	Traquet familier		**R**	
		Cercomela melanura	Blackstart	Traquet à queue noire			**V**
		Myrmecocichla aethiops	Northern Anteater Chat	Traquet brun		**R**	**R**
		Myrmecocichla nigra	Sooty Chat	Traquet commandeur		**R?**	
		Myrmecocichla albifrons	White-fronted Black Chat	Traquet à front blanc		**R**	**R**
		Thamnolaea cinnamomeiventris	Mocking Cliff Chat	Traquet à ventre roux		**R**	
		Monticola saxatilis	Common Rock Thrush	Monticole de roche		**P**	**P**
		Monticola solitarius	Blue Rock Thrush	Monticole bleu		**P**	**P**
		Turdus pelios	African Thrush	Merle africain		**R**	**R**
		Turdus philomelos	Song Thrush	Grive musicienne		**V**	

Sen	Gam	Scientific name	English name	French name	IUCN	Status	
						Sen	Gam
Warblers SYLVIIDAE							
		Melocichla mentalis	African Moustached Warbler	Mélocichle à moustaches		**R**	**§**
		Locustella naevia	Grasshopper Warbler	Locustelle tachetée		**P**	**V**
		Locustella luscinioides	Savi's Warbler	Locustelle luscinoïde		**P**	
		Acrocephalus paludicola	Aquatic Warbler	Phragmite aquatique	**VU**	**P**	
		Acrocephalus schoenobaenus	Sedge Warbler	Phragmite des joncs		**P**	**P**
		Acrocephalus scirpaceus	European Reed Warbler	Rousserolle effarvatte		**P**	**P**
		A. (scirpaceus) baeticatus	African Reed Warbler	Rousserolle africaine		**R**	**R**
		Acrocephalus palustris	Marsh Warbler	Rousserolle verderolle		**V**	
		Acrocephalus arundinaceus	Great Reed Warbler	Rousserolle turdoïde		**P**	**V**
		Acrocephalus rufescens	Greater Swamp Warbler	Rousserolle des cannes		**R**	**R**
		Hippolais pallida	Eastern Olivaceous Warbler	Hypolaïs pâle		**P**	**V**
		Hippolais (pallida) opaca	Western Olivaceous Warbler	Hypolaïs obscure		**P**	**P**
		Hippolais polyglotta	Melodious Warbler	Hypolaïs polyglotte		**P**	**P**
		Hippolais icterina	Icterine Warbler	Hypolaïs ictérine		**V**	**V**
		Eremomela icteropygialis	Yellow-bellied Eremomela	Érémomèle à croupion jaune		**R**	
		Eremomela pusilla	Senegal Eremomela	Érémomèle à dos vert		**R**	**R**
		Sylvietta brachyura	Northern Crombec	Crombec sitelle		**R**	**R**
		Sylvietta virens	Green Crombec	Crombec vert		**R**	**R**
		Phylloscopus trochilus	Willow Warbler	Pouillot fitis		**P**	**P**
		Phylloscopus collybita	Common Chiffchaff	Pouillot véloce		**P**	**P**
		Phylloscopus (collybita) ibericus	Iberian Chiffchaff	Pouillot ibérique		**P**	
		Phylloscopus sibilatrix	Wood Warbler	Pouillot siffleur		**V**	**V**
		Phylloscopus bonelli	Western Bonelli's Warbler	Pouillot de Bonelli		**P**	**P**
		Phylloscopus inornatus	Yellow-browed Warbler	Pouillot à grands sourcils		**V**	**V**
		Sylvia nisoria	Barred Warbler	Fauvette épervière		**V**	**V**
		Sylvia hortensis	Western Orphean Warbler	Fauvette orphée		**P**	**P**
		Sylvia borin	Garden Warbler	Fauvette des jardins		**P**	**P**
		Sylvia atricapilla	Blackcap	Fauvette à tête noire		**P**	**P**
		Sylvia communis	Common Whitethroat	Fauvette grisette		**P**	**P**
		Sylvia curruca	Lesser Whitethroat	Fauvette babillarde		**P**	**V**
		Sylvia rueppelli	Rüppell's Warbler	Fauvette de Rüppell			**V**
		Sylvia melanocephala	Sardinian Warbler	Fauvette mélanocéphale		**P**	
		Sylvia cantillans	Subalpine Warbler	Fauvette passerinette		**P**	**P**
		Sylvia conspicillata	Spectacled Warbler	Fauvette à lunettes		**P**	**V**
		Hyliota flavigaster	Yellow-bellied Hyliota	Hyliote à ventre jaune		**R**	**R**
		Hylia prasina	Green Hylia	Hylia verte		**R**	**R**
Cisticolas and allies CISTICOLIDAE							
		Cisticola erythrops	Red-faced Cisticola	Cisticole à face rousse		**R**	**R**
		Cisticola cantans	Singing Cisticola	Cisticole chanteuse		**R**	**R**
		Cisticola lateralis	Whistling Cisticola	Cisticole siffleuse		**R**	**R**
		Cisticola galactotes	Winding Cisticola	Cisticole roussâtre		**R**	**R**
		Cisticola natalensis	Croaking Cisticola	Cisticole striée		**R**	**R**
		Cisticola guinea	Dorst's Cisticola	Cisticole de Dorst		**R**	**R**

Sen	Gam	Scientific name	English name	French name	IUCN	Status Sen	Status Gam
		Cisticola brachypterus	Short-winged Cisticola	Cisticole à ailes courtes		R	R
		Cisticola rufus	Rufous Cisticola	Cisticole rousse		R	R
		Cisticola juncidis	Zitting Cisticola	Cisticole des joncs		R	R
		Cisticola aridulus	Desert Cisticola	Cisticole du désert		R	
		Cisticola eximius	Black-backed Cisticola	Cisticole à dos noir		R	R
		Prinia subflava	Tawny-flanked Prinia	Prinia modeste		R	R
		Prinia fluviatilis	River Prinia	Prinia aquatique		R	R?
		Heliolais erythropterus	Red-winged Warbler	Prinia à ailes rousses		R	R
		Spiloptila clamans	Cricket Warbler	Prinia à front écailleux		R	
		Apalis flavida	Yellow-breasted Apalis	Apalis à gorge jaune		R	R
		Camaroptera brachyura	Grey-backed Camaroptera	Camaroptère à tête grise		R	R
		Camaroptera chloronota	Olive-green Camaroptera	Camaroptère à dos vert		R	§
		Hypergerus atriceps	Oriole Warbler	Noircap loriot		R	R
Flycatchers MUSCICAPIDAE							
		Fraseria cinerascens	White-browed Forest Flycatcher	Gobemouche à sourcils blancs		R	§
		Melaenornis edolioides	Northern Black Flycatcher	Gobemouche drongo		R	R
		Bradornis pallidus	Pale Flycatcher	Gobemouche pâle		R	R
		Muscicapa striata	Spotted Flycatcher	Gobemouche gris		P	P
		Muscicapa aquatica	Swamp Flycatcher	Gobemouche des marais		R	R
		Myioparus plumbeus	Lead-coloured Flycatcher	Gobemouche mésange		R	R
		Ficedula hypoleuca	Pied Flycatcher	Gobemouche noir		P	P
		Ficedula (hypoleuca) speculigera	Atlas Flycatcher	Gobemouche de l'Atlas		V	
		Ficedula parva	Red-breasted Flycatcher	Gobemouche nain		V	
Monarchs MONARCHIDAE							
		Elminia longicauda	African Blue Flycatcher	Tchitrec bleu		R	R
		Terpsiphone viridis	African Paradise Flycatcher	Tchitrec d'Afrique		R	R
		Terpsiphone rufiventer	Red-bellied Paradise Flycatcher	Tchitrec à ventre roux		R	R
Wattles-eyes, batises and allies PLATYSTEIRIDAE							
		Bias musicus	Black-and-white Flycatcher	Bias musicien			V
		Platysteira cyanea	Common Wattle-eye	Pririt à collier		R	R
		Batis senegalensis	Senegal Batis	Pririt du Sénégal		R	R
Babblers TIMALIIDAE							
		Illadopsis fulvescens	Brown Illadopsis	Akalat brun		R	
		Illadopsis puveli	Puvel's Illadopsis	Akalat de Puvel		R	R
		Turdoides plebejus	Brown Babbler	Cratérope brun		R	R
		Turdoides reinwardtii	Blackcap Babbler	Cratérope à tête noire		R	R
		Turdoides fulva	Fulvous Babbler	Cratérope fauve		R	
		Phyllanthus atripennis	Capuchin Babbler	Phyllanthe à gorge blanche		R	R
Tits PARIDAE							
		Parus (leucomelas) guineensis	White-shouldered Black Tit	Mésange gallonée		R	R
Penduline tits REMIZIDAE							
		Anthoscopus parvulus	Yellow Penduline Tit	Rémiz à ventre jaune		R	R
		Anthoscopus punctifrons	Sennar Penduline Tit	Rémiz du Soudan		R	

Sen	Gam	Scientific name	English name	French name	IUCN	Status Sen	Status Gam
Spotted Creeper SALPORNITHIDAE							
		Salpornis spilonotus	Spotted Creeper	Grimpereau tacheté		**V**	**V**
Sunbirds NECTARINIIDAE							
		Anthreptes longuemarei	Western Violet-backed Sunbird	Souimanga violet		**R**	**R**
		Anthreptes gabonicus	Brown Sunbird	Souimanga brun		**R**	**R**
		Cyanomitra verticalis	Green-headed Sunbird	Souimanga à tête verte		**R**	**R**
		Cyanomitra olivacea	Olive Sunbird	Souimanga olivâtre		**R**	
		Chalcomitra senegalensis	Scarlet-chested Sunbird	Souimanga à poitrine rouge		**R**	**R**
		Hedydipna collaris	Collared Sunbird	Souimanga à collier		**R**	**R**
		Hedydipna platura	Pygmy Sunbird	Souimanga pygmée		**R/M**	**R/M**
		Cinnyris chloropygius	Olive-bellied Sunbird	Souimanga à ventre olive		**R**	**V**
		Cinnyris pulchellus	Beautiful Sunbird	Souimanga à longue queue		**R**	**R**
		Cinnyris venustus	Variable Sunbird	Souimanga à ventre jaune		**R**	**R**
		Cinnyris coccinigastrus	Splendid Sunbird	Souimanga éclatant		**R**	**R**
		Cinnyris cupreus	Copper Sunbird	Souimanga cuivré		**R**	**R**
White-eyes ZOSTEROPIDAE							
		Zosterops senegalensis	Yellow White-eye	Zostérops jaune		**R**	**R**
Shrikes LANIIDAE							
		Lanius meridionalis	Southern Grey Shrike	Pie-grièche méridionale		**R**	**M**
		Lanius isabellinus	Isabelline Shrike	Pie-grièche isabelle		**P**	**V**
		Lanius senator	Woodchat Shrike	Pie-grièche à tête rousse		**P**	**P**
		Corvinella corvina	Yellow-billed Shrike	Corvinelle à bec jaune		**R**	**R**
Bush-shrikes MALACONOTIDAE							
		Malaconotus blanchoti	Grey-headed Bush-shrike	Gladiateur de Blanchot		**R**	**R**
		Chlorophoneus sulfureopectus	Sulphur-breasted Bush-shrike	Gladiateur soufré		**R**	**R**
		Tchagra senegalus	Black-crowned Tchagra	Tchagra à tête noire		**R**	**R**
		Dryoscopus gambensis	Northern Puffback	Cubla de Gambie		**R**	**R**
		Laniarius barbarus	Yellow-crowned Gonolek	Gonolek de Barbarie		**R**	**R**
		Nilaus afer	Brubru	Brubru africain		**R**	**R**
Helmetshrikes PRIONOPIDAE							
		Prionops plumatus	White Helmetshrike	Bagadais casqué		**R**	**R**
Orioles ORIOLIDAE							
		Oriolus auratus	African Golden Oriole	Loriot doré		**R/M**	**R**
		Oriolus oriolus	Eurasian Golden Oriole	Loriot d'Europe		**P**	**P**
Drongos DICRURIDAE							
		Dicrurus ludwigii	Square-tailed Drongo	Drongo de Ludwig		**R**	**R**
		Dicrurus adsimilis	Fork-tailed Drongo	Drongo brilliant		**R**	**R**
Crows CORVIDAE							
		Corvus ruficollis	Brown-necked Raven	Corbeau brun		**M**	**V**
		Corvus albus	Pied Crow	Corbeau pie		**R**	**R**
		Ptilostomus afer	Piapiac	Piapiac africain		**R**	**R**

Sen	Gam	Scientific name	English name	French name	IUCN	Status Sen	Status Gam
Starlings STURNIDAE							
		Onychognathus neumanni	Neumann's Starling	Rufipenne de Neumann		**R**	
		Lamprotornis purpureus	Purple Glossy Starling	Choucador pourprée		**R**	**R**
		Lamprotornis chalcurus	Bronze-tailed Glossy Starling	Choucador à queue violette		**R**	**R**
		Lamprotornis chalybaeus	Greater Blue-eared Starling	Choucador à oreillons bleus		**R**	**R**
		Lamprotornis chloropterus	Lesser Blue-eared Starling	Choucador de Swainson		**R**	**R**
		Lamprotornis splendidus	Splendid Glossy Starling	Choucador splendide		**R**	**M/R?**
		Lamprotornis caudatus	Long-tailed Glossy Starling	Choucador à longue queue		**R**	**R**
		Lamprotornis pulcher	Chestnut-bellied Starling	Choucador à ventre roux		**R**	**R**
		Cinnyricinclus leucogaster	Violet-backed Starling	Spréo améthyste		**M**	**M**
		Creatophora cinerea	Wattled Starling	Étourneau caronculé			**V**
		Sturnus vulgaris	Common Starling	Étourneau sansonnet			**V?**
Oxpeckers BUPHAGIDAE							
		Buphagus africanus	Yellow-billed Oxpecker	Piquebœuf à bec jaune		**R**	**R**
Sparrows PASSERIDAE							
		Passer griseus	Northern Grey-headed Sparrow	Moineau gris		**R**	**R**
		Passer domesticus	House Sparrow	Moineau domestique		**R**	**R**
		Passer simplex	Desert Sparrow	Moineau blanc			**V**
		Passer luteus	Sudan Golden Sparrow	Moineau doré		**R**	**M**
		Petronia dentata	Bush Petronia	Petit Moineau		**R**	**R**
		Petronia pyrgita	Yellow-spotted Petronia	Moineau à point jaune		**R**	
Weavers PLOCEIDAE							
		Bubalornis albirostris	White-billed Buffalo Weaver	Alecto à bec blanc		**R**	**R**
		Plocepasser superciliosus	Chestnut-crowned Sparrow Weaver	Mahali à calotte marron		**R**	**R**
		Sporopipes frontalis	Speckle-fronted Weaver	Sporopipe quadrillé		**R**	**R**
		Malimbus nitens	Blue-billed Malimbe	Malimbe à bec bleu		**R**	**§**
		Anaplectes rubriceps	Red-headed Weaver	Tisserin écarlate		**R**	**R**
		Ploceus luteolus	Little Weaver	Tisserin minule		**R**	**R**
		Ploceus nigricollis	Black-necked Weaver	Tisserin à cou noir		**R**	**R**
		Ploceus vitellinus	Vitelline Masked Weaver	Tisserin vitellin		**R**	**R**
		Ploceus heuglini	Heuglin's Masked Weaver	Tisserin masqué		**R**	**R**
		Ploceus nigerrimus	Vieillot's Black Weaver	Tisserin noir		**§**	**V**
		Ploceus cucullatus	Village Weaver	Tisserin gendarme		**R**	**R**
		Ploceus melanocephalus	Black-headed Weaver	Tisserin à tête noire		**R**	**R**
		Pachyphanthes superciliosus	Compact Weaver	Tisserin gros-bec		**R**	
		Amblyospiza albifrons	Grosbeak Weaver	Amblyospize à front blanc		**V**	
		Quelea erythrops	Red-headed Quelea	Travailleur à tête rouge		**M/R?**	**M/R?**
		Quelea quelea	Red-billed Quelea	Travailleur à bec rouge		**R/M**	**M**
		Euplectes franciscanus	Northern Red Bishop	Euplecte franciscain		**R**	**R**
		Euplectes hordeaceus	Black-winged Bishop	Euplecte monseigneur		**R**	**R**
		Euplectes afer	Yellow-crowned Bishop	Euplecte vorabé		**R**	**R**
		Euplectes macroura	Yellow-mantled Widowbird	Euplecte à dos d'or		**R**	**R**
		Euplectes ardens	Red-collared Widowbird	Euplecte veuve-noire		**§**	**V**

Sen	Gam	Scientific name	English name	French name	IUCN	Status Sen	Status Gam
Estrildid finches ESTRILDIDAE							
		Nigrita bicolor	Chestnut-breasted Negrofinch	Nigrette à ventre roux		**R**	**V**
		Nesocharis capistrata	Grey-headed Oliveback	Dos-vert à joues blanches			**V**
		Estrilda caerulescens	Lavender Waxbill	Astrild queue-de-vinaigre		**R**	**R**
		Estrilda melpoda	Orange-cheeked Waxbill	Astrild à joues oranges		**R**	**R**
		Estrilda troglodytes	Black-rumped Waxbill	Astrild cendré		**R**	**R**
		Spermophaga haematina	Western Bluebill	Sénégali sanguin		**R**	**R**
		Pyrenestes sanguineus	Crimson Seedcracker	Pyréneste gros-bec		**R**	**R**
		Uraeginthus bengalus	Red-cheeked Cordon-bleu	Cordonbleu à joues rouges		**R**	**R**
		Euschistospiza dybowskii	Dybowski's Twinspot	Sénégali à ventre noir		**R**	
		Pytilia phoenicoptera	Red-winged Pytilia	Beaumarquet aurore		**R**	**R**
		Pytilia melba	Green-winged Pytilia	Beaumarquet melba		**R**	**R**
		Lagonosticta senegala	Red-billed Firefinch	Amarante du Sénégal		**R**	**R**
		Lagonosticta rufopicta	Bar-breasted Firefinch	Amarante pointé		**R**	**R**
		Lagonosticta larvata	Black-faced Firefinch	Amarante masqué		**R**	**R**
		Lagonosticta rara	Black-bellied Firefinch	Amarante à ventre noir		**R**	
		Lagonosticta virata	Kulikoro (Mali) Firefinch	Amarante de Kulikoro		**R**	
		Amadina fasciata	Cut-throat Finch	Amadine cou-coupé		**R/M**	**M**
		Ortygospiza atricollis	Black-faced Quailfinch	Astrild-caille à lunettes		**R**	**R**
		Sporaeginthus subflavus	Zebra Waxbill	Bengali zébré		**R**	**M**
		Spermestes cucullatus	Bronze Mannikin	Capucin nonnette		**R**	**R**
		Spermestes fringilloides	Magpie Mannikin	Capucin pie		**R**	**V**
		Euodice cantans	African Silverbill	Capucin bec-d'argent		**R**	**R**
Whydahs and indigobirds VIDUIDAE							
		Anomalospiza imberbis	Cuckoo Finch	Anomalospize parasite			**V**
		Vidua macroura	Pin-tailed Whydah	Veuve dominicaine		**R**	**R**
		Vidua orientalis	Sahel Paradise Whydah	Veuve à collier d'or		**R**	**R**
		Vidua interjecta	Exclamatory Paradise Whydah	Veuve nigérienne		**R**	**R**
		Vidua chalybeata	Village Indigobird	Combassou du Sénégal		**R**	**R**
		Vidua wilsoni	Wilson's Indigobird	Combassou de Wilson		**R**	
		Vidua nigeriae	Quailfinch Indigobird	Combassou du Nigeria			**R**
True finches FRINGILLIDAE							
		Serinus leucopygius	White-rumped Seedeater	Serin à croupion blanc		**R**	**R**
		Serinus mozambicus	Yellow-fronted Canary	Serin du Mozambique		**R**	**R**
		Carduelis cannabina	Common Linnet	Linotte mélodieuse		**V**	
Buntings EMBERIZIDAE							
		Emberiza flaviventris	Golden-breasted Bunting	Bruant à poitrine dorée		**R**	
		Emberiza affinis	Brown-rumped Bunting	Bruant à ventre jaune		**R**	**R**
		Emberiza tahapisi	Cinnamon-breasted Rock Bunting	Bruant cannelle		**M/R?**	**M/R?**
		Emberiza (striolata) sahari	House Bunting	Bruant du Sahara		**V**	**V**
		Emberiza hortulana	Ortolan Bunting	Bruant ortolan		**P**	**V**
		Emberiza calandra	Corn Bunting	Bruant proyer		**V**	

APPENDIX
SPECIES NOT ACCEPTED ON THE SENEGAL AND GAMBIA LISTS

The claimed occurrence in Senegal and/or The Gambia of the following species needs confirmation or is erroneous. Species marked * are described and illustrated in this guide.

Calidris fuscicollis	**White-rumped Sandpiper**	Bécasseau de Bonaparte	(The Gambia)
Limicola falcinellus	**Broad-billed Sandpiper**	Bécasseau falcinelle	(Senegal)
**Macrodipteryx vexillarius*	**Pennant-winged Nightjar**	Engoulevent porte-étendard	(The Gambia)
Eurystomus gularis	**Blue-throated Roller**	Rolle à gorge bleue	(Senegal)
Phoeniculus bollei	**White-headed Wood-hoopoe**	Irrisor à tête blanche	(Senegal)
**Bycanistes subcylindricus*	**Black-and-white-casqued Hornbill**	Calao à joues grises	(The Gambia)
**Pogoniulus atroflavus*	**Red-rumped Tinkerbird**	Barbion à croupion rouge	(Senegal)
Pogoniulus subsulphureus	**Yellow-throated Tinkerbird**	Barbion à gorge jaune	(Senegal)
**Tricholaema hirsuta*	**Hairy-breasted Barbet**	Barbican hérissé	(Senegal)
**Prodotiscus insignis*	**Cassin's Honeybird**	Indicateur pygmée	(Senegal)
**Pinarocorys erythropygia*	**Rufous-rumped Lark**	Alouette à queue rousse	(The Gambia)
**Sylvietta denti*	**Lemon-bellied Crombec**	Crombec à gorge tachetée	(The Gambia)
**Camaroptera chloronota*	**Olive-green Camaraoptera**	Camaroptère à dos vert	(Senegal & The Gambia)
Muscicapa ussheri	**Ussher's Flycatcher**	Gobemouche d'Ussher	(Senegal)
Ficedula albicollis	**Collared Flycatcher**	Gobemouche à collier	(Senegal)
**Megabyas flammulatus*	**Shrike Flycatcher**	Bias écorcheur	(The Gambia)
Illadopsis rufescens	**Rufous-winged Illadopsis**	Akalat à ailes rousses	(Senegal)
**Laniarius (aethiopicus) major*	**Tropical Boubou**	Gonolek d'Abyssinie	(Senegal)
**Lagonosticta rubricata*	**Blue-billed Firefinch**	Amarante flambé	(Senegal)
**Spermestes bicolor*	**Black-and-white Mannikin**	Capucin bicolore	(Senegal & The Gambia)
**Vidua larvaticola*	**Barka Indigobird**	Combassou de Barka	(Senegal & The Gambia)
Fringilla montifringilla	**Brambling**	Pinson du nord	(Senegal)

REFERENCES

1. GENERAL AND REGIONAL REFERENCES

Alström, P., Colston, P. & Lewington, I. (1991) *A Field Guide to the Rare Birds of Britain and Europe.* HarperCollins, London.

Bannerman, D. A. (1930–1951) *The Birds of Tropical West Africa,* Vols. 1–8. Crown Agents, London.

Bannerman, D. A. (1953) *The Birds of West and Equatorial Africa.* 2 vols. Oliver & Boyd, Edinburgh & London.

BirdLife International (2000) *Threatened Birds of the World.* Lynx Edicions, Barcelona & BirdLife International, Cambridge.

BirdLife International (2011) Species factsheets. http://www.birdlife.org (accessed on 1 June 2011).

Borrow, N. & Demey, R. (2001) *Birds of Western Africa.* Christopher Helm, London.

Borrow, N. & Demey, R. (2004) *Field Guide to the Birds of Western Africa.* Christopher Helm, London.

Brown, L. H., Urban, E. K. & Newman, K. (1982) *The Birds of Africa,* Vol. 1. Academic Press, London.

Cramp, S. (ed.) (1985–1992) *The Birds of the Western Palearctic,* Vols. 4–6. Oxford University Press, Oxford.

Cramp, S. & Perrins, C. M. (eds.) (1993–1994) *The Birds of the Western Palearctic,* Vols. 7–9. Oxford University Press, Oxford.

Cramp, S. & Simmons, K. E. L. (eds.) (1977–1983) *The Birds of the Western Palearctic,* Vols. 1–3. Oxford University Press, Oxford.

David, N. & Gosselin, M. (2002a) Gender agreement of avian species names. *Bull. Brit. Orn. Club* 122: 14–49.

David, N. & Gosselin, M. (2002b) The grammatical gender of avian genera. *Bull. Brit. Orn. Club* 122: 257–282.

David, N. & Gosselin, M. (2011) Gender agreement of avian species-group names under Article 31.2.2 of th ICZN Code. *Bull. Brit. Orn. Club* 131: 103–115.

Delany, S., Scott, D., Dodman, T. & Stroud, D. (2009) *An Atlas of Wader Populations in Africa and Western Eurasia.* Wetlands International, Wageningen.

Dickinson, E. C. (ed.) (2003) *The Howard & Moore Complete Checklist of the Birds of the World.* Christopher Helm, London.

Dowsett, R. J. & Forbes-Watson, A. D. (1993) *Checklist of Birds of the Afrotropical and Malagasy Regions. Volume 1: Species limits and distribution.* Tauraco Press, Liège.

Dowsett, R. J. & Dowsett-Lemaire, F. (eds.) (1993) *A Contribution to the Distribution and Taxonomy of Afrotropical and Malagasy Birds.* Tauraco Res. Rep. 5. Tauraco Press, Liège.

Fishpool, L. D. C. & Evans, M. I. (eds.) (2001) *Important Bird Areas in Africa and Associated Islands: Priority Sites for Conservation.* BirdLife Conservation Series No. 11. Pisces Publications, Newbury & BirdLife International, Cambridge, UK.

Fry, C. H., Keith, S. & Urban, E. K. (eds.) (1988, 2000) *The Birds of Africa,* Vols. 3 & 6. Academic Press, London.

Gill, F. & Donsker, D. (eds.) (2011) IOC World Bird Names (version 2.8). http://www.worldbirdnames.org/ (accessed on 1 June 2011).

Hall, B. P. & Moreau, R. E. (1970) *An Atlas of Speciation in African Passerine Birds.* British Museum (Natural History), London.

Helbig, A.J., Kocum, A., Seibold, I. & Braun, M. J. (2005) A multi-gene phylogeny of aquiline eagles (Aves: Accipitriformes) reveals extensive paraphyly at the genus level. *Molec. Phylogen. Evol.* 35: 147–164.

del Hoyo, J., Elliott, A. & Christie, D. (eds.) (2003–2010) *Handbook of the Birds of the World,* Vols. 8–15. Lynx Edicions, Barcelona.

del Hoyo, J., Elliott, A. & Sargatal, J. (eds.) (1992–2002) *Handbook of the Birds of the World*, Vols. 1–7. Lynx Edicions, Barcelona.

Keith, S., Urban, E. K. & Fry, C. H. (eds.) (1992) *The Birds of Africa*, Vol. 4. Academic Press, London.

Knox, A. G., Collinson, M., Helbig, A. J., Parkin, D. T. & Sangster, G. (2002) Taxonomic recommendations for British birds. *Ibis* 144: 707–710.

Mackworth-Praed, C. W. & Grant, C. H. B. (1970–1973) *Birds of West Central and Western Africa*. 2 vols. Longmans, London.

Malling Olson, K. & Larsson, H. (2003) *Gulls of Europe, Asia and North America*. Christopher Helm, London.

Sangster, G., Collinson, J. M., Helbig, A. J., Knox, A. G. & Parkin, D. T. (2004) Taxonomic recommendations for British birds: second report. *Ibis* 146: 153–157.

Scott, D. A. & Rose, P. M. (1996) *Atlas of Anatidae Populations in Africa and Western Eurasia*. Wetlands International Publication No. 41. Wetlands International, Wageningen.

Snow, D. W. (1978) *An Atlas of Speciation in African Non-Passerine Birds*. British Museum (Natural History), London.

Snow, D. & Perrins, C. (1998) *The Birds of the Western Palearctic. Concise Edition*. 2 vols. Oxford University Press, Oxford.

Stattersfield, A. J., Crosby, M. J., Long, A. J. & Wege, D. C. (1998) *Endemic Bird Areas of the World: Priorities for Biodiversity Conservation*. BirdLife International, Cambridge.

Svensson, L., Grant, P. J., Mullarney, K. & Zetterström, D. (1999) *Collins Bird Guide*. HarperCollins, London.

Urban, E. K., Fry, C. H. & Keith, S. (eds.) (1986, 1997) *The Birds of Africa*, Vols. 2 & 5. Academic Press, London.

Zwarts, L., Bijlsma, R. G., van der Kamp, J. & Wymenga, E. (2009) *Living on the Edge: Wetlands and Birds in a Changing Sahel*. KNNV Publishing, Zeist.

2. COUNTRY REFERENCES

THE GAMBIA

Baillon, F. & Dubois, P. J. (1992) Nearctic gull species in Senegal and The Gambia. *Dutch Birding* 14: 49–50.

Ballantyne, S. (2006) Bird diversity in Nyassang Forest Park, The Gambia. *Malimbus* 28: 134–142.

Ballantyne, S. (2007) Bird diversity in Nyassang Forest Park, The Gambia (*Malimbus* 28: 134–142): corrections and further information. *Malimbus* 29: 45–46.

Barlow, C. R. (2002) First nest record for Bronze-winged Courser *Cursorius chalcopterus* in Senegambia. *Bull. Afr. Bird Club* 9: 134–135.

Barlow, C. R. (2003) First conclusive evidence of breeding in Senegambia and parental behaviour of Black Coucal *Centropus grillii*. *Bull. Afr. Bird Club* 10: 53–54.

Barlow, C. R. (2007a) First records of Icterine Warbler *Hippolais icterina* for The Gambia and Senegal. *Bull. Afr. Bird Club* 14: 72–74.

Barlow, C. R. (2007b) First record of Yellow-browed Warbler *Phylloscopus inornatus* for The Gambia. *Bull. Afr. Bird Club* 14: 74–75.

Barlow, C. R. (2009) Three records of Solitary Sandpiper *Tringa solitaria* in The Gambia. *Bull. Afr. Bird Club* 16: 209–210.

Barlow, C. R. & Dodman, T. (2011) A record of Black-necked Grebe *Podiceps nigricollis* in The Gambia in 1989 constitutes an addition to the national list. *Malimbus* 33: 55–57.

Barlow, C., Wacher, T. & Disley, T. (1997) *A Field Guide to Birds of The Gambia and Senegal*. Pica Press, Mountfield.

Barlow, C. R. & Gale, G. (1999) Information obtained from nine road-killed Red-necked Nightjar *Caprimulgus ruficollis* in The Gambia, in winters 1990–1997. *Bull. Afr. Bird Club* 6: 48–51.

Barnett, L. K. & Emms, C. (2001) New species and breeding records for The Gambia. *Bull. Afr. Bird Club* 8: 44-45.

Barnett, L. K., Emms, C. & Camara, A. (2001) The birds of Bijol Island, Tanji River (Karinti) Bird Reserve, The Gambia. *Bull. Afr. Bird Club* 8: 39-43,

Borrow, N. (1997) Red-crested Bustard *Eupodotis ruficrista* and Adamawa Turtle Dove *Streptopelia hypopyrrha*, new to The Gambia, and sightings of Great Snipe *Gallinago media*. *Malimbus* 19: 36–38.

Brown, R. G. B. (1979) Seabirds of the Senegal upwelling and adjacent waters. *Ibis* 121: 283–292.

Clark, W. S. & Barlow, C. (2004) Status of Short-toed Snake Eagle *Circaetus gallicus* and Beaudouin's Snake Eagle *C. beaudouini* in The Gambia. *Bull. Afr. Bird Club* 11: 27–29.

Crewe, M. D. (2001) Selected observations from The Gambia, 1997–1999, with comments on the identification of a number of species. *Bull. Afr. Bird Club* 8: 113–116.

Crewe, M. D., Crewe, M. A. & Sanyang, T. (2008) First record of Rüppell's Warbler *Sylvia rueppelli* for The Gambia. *Bull. Afr. Bird Club* 15: 91–92.

Crewe, M. D. & Small, B. J. (2002) Temminck's Horned Lark *Eremophila bilopha* – a new species for The Gambia. *Bull. Afr. Bird Club* 9: 137–138.

Crouzier, P., Dubois, P. J., Frémont, J.-Y. & Verneau, A. (2011) Black Guillemot at N'Gor, Senegal, in October 2008. *Dutch Birding* 33: 43–44.

Geeson, J. & Geeson, J. (1990) First Red Kite record for The Gambia. *Malimbus* 11: 144.

Gore, M. E. J. (1990) *Birds of The Gambia* (2nd edn.). Check-list No. 3. British Ornithologists' Union, Tring.

High, J. (2006) First record of Bridled Tern *Sterna anaethetus* for The Gambia. *Bull. Afr. Bird Club* 13: 77–78.

Jones, R. M. (1991) The status of larks in The Gambia, including first records of Sun Lark *Galerida modesta*. *Malimbus* 13: 67–73.

Jones, R. M. (1992) European Crag Martin *Ptyonoprogne rupestris* in The Gambia. *Malimbus* 14: 21–23.

King, J. M. B. (2003) Baillon's Crake *Porzana pusilla*, new to The Gambia, with notes on seven other species. *Malimbus* 25: 59–61.

King, M. (2000) Noteworthy records from Ginak Island, The Gambia. *Malimbus* 22: 77–85.

Kirk, G. & Barlow, C. R. (2002) Second confirmed record of Forbes's Plover *Charadrius forbesi* for The Gambia. *Bull. Afr. Bird Club* 9: 138–139.

Kirtland, C. A. E. & Rogers, E. P. (1997) First record and nesting of White-crested Tiger-Heron *Tigriornis leucolophus* in The Gambia. *Bull. Afr. Bird Club* 4: 105–106.

Mikkola, A. & Mikkola, H. (2002) First record of Red-footed Falcon *Falco vespertinus* in The Gambia. *Bull. Afr. Bird Club* 9: 45. [Corrigendum 9: 93.]

Morel, G. J. & Morel, M.-Y. (1990) *Les Oiseaux de Sénégambie*. ORSTOM, Paris.

Payne, R. B., Payne, L. L. & Barlow, C. R. (1997) Observation of Savile's Bustard *Eupodotis savilei* in The Gambia. *Malimbus* 19: 97–99.

Payne, R. B., Barlow, C. R. & Wacher, T. (2000) Adamawa Turtle Dove *Streptopelia hypopyrrha* in The Gambia, with comparison of its calls in The Gambia and Nigeria. *Malimbus* 22: 37–40.

Ranner, A., Tebb, G. & Craig, M. (2000) First record of Little Crake *Porzana parva* in The Gambia. *Bull. Afr. Bird Club* 7: 51–52.

Robel, D. (1998) An observation of Ayres' Hawk-Eagle *Hieraaetus dubius* in The Gambia. *Bull Afr. Bird Club* 5: 128–129.

Robinson, P. (2001) The Gambia. *In:* Fishpool, L. D. C. & Evans, M. I. (eds.) *Important Bird Areas in Africa and Associated Islands: Priority Sites for Conservation.* Pisces Publications, Newbury & BirdLife International, Cambridge.

Roy, K. (2009) First record of Wahlberg's Honeyguide *Prodotiscus regulus* for The Gambia. *Bull. Afr. Bird Club* 16: 90–91.

Rumsey, S. J. R. (2001) Recoveries of birds ringed in Sénégal and The Gambia during recent years. *Bull. Afr. Bird Club* 8: 46-47.

Smalley, M. E. (1979) Dowitcher in The Gambia. *Malimbus* 1: 68.

Van Welie, L. (2003) Franklin's Gull *Larus pipixcan* in The Gambia. *Malimbus* 25: 97–99.

Wacher, T. (1993) Some new observations of forest birds in The Gambia. *Malimbus* 15: 24–37.

SENEGAL

Aransay, N., Pacheco, L. & Zabala, I. (in press) First records of Narina's Trogon *Apaloderma narina* for Senegal. *Bull Afr. Bird Club*

Bailleul, P. & Le Gal, P.-Y. (1991) Observation du Grèbe à cou noir au Sénégal. *Malimbus* 13: 40–41.

Baillon, F. (1989) Découverte d'une colonie nicheuse de Héron cendré (*Ardea cinerea* L.) au Sénégal. *Alauda* 57: 233.

Baillon, F. (1991) Une nouvelle espèce pour le Sénégal: le Goéland à bec cerclé *Larus delawarensis*. *Alauda* 59: 113.

Baillon, F. (1992) *Streptopelia cf. hypopyrrha*, nouvelle espèce de tourterelle pour le Sénégal. *Oiseau & R.F.O.* 62: 320–334.

Baillon, F. & Cormier, J. P. (1993) Variation d'abondance de *Circus pygargus* (L.) dans quelques sites du Sénégal entre les hivers 1988–1989 et 1989–1990. *Oiseau & R.F.O.* 63: 66–70.

Baillon, F. & Dubois, P. (1991) Seawatching from Cape Verde, Senegal. *Birding World* 4: 440–442.

Baillon, F. & Dubois, P. J. (1992) Nearctic gull species in Senegal and The Gambia. *Dutch Birding* 14: 49–50.

Baillon, F. & Sylla, S. I. (1991) Nouvelles données sur la reproduction de l'Échasse blanche (*Himantopus himantopus himantopus*, Linnaeus 1758) au Sénégal. *Oiseau & R.F.O.* 61: 62–65.

Barlow, C. R. (2007a) First records of Icterine Warbler *Hippolais icterina* for The Gambia and Senegal. *Bull. Afr. Bird Club* 14: 72–74.

Barlow, C., Wacher, T. & Disley, T. (1997) *A Field Guide to Birds of The Gambia and Senegal*. Pica Press, Mountfield.

Bengtsson, K. (1995) More observations of Audouin's Gulls *Larus audouinii* in Senegal. *Malimbus* 17: 102.

Bengtsson, K. (1997) Some interesting bird observations in Mauritania and Senegal. *Malimbus* 19: 96–97.

Brasseur, R. E. (2000) African Swallow-tailed Kite *Chelictinia riocourii* breeding in the Saloum Delta, Sénégal. *Bull. Afr. Bird Club* 7: 134.

Brown, R. G. B. (1979) Seabirds of the Senegal upwelling and adjacent waters. *Ibis* 121: 283–292.

Colston, P. R. & Morel, G. J. (1984) A new subspecies of the African Reed Warbler *Acrocephalus baeticatus* from Senegal. *Bull. Brit. Orn. Club* 104: 3–5.

Colston, P. R. & Morel, G. J. (1985) A new subspecies of the Rufous Swamp Warbler *Acrocephalus rufescens* from Senegal. *Malimbus* 7: 61–62.

Cormier, J. P. & Baillon, F. (1991) Concentration de Busards cendrés *Circus pygargus* (L.) dans la région de Mbour (Sénégal) durant l'hiver 1988–1989: utilisation du milieu et régime alimentaire. *Alauda* 59: 163–168.

Coulthard, N. D. (2001) Senegal. *In:* Fishpool, L. D. C. & Evans, M. I. (eds.) *Important Bird Areas in Africa and Associated Islands: Priority Sites for Conservation*. Pisces Publications, Newbury & BirdLife International, Cambridge.

Clark, J. A., Wernham, C. V., Balmer, D. E., Adams, S. Y., Griffin, B. M., Blackburn, J. R., Anning, D. & Milne, L. J. (2001) Bird Ringing in Britain and Ireland in 1999. *Ringing & Migr.* 20: 239–288.

Condamin, M. (1987) Le Pluvier de Leschenault (*Charadrius leschenaultii*), espèce nouvelle pour le Sénégal. *Malimbus* 9: 131–133.

Cruse, R. (2004) Yellow-browed Wabler *Phylloscopus inornatus* in Senegal in December 2003. *Bull. Afr. Bird Club* 11: 147–148.

Delaporte, P. & Dubois, P. J. (1990) Premier recensement de laridés sur les côtes sénégambiennes. *Alauda* 58: 163–172.

Devisse, R. (1992) Première observation au Sénégal du Martinet marbré *Tachymarptis aequatorialis*. *Malimbus* 14: 16.

Dubois, P. J. (1990) Audouin's Gulls in Senegal. *Dutch Birding* 12: 25–26.

Dubois, P. J., Holmström, N. & Verneau, A. (2009) La péninsule du Cap-Vert à Dakar, Sénégal, est-elle la "Mecque" du seawatching? *Ornithos* 16: 220–232.

Erard, C. & Morel, G. J. (1994) La sous-espèce du Cochevis modeste *Galerida modesta* en Sénégambie. *Malimbus* 16: 56–57.

Ericsson, S. (1989) Notes on birds observed in Gambia and Senegal in November 1984. *Malimbus* 11: 88–94.

Gruwier, C., Claerbout, S. & Portier, B. (2001) Bergeronnette citrine *Motacilla citreola* à Dakar: première mention pour le Sénégal. *Bull. Afr. Bird Club* 8: 139-140.

Guillou, J.-J. (1987) Nidification de *Thalassornis leuconotus* au Sénégal. *Alauda* 55: 149.

Haass, N. & Engelhardt, C. (2002) *Calonectris* shearwaters off West Africa. *Birding World* 14: 514.

King, M. (2000) Breeding of Swallow-tailed Kite in Senegal. *Malimbus* 22: 90–91.

Lambert, K. (1978) Feldsperling, *Passer montanus*, und Bergfink, *Fringilla montifringilla*, in den Tropen. *Beitr. Vogelkd*. 24: 103.

La Rédaction (1988) [Rejection of *Larus belcheri* in Senegal]. *Alauda* 56: 68.

Mackrill, E. J. (1989) Audouin's Gulls in Senegal in January 1989. *Dutch Birding* 11: 122–123.

Marr, T., Newell, D. & Porter, R. (1998) Seabirds off Senegal, West Africa. *Bull. Afr. Bird Club* 5: 22–28.

Marr, T. & Porter, R. (1992) Spring seabird passage off Senegal. *Birding World* 5: 391–394.

Meeth, P. (1981) Baird's Sandpiper in Senegal in December 1965. *Dutch Birding* 3: 51.

Morel, G. J. & Morel, M.-Y. (1989) Une héronnière mixte sur le lac de Guier (Sénégal) avec référence spéciale à *Ixobrychus m. minutus* et *Platalea leucorodia*. *Oiseau & R.F.O.* 59: 290–295.

Morel, G. J. & Morel, M.-Y. (1990) *Les Oiseaux de Sénégambie*. ORSTOM, Paris.

Mullié, W. C. & Mullié-Mbézellé, G. (2001) First breeding record of Hadada *Bostrychia hagedash* from Senegal. *Malimbus* 23: 59-62.

Payne, R. B. (1997) The Mali Firefinch *Lagonosticta virata* in Senegal. *Malimbus* 19: 39–41.

Petersen, B. S., Christensen, K. D. & Jensen, F. P. (2007) Bird population densities along two precipitation gradients in Senegal and Niger. *Malimbus* 29: 101–121.

Printemps, T., Rouillon, Y. & Morel, G. J. (1999) Observation de la Bernache cravant *Branta bernicla* au Sénégal. *Malimbus* 21: 114–115.

Riddiford, N. (1990) Collared Flycatcher *Ficedula albicollis* in Senegal. *Malimbus* 11: 149–150.

Rodwell, S. P., Sauvage, A., Rumsey, S. J. R. & Bräunlich, A. (1996) An annotated check-list of birds occurring at the Parc National des Oiseaux du Djoudj in Senegal, 1984–1994. *Malimbus* 18: 74–111.

Rouchouse, C (1985) Sédentarisation de *Monticola solitarius* au Cap de Naze, Sénégal. *Malimbus* 7: 91–94.

Rumsey, S. (1992) More news from Senegal. *B.T.O. News* 183: 13.

Rumsey, S. J. R. (2001) Recoveries of birds ringed in Sénégal and The Gambia during recent years. *Bull. Afr. Bird Club* 8: 46-47.

Salewski, V. (2008) Eastern Olivaceous Warbler *Hippolais pallida reiseri* wintering in the Senegal valley. *Malimbus* 30: 172–175.

Salewski, V., Bargain, B., Diop, I. & Flade, M. (2009) Quest for a phantom—the search for the winter quarters of the Aquatic Warbler *Acrocephalus paludicola*. *Bull. Afr. Bird Club* 16: 61–66.

Salewski, V. & Becker, P. (2009) A Stone-curlew *Burhinus oedicnemus* in a flock of Senegal Thick-knees *B. senegalensis* in Senegal. *Malimbus* 31: 118–120.

Sauvage, A. & Rodwell, S. P. (1998) Notable observations of birds in Senegal (excluding Parc National des Oiseaux du Djoudj), 1984–1994. *Malimbus* 20: 75–122.

Sauvage, A., Rumsey, S. & Rodwell, S. (1998) Recurrence of Palaearctic birds in the lower Senegal river valley. *Malimbus* 20: 33–53.

Schricke, V., Triplet, P. & Leray, G. (2001) La Foulque macroule *Fulica atra*, une nouvelle espèce nicheuse au Sénégal. *Alauda* 69 : 328.

Strandberg, R. & Olofsson, P. (2007) Seabird counts at N'Gor, Senegal, in November 2006. *Malimbus* 29: 128–130.

Talamelli, A. (2007) First record of Eurasian Black Vulture *Aegypius monachus* for Senegal. *Bull. Afr. Bird Club* 14: 203–204.

Tréca, B. & Sakho, M. (1995) Confirmation de la présence du Martinet alpin *Apus melba* au Sénégal. *Malimbus* 17: 100–101.

Tréca, B. (1996) Fréquentation de bassins de lagunage par le Grèbe castagneux *Tachybaptus ruficollis* et le Grèbe à cou noir *Podiceps nigricollis* au Sénégal. *Alauda* 64: 421–428.

Tréca, B. (1997) Les Chevaliers combattants *Philomachus pugnax* L. dans le nord du Sénégal. Composition des populations par sexe et classes d'âge. *Alauda* 65: 161–166.

Triplet, P., Schricke, V. & Tréca, B. (1995) L'exploitation de la basse vallée du Sénégal par les anatidés paléarctiques. Une actualisation des données. *Alauda* 63: 15–24.

Triplet, P. & Schricke, V. (2008) Conséquences de l'ouverture d'une brèche dans la Langue de Barbarie du delta du Sénégal. *Alauda* 76: 157–159.

Triplet, P. & Yésou, P. (1999) La Spatule blanche *Platalea leucorodia* hivernant dans le delta du Fleuve Sénégal. *Malimbus* 21: 77–81.

Trolliet, B., Fouquet, M., Triplet, P., Girard, O. & Yésou, P. (1995) A propos de l'hivernage de la barge à queue noire *Limosa limosa* dans le delta du Sénégal. *Alauda* 63: 246–247.

Van der Have, T. M. & van der Hoop, C. (2002) Spotted Sandpiper at Île de Gorée, Senegal, in February 2001. *Dutch Birding* 24: 156–157.

Verschuren, J. & Dupuy, A. (1987) Notes sur les oiseaux des Parcs nationaux littoraux du Sénégal. *Gerfaut* 77: 405–442.

Yésou, P. & Triplet, P. (1995) The Common Gull *Larus canus* in Senegal. *Malimbus* 17: 26–27.

Yésou, P. & Triplet, P. (1995) La Mouette atricille *Larus atricilla* au Sénégal. *Alauda* 63: 335.

3. ACOUSTIC REFERENCES

Barlow, C., Hammick, J. & Sellar, P. (2002) *Bird Song of The Gambia & Senegal—An Aid to Identification.* 3 CDs. Mancarin Productions, Wimborne.

Chappuis, C. (2000) *African Bird Sounds: Birds of North, West and Central Africa and Neighbouring Atlantic Islands.* 15 CDs. Société d'Études Ornithologiques de France, Paris & British Library, London.

INDEX

D

E

L

M

N

O

P

Q

R

S

X

Y

Z

QUICK INDEX TO THE MAIN GROUPS OF BIRDS

Figures in **bold** refer to plate numbers